Biophysical Research for Asian Agroforestry

BIOPHYSICAL RESEARCH FOR ASIAN AGROFORESTRY

Editors

MARTHA E. AVERY
Forester, Argonne National Laboratory, USA

MELVIN G.R. CANNELL
Head of Station, Institute of Terrestrial Ecology, Scotland

CHIN K. ONG
Agronomist, ICRISAT, India

WINROCK INTERNATIONAL, USA/1991
and
OXFORD & IBH PUBLISHING CO. PVT. LTD.
New Delhi Bombay Calcutta

This volume was sponsored by the Forestry/Fuelwood Research and Development (F/FRED) Project for which Winrock International is the principal contractor. Funded by the U.S. Agency for International Development, the F/FRED Project is designed to help scientists in Asia address the needs of small-scale farmers for fuelwood and other tree products. F/FRED provides a network through which biological and social scientists exchange research plans, methods, and results. The books in this series are designed to further stimulate international efforts to improve the productivity and effectiveness of agroforestry research in meeting increasing human and environmental needs.

To facilitate access to literature cited in this book, copies of most of the articles (but not books) are available on request from:

Research Network Secretary
P.O. Box 1038
Kasetsart Post Office
Bangkok, 10903, Thailand

ISBN 81-204-0569-2

Published in India, in collaboration with South Asia Books, USA, by Mohan Primlani, Oxford & IBH Publishing Co. Pvt. Ltd., 66 Janpath, New Delhi 110 001, for the Winrock - Oxford & IBH Series, phototypeset by Indira Printers and printed at Rekha Printers Pvt. Ltd., A-102/1 Okhla Industrial Area, Phase II, New Delhi 110 020.

1-Y1-20

Foreword

Winrock International Institute for Agricultural Development and the Oxford and IBH Publishing Company began publishing a book series on agroforestry research and practice in 1990. The first volume, *A Handbook for Managing Agroforestry Research*, was written by Dr. John Gordon, Dean of the Yale School of Forestry and Environmental Studies, and myself.

This is the second volume in the series. The editors, Dr. Martha Avery, Dr. Melvin, Cannell and Dr. Chin Ong, have organized our current understanding of agroforestry systems in terms of biological concepts. We believe this conceptual structure will help research specialists and practitioners in framing clearer, more readily testable, and ultimately more useful hypotheses about how agroforestry systems work and how they might be improved.

Other volumes in this series will integrate the social sciences into our understanding of agroforestry and develop more locational specific understanding of particular systems. At present we plan on publishing ten or more volumes on agroforestry over the next three years.

We have three goals in this series. We want each volume to be relevant to the problems faced by researchers and practitioners, but we are especially concerned that the volumes lead toward results of benefit to rural poor people throughout Asia. Our authors and editors are working hard on the quality of their expression as well as the quality of their ideas, so each volume can be easily understood and translated into practice.

Finally, we want the price of each volume to be within reach of students, researchers, and practitioners. We are most grateful to the U.S. Agency for International Development for supporting the first three volumes of this series through its F/FRED Multipurpose Tree Species Network Project being managed by Winrock International. The Ford Foundation and other donors are supporting future volumes, and we are searching for additional support.

General Series Editor
April 1991

WILLIAM R. BENTLEY

Preface

The purpose of this volume is to provide a framework for examining some of the biophysical principles underlying agroforestry. Arguments in favor of growing trees with agricultural crops are often based on untested assumptions, and many of the critical biological issues have not been addressed previously. This volume differs from others in that it focuses on evaluating rather than describing agroforestry systems.

The chapter authors were requested to identify the critical research questions in their fields and then to outline ways for researchers to conduct experiments to address those questions. Some authors responded by restating the questions as testable hypotheses and including sections on the client problem, hypothesis, materials and methods, and critical review. Other authors responded more generally, addressing the critical questions in terms of basic principles. Whichever approach was used, the charge was to assess which theories have actually been proven and which remain as assumptions, then to offer direction for future research. The specific research examples in this volume illustrate many of the concepts presented in the companion volume (volume 1) on research methods.

Conceived essentially as a handbook for new researchers and practitioners in the field of agroforestry, this volume includes chapters on experimental design specifically for agroforestry research; design elements of agroforestry systems; and chapters investigating plant, soil, and livestock interactions. The chapters on allelopathy and biological control can draw on little specific agroforestry research and thus anticipate areas of increasing interest.

Several chapters were solicited to present less common perspectives on agroforestry research and application. The social implications of biophysical research are addressed in the chapter by Hocking, who contends that social objectives are influenced by the specific biophysical research conducted. His chapter reinforces the relationship between the biological and social sciences and serves as a bridge to the companion volume (volume 3) on sociocultural research. The chapter by Jones and Lowry is unusual in that it approaches the objective of increasing fodder availability by improving fodder quality and digestibility, not by increasing the biomass yield from agroforestry systems. The chapter by Briscoe goes beyond small farm agroforestry systems to explore silvicultural applications for industrial and government lands.

The production of this volume has involved many people in numerous countries. We would like to thank the chapter authors for their willingness to address agroforestry research from our particular perspective and the many colleauges who reviewed and commented on the various chapters. We would like to formally acknowledge the institutional support provided by Winrock International and the individual efforts of William R. Bentley, managing editor of the handbook series, Karen Seckler, editorial consultant for this volume, and Barbara Scott for her tireless secretarial support.

MARTHA E. AVERY
MELVIN G.R. CANNELL
CHIN K. ONG

Contents

Contributing Authors

AVERY, MARTHA E., Environmental Research Division, Argonne National Laboratory, 9700 South Cass Avenue, Argonne, Illinois, 60439.

BRISCOE, C.B., Winrock International, 1611 N. Kent Street, Suite 600, Arlington, Virginia, 22209.

CANNELL, MELVIN G.R., Institute of Terrestrial Ecology, Bush Estate, Penicuik, Midlothian EH26 0QB, Scotland.

CHUNTANAPARB, LERT, Faculty of Forestry, Kasetsart University, Bangkok, Thailand.

HOCKING, DRAKE, Swiss Development Cooperation/Nepal, P.O. Box 113, Kathmandu, Nepal.

HORSLEY, STEPHEN B., USDA Forest Service, Northeastern Forest Experiment Station, P.O. Box 928, Warren, Pennsylvania, 16365.

JONES, RAYMOND J., Division of Tropical Crops and Pastures, Davies Laboratory, CSIRO, Private Mail Bag, Post Office, Aitkenvale, Townsville, Queensland, 4814 Australia.

LOWRY, J. BRIAN, Division of Tropical Animal Production, Davies Laboratory, CSIRO Private Mail Bag, Post Office, Aitkenvale, Townsville, Queensland, 4814 Australia.

MACDICKEN, KENNETH G., Winrock International F/FRED, P.O. Box 1038, Kasetsart Post Office, Bangkok 10903, Thailand.

MEAD, ROGER, Department of Applied Statistics, The University of Reading, Whiteknights, P.O. Box 217, Reading, RG6 2AN, U.K.

MILLER, I.M., Department of Botany, University of Glasgow, Glasgow, G12 8QQ, Scotland.

NARAYANAN, R., Department of Microbiology, Tamil Nadu Agricultural University, Coimbatore 641 003, India.

ONG, CHIN K., Resource Management Program, ICRISAT, Patancheru P.O., Andhra Pradesh 502 324, India.

PURUSHOTHAMAN, D., Department of Microbiology, Tamil Nadu Agricultural University, Coimbatore 641 003, India.

REYNOLDS, L., International Livestock Centre for Africa, c/o IITA, PMB 5320, Ibadan, Nigeria.

WHEELER, CHRISTOPHER T. Department of Botany, University of Glasgow, Glasgow G12 8QQ, Scotland.

WIERSUM, K.F., Department of Forest Management, Wageningen Agricultural University, P.O. Box 342, 6700 AH Wageningen, The Netherlands.

YOUNG, ANTHONY, ICRAF, P.O. Box, 30677, Nairobi, Kenya.

ZHAO, LIPING, Department of Plant Protection Nanjing Agricultural University, Nanjing, 210014, People's Republic of China.

PART ONE

Approaching Agroforestry

CHAPTER 1

Designing Experiments for Agroforestry Research

R. Mead

Is it necessary to design experiments? It is, if unambiguous, precise information that is relevant to practical situations is the desired outcome. It may be necessary, however, to apply different design concepts to research station and on-farm experiments in agroforestry because the objectives and constraints of these two situations are different.

In experiments on research stations, it should be possible to produce a high level of control of environmental and plant variation. This control enables the experimenter to ask questions about the form of causative mechanisms as well as more agronomic questions about performance. Research station experiments inevitably have an element of artificiality due to the level of resources that can be applied. Given those resources, the statistical concepts of experimental design should be employed to control the variability of experimental units and thus to maximize precision. That the scope for using standard statistical design concepts is considerable can be seen in the discussion of research proposals for agroforestry in southern Africa (Huxley et al. 1987).

It might appear that there is less scope for statistical design concepts in on-farm experimentation because on-farm experimentation tends to use fewer, but larger, experimental plots, has less choice of plots, and often has fewer resources available for characterizing and assessing the plots. However, the reduced information available makes it even more important to use statistical concepts of design, although these may differ from the concepts used for research station experimentation. Instead of using the geographical blocking systems normally employed on research stations, plots in on-farm experiments can be characterized in a manner similar to patients in medical research experiments, where potential differences of age, sex, size, and history may well be recognized and used to adjust measurements of performance.

GENERAL STATISTICAL PRINCIPLES OF DESIGN RELEVANT TO AGROFORESTRY RESEARCH

The underlying principles of design are, inevitably, no different for agroforestry experiments than for any other area of experimentation. For the purposes of this section, a single experimental unit is defined to include one or more trees or lengths of hedge plus one or more areas of agricultural crop.

The Stages of Experimental Design

The design of any agroforestry experiment consists of three stages:

1) Appropriate experimental units must be identified, and inherent characteristics of those units that are likely to lead to predictable patterns of performance must be recognized. By using the experimenter's specialized knowledge about the agroforestry units, it should be possible to control the level of variability in the set of units through systems of blocking.

2) The objectives of the experiment must be identified and formulated as specific questions, and treatments to provide answers to these questions must be selected. The statistical concepts of treatment structure may be employed to provide more information for each question.

3) The chosen treatments must be compared within the structured set of units. Stages 1 and 2 should be considered independently, any apparent incompatibility, being overcome in stage 3, which involves allocating treatments to particular units within the overall recognized structure of units. Particular facets of the treatment set sometimes require particular patterns of treatment allocation.

Design Resources

In an experimental design, there are n experimental units, and the resources, represented by $n - 1$ degrees of freedom (df), are used in three ways: (1) blocking, or variation control (including covariance), (2) estimation of variance (σ^2), and (3) answering treatment questions.

In an experiment with blocking control, the block size typically will be between 4 and 12 units, and therefore the df required for (1) will usually be between $n/12$ and $n/4$. The minimum requirement of df for (2) is about 10; the maximum df that should be allocated to (3) is about 20. The remaining df are allocated to (3).

An experiment may be inefficient in many ways:

not enough df for (1)
not enough df for (2)
too many df for (2)
not enough treatments to use the df available for (3)
not using other methods of controlling σ^2

Principal Design Concepts

Blocking

With relatively large plots, it is often inappropriate to use randomized complete block designs. The concept of blocking for agroforestry with large plots should almost always indicate the use of incomplete blocks. Note that the construction of incomplete block designs is easy and does not require classical balanced incomplete block design theory. (See Mead 1988, chapter 7, for a general discussion.)

Covariates

It is extremely important to record all information about particular influences on the plots during the experiment in such a way that the information can be used as covariates if that proves appropriate.

Factorials

The concept of factorial structure is insufficiently utilized in agricultural experimentation in general, but is particularly relevant to agroforestry experimentation. With two components, as in intercropping, the number of factors that should be considered is large, particularly the number of two-factor interactions that need to be investigated. At the early stages of an agroforestry research program, it is clear that factorial structure must be extensively used.

Confounding

Using factorial structure in small blocks leads inevitably to confounding. It must be emphasized that confounding is extremely simple and efficient. The computational facilities necessary to analyze data from confounded experiments are widely available, even when the form of the confounding extends beyond that discussed in conventional textbooks. (See Mead 1984 for a discussion of confounding with particular reference to intercropping experiments.)

Split Plot Designs

Because of the disparate size of the two component units in agroforestry experimentation, there may often be situations where a split plot structure is essential for practical reasons. When not practically necessary, split plot designs should be avoided. The advantages claimed for split plot designs in terms of the assessment of interaction are spurious.

Quantitative Factor Levels

In general, three or four levels of a quantitative factor should be adequate for any form of analysis, provided that levels are appropriately chosen. If a straight line relationship is expected, two levels at the extremes of the realistic range provide maximal efficiency. For most curvilinear relationships, three levels (at the two extremes and at a central point) give the most efficient designs.

COMPONENTS OF DESIGN FOR AGROFORESTRY EXPERIMENTS ON RESEARCH STATIONS

The use of design factors specifically relevant to research station experiments in agroforestry revolve around four basic questions.

Size and Form of Experimental Units

What size and form should experimental units take? In any experiment, this choice is crucial, but it is particularly so for agroforestry research experiments because of the two different components, crops and trees, of such a system. The size and shape of an experimental unit must depend on the particular objectives of the experiment.

If the interest of an experiment is in the tree/crop or hedge/crop interface and in understanding and describing the effect of each component on the other, then the ideal form of plot is a single, isolated tree with a surrounding area of crop, or a single length (7 to 10 m) of hedge with crop area on one or both sides. A single tree plot, with sufficient space to examine the effect on the surrounding crop, gives basic information about the component interactions.

It also reflects the increasing recognition in forestry research that single tree plots provide the best precision.

The opposite extreme of the single tree plot, used for basic studies, is the multiple tree or multiple hedge plot used for assessing the relative merits of production systems. Where large experimental units provide information on the costs and profits of a few alternative systems, it is essential to have adequate replication, often using replication across sites. Typically, large units might include eight or nine trees in a plot sufficiently large to avoid problems of shading or root infiltration into adjacent plots (tree roots can spread very long distances—as much as 20 m in semiarid regions).

Between the extremes of single, isolated trees used for basic studies and substantial multitree plots used for production studies, other plot sizes are possible for experiments that compare a range of management and input treatments. Although such experiments normally use compact groups of adjacent plots, the advantages of single tree plots or single hedge-length plots are still relevant. However, the interrelationship with the agricultural component can make the precise definition of plots difficult. (A novel solution to this problem is suggested below in the section on interdependent tree or hedge units and crop units.)

Three basic forms of experimental units are considered in more detail, the first two of which are illustrated for both trees and hedges in figure 1.1.

Multitree or multihedge plots

This type of plot is used for substantial investigations into the performance of different systems and when large areas of trees are available prior to experimentation. Statistical considerations always point to the use of more and smaller plots rather than fewer and larger plots. Experimental plots, therefore, usually include 3 × 3 or 2 × 4 arrays of trees or two or three 10-meter lengths of hedge (figure 1.1a). Larger plots should be avoided unless there are very good practical reasons for their use (such as the sphere of influence of trees) and if sufficient replication of the large plots can be achieved.

The need for guard, or discard, areas of multitree plots arises either because of possible interference or influence between the tree components of adjacent plots or because of the difficulty of defining a typical area of crop for harvesting. The inclusion of guard trees or hedges considerably increases the total plot area; they should be avoided unless the competition effects are likely to be large. If only small competition effects are anticipted, then the loss of precision due to the greater variability of harvested areas resulting from more widely spread, larger plots is probably more of a disadvantage than the small bias introduced by the competition between plots. Guard areas should not be used to attempt to eliminate bias totally—this is never realistic—and the consideration of bias must recognize that experimentation inherently involves unrealistic situations that introduce bias.

Just as it is usually desirable to avoid the use of guard trees or hedges, it is also often important to discard substantial areas of crop that do not lie wholly within the set of trees or hedges in each plot. Crop areas between trees or hedges of different plots should not be recorded as part of the plot yields. This should not cause any difficulty because the total crop area included within a

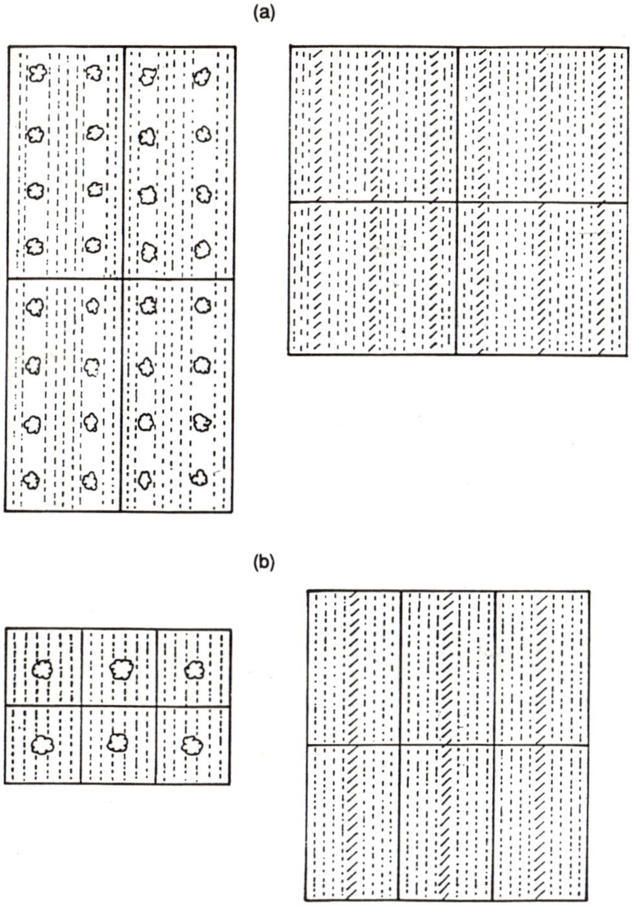

Figure 1.1. Arrangement of trees, or hedges, with crop areas for (a) multitree or multihedge plots and (b) single tree or single hedge plots.

multitree or multihedge plot is usually considerably larger than a typical plot size for sole crop experiments.

A typical plot size for a hedgerow and alley intercropping trial includes two hedges (1 m to 2 m wide), the crop alley between them (3 m to 5 m), and crop areas outside the hedges (2 m to 3 m), giving a total plot width of 9 m to 15 m. Plot length should be about 10 m, with a length of between 6 m and 8 m of the two hedges and the crop area between the hedges used for harvest records.

Single tree or single hedge plots

Much of the information about tree/crop interface effects can be best obtained from plots containing a single tree surrounded by a crop area, or a single hedge length with crop rows on one or both sides (figure 1.1b). Essentially, the single trees or hedge lengths are isolated, and the response of the crop at varying distances and the response of the tree or hedge to varying crop densities can be examined.

A practical rule that has been developed for fundamental research studies in India and southern Africa is that the plot size to surround a single isolated tree should be a square whose sides are three times the height that the tree is expected to reach during the experiment. For hedges, recommended plot sizes are 7 m to 10 m lengths of hedge with 3 m to 4 m of crop area on either side of the hedge.

Where single tree or single hedge plots are used for screening large numbers of tree or hedge genotypes or provenances, or for trials involving management treatments, smaller plots are appropriate so that the distance of the tree or hedge in the next plot is similar to that which would be expected in a production system.

Interdependent tree or hedge units and crop units

When experimental units of single trees or single lengths of hedge are used, researchers often encounter problems of unit definition. Consider an example from India where the experimental treatments included the between-hedge spacing. Three spacings (6 m, 8 m, and 10 m) were tried. To keep plot size small, the experimental unit was defined as one length (10 m) of hedge with half the space (3 m, 4 m, and 5 m) on either side of the central hedge. From the pattern of plots after randomization of spacing treatments (figure 1.2), it can be seen that the actual spaces between hedges, inevitably, were not what was intended. A plot intended for a 6 m spacing next to a plot intended for a 10 m spacing actually results in an 8 m (3 m + 5 m) gap between the hedges on the two plots.

One approach to this problem is to discard the concept of a single plot type and to consider instead a set of single tree units (or single hedge-length units) and a set of crop units. For the simpler hedge and alley-cropping system, there would then be an alternating sequence of hedge units and alley units with different hedge treatments and alley crop treatments. In a possible pattern for three hedge treatments and two alley crop treatments (figure 1.3), each hedge

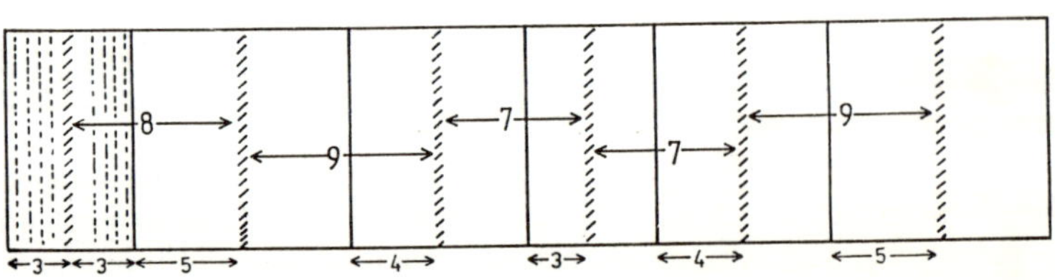

Figure 1.2. Illustration of difficulties implicit in varying alley widths using single hedge plots (distances in meters).

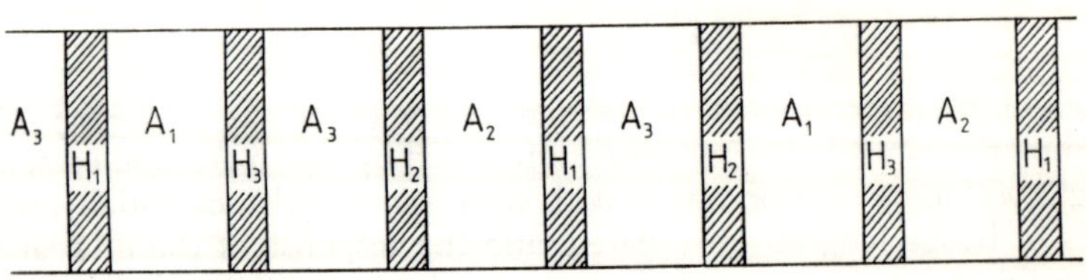

Figure 1.3. Typical arrangements tor experiments using two types of unit—hedge units (H) and alley units (A)—which are assessed interdependently.

unit treatment occurs beside each alley crop treatment and vice versa. With multiple replication involving restricted randomization of treatment allocation, it is possible to arrange experiments so that all pairs of hedge and alley treatment can be included adjacently.

The use of two interdependent sets of experimental units offers numerous opportunities for efficient experimentation using small experimental units. The basic analysis of data from such experiments involves separate analyses of data from each set of experimental units, including the effects of the treatments applied to the adjacent units of the other set. Thus if we consider the model for yields from alley units, the yield will be influenced by the treatment applied directly to the alley, by the blocking factors utilized in the overall structure of units, by both the treatments applied to the adjacent hedge units, and by the interaction of these two treatments with the treatment applied directly to the alley. In addition, there may be treatments applied to larger areas, including several hedge units and several alley units (such as fertilizer levels). Conceptually, the model could be written as follows:

Yield (alley) = block effects +
alley treatment effect +
average effect of two hedge treatments +
alley treatment and hedge treatment interaction +
large plot areal treatment

This form of analysis can be completed simply using multiple regression analysis.

For research involving single trees rather than hedges, the arrangement of tree units and crop plot units involves two dimensions rather than one, and this may lead to more complex design structures. However, the principle illustrated in figure 1.4 is unchanged. Each tree is surrounded by four crop plots and each crop plot by four trees.

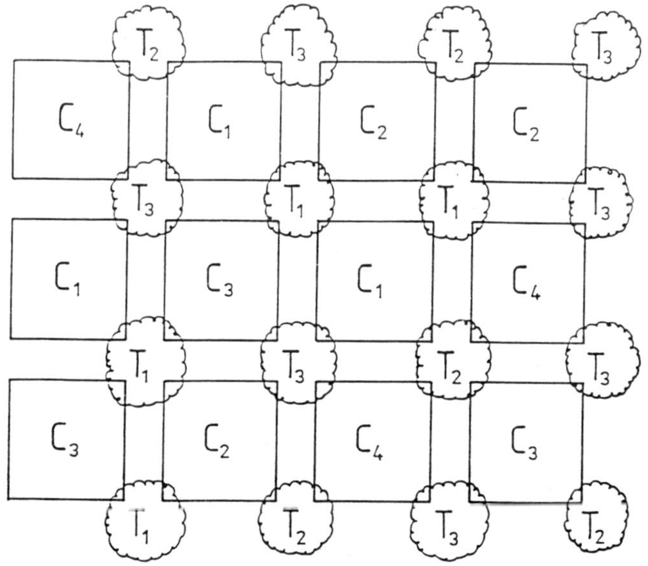

Figure 1.4. Typical arrangement for tree-crop experiments using two types of units, tree and crop, interdependently.

Organization and Structure of Units

What pattern of organization and structure of units is optimal? When large units are employed, using several trees or hedge lengths with crop areas, it is difficult to control variation between units just because the sizes of the units ensure that groups of units cannot be close together. Apart from using blocks of only a small number of plots, say a maximum of six plots per block, and choosing the blocking system with care, there is nothing much that can be done to improve the precision of treatment comparisons.

Smaller plots (single tree, single hedge, or interdependent sets of units) make it possible to choose blocks of homogeneous plots. With hedge and alley systems, it is often appropriate to use hedge units running along contours. When the land is relatively flat, hedge units with a common orientation in a row may form a sensible block. For single tree plots, blocks should usually be rectangular and as nearly square as possible. However, when isolated mature trees are used as plots, they are likely to be dispersed over the experimental area; blocks are then not compact areas but rather are sets of similarly developed and similarly situated trees, possibly scattered quite widely. In fact, a block is a set of homogeneous plots.

A complication can arise with the use of interdependent hedge and alley units when the land is not flat. It is sensible to use hedges running along contours and to regard plots along a contour as blocks. It therefore follows that the sequences of hedge-alley-hedge-alley will run across blocks. The analysis in sorting out the various effects of this sequence is not difficult. However, the design includes blocking units along contours and a transverse blocking system formed by the H-A-H-A sequence (figure 1.5). It is desirable that each hedge treatment occur in a block along a contour as well as in each reverse block. This gives the same kind of structure as a Latin Square.

Selection of Treatments

Many agroforestry research programs are relatively new, and the amount of practical experimental work they have produced is, thus far, not very large. On what basis then should treatments be selected? It is appropriate to emphasize that it is essential to employ factorial structures in the early stages of an experimental research program.

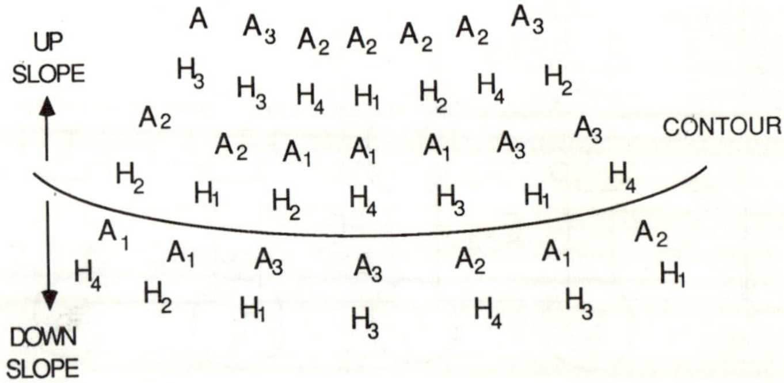

Figure 1.5. Contour and transverse blocking of hedge-alley sequences.

In a complex system such as agroforestry, there are many factors that could be investigated. In a hedge and alley system, the factors that could influence the productivity of the system include

alley width
hedge width
hedge-planting density
hedge-cutting times
hedge-cutting height
alley crop overall density
alley crop row spacing
alley crop genotype
hedge species and provenance
fertilizer factors (several)
alley crop planting time

This list is by no means exhaustive.

Particularly with the various spatial arrangements and timing factors, it is almost inevitable that the effects of these factors will not be independent. It is therefore essential at any early stage of any agroforestry experimental research program to utilize the efficiency provided by factorial structure and to obtain information about the interdependence of the effects of various factors. While this information is directly important, when interactions appear to be negligible, it also may allow future experiments in the program to be planned to take advantage of the assumed absence of interaction. The alternative philosophy of planning very simple experiments involving only a single factor during the early stages of a research program can be very wasteful of resources.

The most useful treatment structures for early stages of a research program are the 2^n factorials in which a substantial number of factors, each at two levels, are included. Normally, fractional replication—using only half or a quarter of the possible factorial combinations—and confounding should be used for grouping the treatment combinations to be allocated to different blocks of units.

For example, suppose an experimental unit consists of two parallel 10 m lengths of 1 m wide hedge with a 6 m wide alley between the two hedges and two 3 m areas on the outside of each hedge (total area 10 m $\times$ 14 m). Suppose also that it is desired to investigate

two hedge densities (factor D)
two hedge species (factor H)
two crop row spacings (factor R)
two hedge-cutting patterns (factor C)
two fertilizer regimes (factor F)
two crop genotypes with very distinct physiological differences (factor G)

A reasonable design might be a half-replicate of the 2^6 using 32 factorial combinations in four blocks of eight units, with confounded effects HF (aliased with DRCG), DFG (aliased with HRC), and DHG (aliased with RCF). It is inevitable that one two-factor interaction should be confounded; the HF interaction should not be substantial. The design shown in figure 1.6 might be suitable, but for an actual experiment, of course, the design would be blocked in the light of the experimenter's knowledge about the available land and after discussion with a statistician.

BLOCK I	BLOCK II
$d_0\ h_0\ r_0\ c_0\ f_0\ g_0$	$d_0\ h_0\ r_0\ c_0\ f_1\ g_1$
$d_1\ h_0\ r_0\ c_0\ f_0\ g_1$	$d_1\ h_0\ r_0\ c_0\ f_1\ g_0$
$d_0\ h_0\ r_1\ c_1\ f_0\ g_0$	$d_0\ h_0\ r_1\ c_1\ f_1\ g_1$
$d_1\ h_0\ r_1\ c_1\ f_0\ g_1$	$d_1\ h_0\ r_1\ c_1\ f_1\ g_0$
$d_0\ h_1\ r_0\ c_1\ f_1\ g_1$	$d_0\ h_1\ r_0\ c_1\ f_0\ g_0$
$d_1\ h_1\ r_1\ c_0\ f_1\ g_0$	$d_1\ h_1\ r_0\ c_1\ f_0\ g_1$
$d_0\ h_1\ r_1\ c_0\ f_1\ g_1$	$d_0\ h_1\ r_1\ c_0\ f_0\ g_0$
$d_1\ h_1\ r_1\ c_0\ f_1\ g_0$	$d_1\ h_1\ r_1\ c_0\ f_0\ g_1$

BLOCK III	BLOCK IV
$d_0\ h_0\ r_0\ c_1\ f_0\ g_1$	$d_0\ h_0\ r_0\ c_1\ f_1\ g_0$
$d_1\ h_0\ r_0\ c_1\ f_0\ g_0$	$d_1\ h_0\ r_0\ c_1\ f_1\ g_1$
$d_0\ h_0\ r_1\ c_0\ f_0\ g_1$	$d_0\ h_0\ r_1\ c_0\ f_1\ g_0$
$d_1\ h_0\ r_1\ c_0\ f_0\ g_0$	$d_1\ h_0\ r_1\ c_0\ f_1\ g_1$
$d_0\ h_1\ r_0\ c_0\ f_1\ g_0$	$d_0\ h_1\ r_0\ c_0\ f_0\ g_1$
$d_1\ h_1\ r_0\ c_0\ f_1\ g_1$	$d_1\ h_1\ r_0\ c_0\ f_0\ g_0$
$d_0\ h_1\ r_1\ c_1\ f_1\ g_0$	$d_0\ h_1\ r_1\ c_1\ f_0\ g_1$
$d_1\ h_1\ r_1\ c_1\ f_1\ g_1$	$d_1\ h_1\ r_1\ c_1\ f_0\ g_0$

Figure 1.6. Treatment allocation for a half replicate of 2^6 in four blocks of 8 plots per block for factors D, H, R, C, F and G (defined in text).

Systematic Spacing Designs

When are systematic spacing designs useful? Some experimental objectives concerned with the investigation of spacing or crop density factors may require using a design that changes density or spacing systematically across each of several main plots. Systematic spacing changes within a main plot can be arranged in one direction or in two perpendicular directions, and they may involve either or both components of the agroforestry system.

Some of the systematic designs suggested for intercropping by Mead and Stern (1980) may be appropriate for tree-crop combinations where the trees are not planted in hedges but may be quite close together (figure 1.7a, b, and c). Chetty and Rao (1981) have a different approach for examining the effect of varying alley width (figure 1.7d).

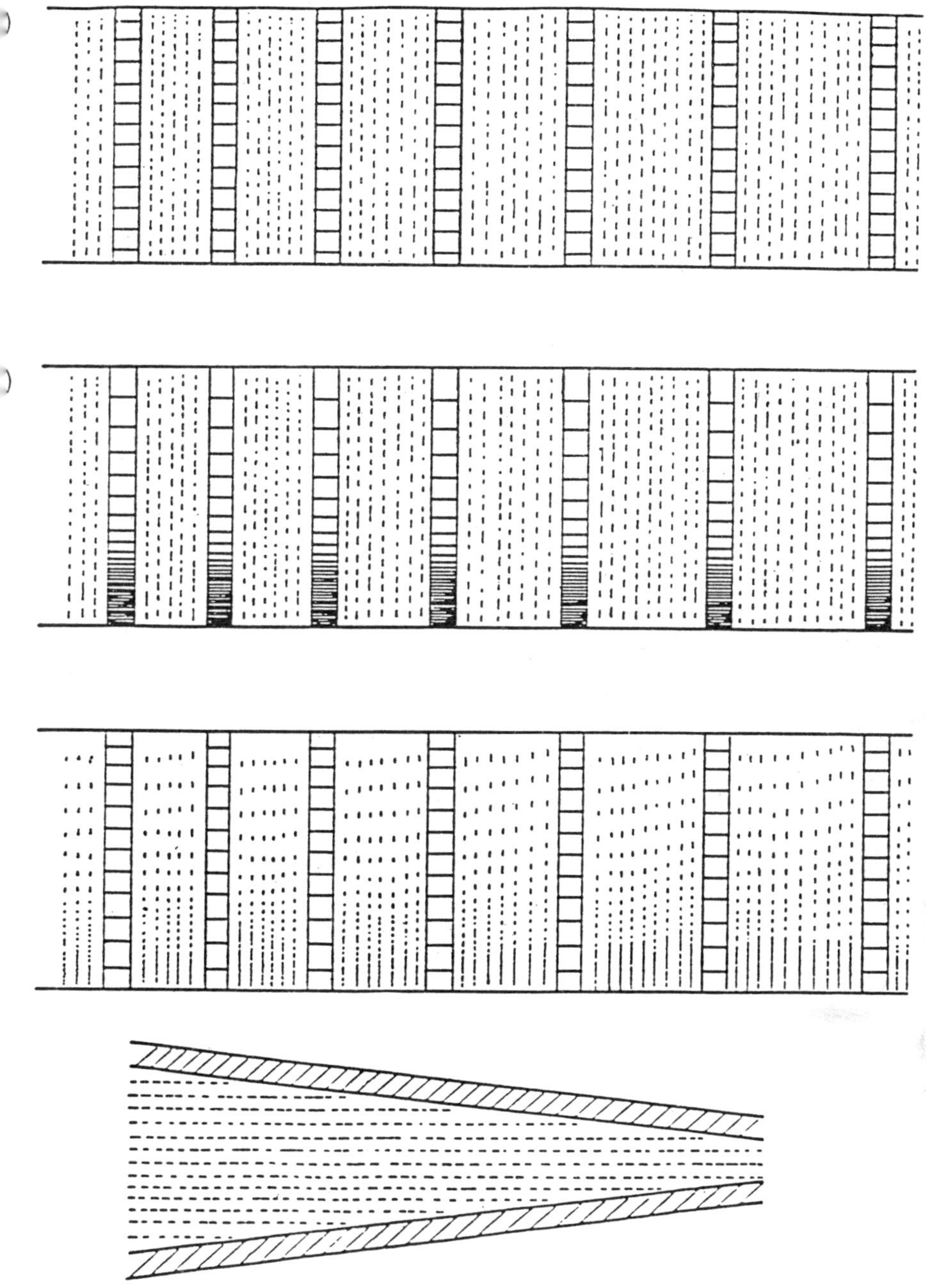

Figure 1.7. Possible systematic design structures for investigating spacing effects.
(a) A one-dimensional variation of alley width.
(b) A two-dimensional variation of alley width and hedge density.
(c) Crop density varied along the alleys with alley width varied between alleys.
(d) Chetty's and Rao's different approach for examining the effect of varying alley width. Source for (d): Chetty and Rao 1981.

Systematic designs can provide useful information about the effects of a wide range of spacings, but there are a number of potential dangers in the use of such designs. These dangers arise from the failure to have proper replication, from being too clever in fitting many systematic components together, and from the size of tree components. Unless a systematic row design is replicated at least three times, it cannot provide analyzable information and must be regarded as a pilot study with a pattern of results that may be caused by factors other than the imposed systematic spatial variation. Replicates should include randomization of the direction of the systematic change.

The arrangement of several systematic spacing or density variation units to form geometric designs with the different units spreading out in different directions is particularly dangerous because the different systematic changes are inevitably confused with other environmental changes. Patterns in the yield results are therefore liable to misinterpretation, which in the absence of genuine, randomized replication cannot be properly assessed.

One particular problem in the use of systematic designs for studying trees is the sheer size of the trees and the consequent breakdown at close spacings of the assumption of the relevance of the spacings within the systematic change. The original systematic designs of Nelder (1962) are intended for vegetable crops where adequate guard areas can be provided. For trees, however, the rate of the systematic change of density must proceed more slowly at the high-density end. The fan design, although attractive, will not work. The corresponding row design with modified patterns of density changes and considerably more guard rows can still provide relevant information, but such a design needs careful thought (particularly about the practical behavior of trees).

Analysis of systematic design results

It is important to emphasize that using a systematic design or a systematic component of a design usually leads to a different form of data analysis than will result from a randomized design. If any analysis and a statistical assessment of precision is to be attempted, it is essential that the main plots, within which spacings are varied systematically, be replicated; normally there should also be several main plot treatments. If there is inadequate replication or no replication, then it is not possible to assess whether the response pattern over the systematically varied treatments shows consistency that can be expressed in terms of statistical significance.

If there is sufficient replication of the main plots within which systematic treatment variation occurs, then the appropriate method of analysis is (1) to summarize the pattern of response to the systematic treatment factor in the form of a fitted response model for each main plot and (2) to analyze the variation of the fitted response models over the main plots. It is extremely important not to use a split plot analysis as if the systematically applied treatments had been randomly allocated. Such an analysis can give a very misleading assessment of the precision of response to the systematic factor.

For example, if the alley width is systematically varied in 10 steps from 2 m to 8 m across each main plot, if the experiment includes five species used to provide the hedge, and if each species is replicated three times, then there will be 15 main plots (3 blocks × 5 species) with systematic variation of alley width across each plot (and the direction of the systematic change chosen randomly for each plot). To examine the effect of alley width on hedge production per unit area (p), an asymptotic response function of the form

$$\log(p) = a - b \text{ (alley width)}$$

might be appropriate. (This could be investigated initially by using graphical plots; other response functions are possible.) This function would be fitted to the data for each of the 15 main plot combinations producing tables of *a* and *b* values.

a values

	B_1	B_2	B_3
Sp1			
Sp2			
Sp3			
Sp4			
Sp5			

b values

	B_1	B_2	B_3
Sp1			
Sp2			
Sp3			
Sp4			
Sp5			

The *a* values represent an upper limit estimate of the achievable hedge production as the alley is gradually eliminated, essentially representing the species productivity as a pure hedge strain. The *b* values represent the rate of reduction in production as the alley gap is increased. An analysis of each set of 15 values would provide information about the pattern and consistency of differences between species in terms of *a* or *b*.

The same form of analysis would be appropriate for measurements on crop performance in the alley.

The split plot analysis of the form

Source	*df*
Blocks	2
Species	4
Error	8
Widths	9
Sp × W	36
Error	90
Total	149

is wrong, as explained earlier, because it assumes random treatment allocation.

Systematic designs have great advantages in the efficient use of space (they do not need guard areas because of the gradual treatment changes), and they give information across a wide range of levels of the systematic factor in a manner particularly appropriate for density variation. They cannot, however, provide very precise assessment of the comparisons between particular alternative densities and, if used without replication, can be very misleading (for example, where the systematic change of density coincides with a slope). They provide a useful, but potentially dangerous, tool in the experimenter's armory.

Measurements and Analysis

It must be recognized that there is not a single form of statistical analysis that is appropriate to all forms of agroforestry data. Even for a single set of experimental data, it is important to use several different forms of analysis. For the two components of an agroforestry system, the data may occur in different structural forms. In general, data structures from agroforestry experiments are complex, with different forms of yield information available for different subsets of experimental units.

Valid comparisons

In considering alternative possibilities for the analysis of data from agroforestry experiments, it is essential that the principle of comparing like with like is obeyed. If yields are measured in different units, or over different time periods, or for different species, then comparisons are generally not valid and should not be attempted. A set of ten treatments (table 1.1) illustrates the difficulties and possibilities. Any actual experiment would be unlikely to include such a diverse set of treatments, though there typically would be several representatives of some of the treatment types illustrated.

Table 1.1. The structure for ten treatments

Hedge/tree crop			Agricultural crop			
Species	Annual yield	Growth increment	Species	Yield	Monetary value	Σ Relative performance
1 I	y_1	–	–	–	r_1	–
2 II	–	z_2	–	–	(r_2)	–
3 –	–	–	C	c_3	r_3	–
4 –	–	–	D	d_4	r_4	–
5 I	y_5	–	C	c_5	r_5	$y_5/y_1 + c_5/c_3$
6 I	y_6	–	C	c_6	r_6	$y_6/y_1 + c_6/c_3$
7 I	y_7	–	D	d_7	r_7	$y_7/y_1 + d_7/d_4$
8 I	y_8	–	D	d_8	r_8	$y_8/y_1 + d_8/d_4$
9 II	–	z_9	C	c_9	(r_9)	$z_9/z_2 + c_9/c_3$
10 II	–	z_{10}	D	d_{10}	(r_{10})	$z_{10}/z_2 + d_{10}/d_4$

A comparison is valid only when the units of measurement are identical. Thus it is valid to investigate the effect of different agricultural crops on hedge productivity (y_1, y_5, y_6, y_7, y_8) or on tree growth (z_2, z_9, z_{10},) and the effect different tree environments on crop yield (c_3, c_5, c_6, c_9, or d_4, d_7, d_8, d_{10},). The effects of different treatment systems on pairs of yields may be assessed by comparing the pair (y_5, c_5,) with (y_6, c_6,) or (y_7, d_7,) with (y_8, d_8,). Particular combinations of the pair of yields may also be compared so that (y_5,/y_1, + c_5,/c_3,) may be compared with $y_6/y_1 + c_6/c_3$. However, it is not valid to compare (y_5,/y_1, + c_5,/c_3,) with (y_7,/y_1, + d_7, d_4,) because the divisors for the second ratio components are different, and the resulting sums of ratios are not comparable quantities (ratios may be large either because the first value is large or because the divisor is small). When these sums of ratios are interpreted as land equivalent ratios (LERs) (Willey 1979, Mead and Riley 1981), they represent the land areas required to produce the equivalent of sole crop yields. However, land areas required to grow crop C are not comparable with land areas needed to grow crop *D*. Comparison of biological efficiency through LERs cannot be valid for different crop combinations.

The only measure by which different component combinations can be compared must be a variable, such as money, to which all component yields can be directly converted and which has a practical meaning. Even using the monetary scale is of dubious validity for treatments 2, 9, and 10 because the growth increments are not simply translated into financial value.

The variety of forms of analysis

The only form of analysis that retains all the available information is multivariate. When the performance of each component crop may be summarized in a single yield, then a bivariate analysis of variance is the most powerful technique available. However, only those experimental units can be included in a bivariate analysis for which both yields may be measured.

The methodology of bivariate analysis has been developed for intercropping by Pearce and Gilliver (1978, 1979) and, in more technical detail, by Dear and Mead (1983, 1984). This approach to analysis can certainly be useful when comparing agroforestry treatment combinations where the two components are assessed in either similar or quite different ways. The use of measurements over different time periods causes no problems with bivariate analysis because the bivariate analysis examines the joint variation of the two measurements. It can interpret whatever pattern of variation occurs, including situations where the two measurements appear to behave almost independently or where the two measurements are highly correlated. (For a more extensive explanation of why bivariate analysis is such a powerful tool, see Mead 1986.)

To compare yields from multiple crop plots with those from single crop plots, a univariate analysis for a single crop yield is necessary. It is important when contemplating an analysis using yields from both single crop plots and multiple crop plots to check that the variability of yields is homogeneous between the two types of plots. It should be expected that variability will probably not be homogeneous.

A PHILOSOPHY FOR ON-FARM EXPERIMENTATION

The history of on-farm experimentation in agricultural research and extension does not seem to be a very happy one. Experiments carried out on farms tend to have been derived as small-scale, poor relations of experiments at research institutes and tend to produce conclusions that (1) are very imprecise, (2) suggest farms are very different from research institutes, (3) do not convince anybody, and (4) are not always valuable as demonstration plots. Since it is clear that there should be benefits from performing experiments under farm conditions, for which the conclusions of research experiments are intended to be predictive, it is important to look again at the opportunities and restrictions of on-farm experiments and at the statistical expertise available for designing them. The ideas presented here have been developed by Huxley and Mead (1988) and have been stimulated by examining the apparent differences of statistical philosophy for designing experiments in agricultural, medical, and industrial contexts.

The crucial requirements for experiments using plots on farms are

controlling variability, while
utilizing environmental variation
allowing treatments to be applied to plots
providing replication
with randomization validity

A plot is defined as an area of land including hedge (or tree) or crop area, or both; it is fairly small (3 to 20 m^2) and is clearly definable in size, shape, and context within the normal activity of the farm. On any farm, there could be very large numbers of potential plots, of which only a small number (15 to 30) might be used in an experiment. A single experiment might include sets of plots from several farms.

The traditional method of controlling variability in agricultural crop experiments is through blocking, which has become identified with the procedure of selecting compact sets of adjacent plots that are recognized as blocks. Many such blocking systems are quite effective in achieving sets of homogeneous plots within blocks, though many blocking systems are set up without any real thought about the purpose of blocking. Successful blocking requires that the units allocated to a block should be expected to perform very similarly, and correspondingly, that units in different blocks be expected to perform very differently. To achieve this objective, it is not necessary that plots in a block be physically adjacent. The blocking principle is used in medical experiments where each experimental unit is a patient. Patients do not occur in geographically compact groups, but it is still possible to determine groups (or blocks) of patients who, because of their physical characteristics and history, might be expected to perform similarly. In the same way, plots on a farm can be thought of as grouped into blocks by slope, altitude, orientation, previous history, and possibly, proximity.

However, some effects of the characteristics that are useful in thinking about controlling variability by blocking may be of interest in themselves. We could use soil fertility as a blocking characteristic for plots, but we may also be interested in differences caused by different soil fertility levels, possibly in combination with applied treatments. Again, the analogy with medical experiments is useful. The fact that men and women are expected to react differently to a particular drug could lead to using sex as a blocking factor, but many medical trials would be interested in precisely how the response to drugs differs between males and females, and sex could then be thought of as a treatment factor. Just as sex cannot be randomly allocated to individual patients, soil fertility is a characteristic of each plot and cannot be randomly chosen. A suitable name for such a treatment is an *existing treatment* (in contrast to an *applied treatment*); others have used the names *ecological treatment* or *environmental treatment*. In addition to the existing treatment, there are usually applied treatments in an on-farm experiment.

The amount of replication that a single farm can afford and whether several farms are necessary to provide adequate overall replication must be considered. It is also important that the actual plots included in an on-farm experiment should be a random sample from some recognizable set of possible plots, in the same sense that plots included in an agricultural survey are randomly selected from a population of possible plots.

In their 1988 ICRAF technical report, Huxley and Mead propose that this approach to on-farm experimentation should include the following steps:

1) Identify a large population of observational units, probably spread over several farms.

2) Classify each unit according to its level of each of several existing treatment factors.

3) Further classify each unit according to its level of each of several blocking factors. These factors, unlike the existing treatment factors, are of no direct relevance to the questions the research is intended to answer.

4) Identify any applied treatments (the type of treatment used in an on-station experiment) to be included in the on-farm experiment.

5) Within blocks of similar units, where blocks are defined by step 3, units are *randomly* selected to represent different existing treatments (step 2) from a population of available quadrats (plots) of that kind. If step 4 is relevant, allocated levels of applied treatments are implimented.

The analysis of data from an on-farm experiment involves separating out the effects of the blocking factors, the existing treatment factors, the applied treatment factors, and the interactions between existing and applied treatment factors. This analysis can be achieved by fitting a general linear model that allows for the inevitable lack of completeness in the occurrence of combinations. (For example, it may not be possible to find plots for each combination of each level of a blocking factor that has each level of an existing treatment factor.)

The classification of plots for the various possible existing treatment factors can require substantial resources from the research team and, to a lesser extent, from the farmer. However, this classification need not require high-precision measurements; broad scoring systems may often be adequate.

Identifying blocking classifications that control the remaining variability, after allowing for the variability due to existing treatment factors, is crucial to the success of the experiments in providing useful information. If blocking is conscientiously performed, the block sizes will vary within an experiment, and the general principle of designing experiments in incomplete blocks (Mead 1988, chapter 7) will be appropriate. It must be emphasized that analysis of data from such non-orthogonal block designs is straightforward, if computationally complex, with modern statistical computing packages.

If the design of on-farm agroforestry experiments is approached within the structure proposed in this section, it is, of course, possible to omit particular components of the structure. The applied treatment factors can be omitted, making the experiment resemble a survey with various stratification factors. Alternatively, if the existing treatment factors are omitted, the experiment will be more similar to traditional experimental structures. But even in the latter case, it is important not to revert to the traditional practice of using miniaturized experimental structures of the research station for on-farm experiments. The crucial factor is in the blocking by plot characteristic rather than by groups of adjacent plots. It is arguable whether the definition of groups of adjacent plots as blocks is appropriate for research station experimentation since the experimenter should often know more about the plots than is implied by the form of blocking. In the less environmentally homogeneous on-farm plots, however, it is essential to think much more carefully about blocking.

ACKNOWLEDGEMENTS

Many of the ideas in this chapter have been stimulated by discussions with Peter Huxley for which I am grateful, even though we disagree about several aspects of experimental design.

LITERATURE CITED

Chetty, C.K.R. and U.M.B. Rao. 1981. Experimental designs for intercropping systems and analysis of data. In: *Proceedings of the International Workshop on Intercropping*, 10-13 January, 1979, ICRISAT, Hyderabad, India.

Dear, K.B.G., and R. Mead. 1983. The use of bivariate analysis techniques for the presentation, analysis and interpretation of data. *Statistics in intercropping, Technical Report 1*. Department of Applied Statistics, University of Reading, Berkshire, UK.

Dear, K.B.G., and R. Mead. 1984. Testing assumptions and other topics in bivariate analysis. *Statistics in intercropping, Technical Report 2*. Department of Applied Statistics, University of Reading, Berkshire, UK.

Huxley, P.A., R. Mead, 1988. An ecological approach to on-farm experimentation. *ICRAF Working Paper 52*. ICRAF, Nairobi.

Huxley, P.A., and R. Mead. and D. Ngugi. 1987. National agroforestry research proposals for southern Africa, AFRENA. ICRAF, Nairobi.

Mead, R. 1984. Confounded experiments are simple, efficient and misunderstood. *Experimental Agriculture* 20:185-201.

Mead, R. 1986. Statistical methods for multiple cropping. In: *Multiple cropping: principles, practice and potential*. (C.A. Francis, ed.). New York: MacMillan. pp 317–350.

Mead, R. 1988. *The design of experiments: statistical principles for practical application*. London: Cambridge University Press.

Mead, R., and J. Riley. 1981. A review of statistical ideas relevant to intercropping research (with discussion). *Journal of the Royal Statistical Society* A 144:462-509.

Mead, R., and R.D. Stern. 1980. Designing experiments for intercropping research. *Experimental Agriculture* 16: 329-342.

Nelder, J.A. 1962. New kinds of systematic designs for spacing experiments. *Biometrics* 18: 283-307.

Pearce, S.C., and B. Gilliver. 1978. The statistical analysis of data from intercropping experiments. *Journal of Agricultural Science* 91:625-632.

Pearce, S.C., and B. Gilliver. 1979. Graphical assessment of intercropping methods. *Journal of Agricultural Science* 93:51-58.

Willey, R.W. 1979. Intercropping—its importance and research needs. Parts I and II. *Field Crop Abstracts* 32: 1-10, 73-85.

CHAPTER 2

Orienting Agroforestry Research Toward Social Objectives

Drake Hocking

Many scientists believe that the work they do is socially and politically neutral, that only the *use* made of their results has social effects. To a certain extent this is true. But ultimately, if the results of *any* scientific research (including agroforestry) are used, there will be some social outcome. Responsible scientists will consider therefore what the social effects of their work might be and how marginal adjustments of intended research programs might change that eventual outcome. This is particularly relevant for agroforestry research because the rural poor are commonly portrayed as being its primary beneficiaries, and research in this field is intended to have a bias toward social equity. Thus it is important to examine *in advance* whether this broad social objective has a good chance of being achieved.

One important caveat should be mentioned here: no matter what the intentions of agroforestry are, the very poorest—those without *any* land—can be reached only indirectly by changes in land-use technology. Only when such changes generate additional employment (such as operation of simple bamboo tubewells in Bangladesh), or reduce consumer prices of basic commodities, will some portion of the benefits flow to the landless. This should not be forgotten when weighing overall priorities for research and investment. While improving technologies is really the desired priority of research in agroforestry, it must be acknowledged that such improvements can only indirectly help the landless.

The object of this chapter is to introduce into the thinking of biophysical scientists a concern for the social and distributional outcomes of research, and to provide a method that researchers can use to evaluate their own proposed research programs. The third book in this series, on applying social science theory and methods to agroforestry research provides the reader with more detail.

First, this chapter defines key terms. Second, two important research topics are evaluated in the context of agroforestry's acknowledged goal of helping the rural poor: (1) species selection and modification and (2) research in agroforestry technologies. Third, a method for assessing the social impact of agroforestry research by researchers themselves is outlined. Fourth, social objectives, both explicit and impicit, are discussed. And finally, two examples demonstrate how these factors fit together to shape agroforestry as it operates in developing countries today.

DEFINITIONS

Confusion exists among practitioners, scientists, and beneficiary populations about the meanings, contents, and objectives of various terms used in the general areas of agroforestry and social forestry, which for want of a general consensus, are here called "generic terms." Misunderstandings occur and much psychic energy is consumed when noninterchangeable terms are used interchangeably. The two definitions below are the key terms used throughout this chapter.

agroforestry A set of land-use systems that combine trees with pasture, arable crops, and/or animal production on the same land unit, either simultaneously or in short sequence. Thus agroforestry is a set of technologies or practices, as distinct from a program or policy. Certain agroforestry technologies find valuable application in programs of social forestry or community forestry.

social forestry The use of trees, and/or tree planting, to pursue social objectives (usually betterment of the poor). Social forestry is a program that may include many elements of agroforestry. It often includes, but is not interchangeable with, *community forestry*, which has a narrower meaning. A broader discussion of the term follows in the chapter section on case studies.

There is also a spectrum of related, but slightly different, terms used in different localities that convey significant differences in meaning. These differences, of course, imply different objectives that need to be considered. Definitions of a range of these related terms are given in an appendix to this chapter.

RESEARCH IS NOT NEUTRAL: TWO EXAMPLES IN AGROFORESTRY

Two examples of agroforestry research and their built-in social implications are briefly considered here. Each example leads to a hypothesis about the social effects of research. The first shows how the selection or modification of a species to improve its usefulness may not produce the intended social benefit. The second questions the impact of research designed to intensify already known agroforestry technologies. Even these seemingly beneficial objectives can be misapplied or have some results opposite to those intended.

Example 1—Species Selection and Modification

The rural poor are more likely to be able to harvest firewood from trees if the tree species are otherwise useless and difficult to manage for other applications.

The species *Prosopis juliflora*, mainly used in rural India for fuelwood, is a good case in point. Its rapid growth and vigorous coppicing makes it harvestable at an early age, and repeated harvesting at small dimensions prevents the emergence of a commercially valuable product. Its vicious thorns deter anyone except the most needy from attempting a casual harvest, but harvesting it for sale as small fuelwood provides a living for those people with low opportunity cost of time, such as the rural poor and underemployed landless.

Research is underway now to select or develop a thornless variety of *P. juliflora*, with the explicit objective of making it easier to manage for the people who harvest it. However, a successful outcome of this research may make the shrub attractive to other social groups previously deterred by the thorniness. If so, the final outcome may very well be that the truly deprived will have less access to it than they have now.

Hypothesis

The choice of tree species for development by selection or breeding, and the direction or objectives of selection or breeding, may be distributionally regressive.

Method of Testing

A literature survey is needed to first list the tree species on which such research has been done, and then to sort the list according to the most likely users of those species: small, subsistence farmers or large, commercial timber or wood users. Second, the actual directions of the research (intended and final outcome) should be sorted similarly. Certain objectives are likely to be intended exclusively for small subsistence farmers (such as enhanced fodder yield or, as in the case of *P. juliflora*, a feature making the tree useless for other users). However, some objectives (such as fast growth of straight trees) could be desirable for both large and small users, so the separation would be less distinct.

Example 2—Technology Intensification

As defined above, agroforestry is a technology, a land-use system. Traditionally, systems of agroforestry may be either extensive or intensive. Extensive systems include large tracts of land of widely scattered trees (*Butyrospermum paradoxum* in Mali, *Acacia albida* in much of the Sahel, *Prosopis cineraria* in Rajasthan, India) with seasonal undercrops of food grains or pasture. These systems are common in semiarid regions where productivity is low due to limitations of soil and/or climate. Intensive systems are more typical of productive areas with fertile soils and abundant rainfall. The multistoried home gardens of Kerala, India, or the Chagga coffee/food crop systems on Kilimanjaro in Tanzania are good examples.

The concept of agroforestry that is widespread among thinking politicians, land-use managers, and technocrats, and is accepted even by some researchers, is that it is a land-use system particularly suitable for resource-poor marginal land, even for wasteland. Because such land is usually owned and cultivated by poor and small farmers, research in agroforestry has been much promoted as a way to improve social equity by increasing productivity of poor lands.

It is thus argued that agroforestry (and therefore agroforestry research) is inherently biased toward the poor. In the context of general policies favoring greater social equity, this has helped to divert very significant resources toward agroforestry research. How are these funds used and with what results?

There are specific agroforestry systems that may be adapted to poor quality land owned by poor people, but research in this area seldom attracts the interest of researchers or the funds of research agencies. On the other hand, research on *intensification* of agroforestry technology is well funded. This is a problem because research on intensification is biased away from the very poor: it places greater demands on resources (soil fertility, water) and so is inherently better suited to *fertile* sites (and the rich farmers who own them) than to marginal ones or to wastelands. Contrary to the rhetoric, the most likely beneficiaries of this agroforestry research would be the rich!

Who then has benefited from the heavy promotion of agroforestry research as a technology for marginal lands? Mainly the research establishment. Who will benefit from the research in the long run depends very strongly on *what* research is eventually done when the research infrastructure has been built up. But unless deliberate and conscious efforts are made to direct research toward systems suitable to marginal land and wastelands, the inherent tendency is to work on better, more fertile, well-endowed land. Such land has a greater inherent potential for complex, intensified systems and will also yield research results more quickly (always an attraction for scientists who are assessed on their productivity).

Thus research in agroforestry is *not* socially or politically neutral—not for the researcher, not for the funding agencies, and certainly not for the beneficiaries. Who benefits from it depends very much on what system is being researched.

Hypothesis

Research in agroforestry leans toward those systems that intensify land use.

Method of Testing

The stated and unstated objectives (see section on social objectives, explicit and implicit) and probable outcomes of the array of experiments underway at various research institutions involved in agroforestry research must be analyzed. A method for doing so is presented in the following social audit section.

Such an analysis of on-farm and on-station research would classify experiments into those tending to intensify technology suitable for better sites and those tending to extend technology suitable for marginal lands. This distribution data could be compared within and across institutions, ecological zones, and countries. The result would be illuminating.

METHODS OF SOCIAL AUDIT FOR AGROFORESTRY RESEARCH

These two examples of agroforestry research, though brief, demonstrate that research itself can have social implications and therefore why the scientist must be concerned. What scientists need is a framework with which to evaluate their research proposals so outcomes can be predicted and problems can be anticipated and avoided wherever possible. Such a framework is provided in the *social audit*, a term first applied generally to analysis of rural development by Robert Chambers (1983) in his book *Rural Development: Putting the Last First*. The main emphasis of Chambers' book, like that of this chapter, is to help development workers, including agroforestry research scientists, perceive clearly the likely distributional implications of their work and to stimulate them to take social objectives seriously. Who will win? Who will lose? are the questions all researchers, as well as people working in applied fields, must ask.

Chambers proposed a simple matrix table that sets out a number of possible combinations of winners and losers in order to show that, although one might wish to target a project's benefits exclusively to the rural poor, it is essential to also consider what side effects there might be in the overall social context.

Principal Factors

The following are principal factors that require examination in a social audit:

1. Perceptual biases
2. Socioecological environments
3. Traditional and statutory tree and land ownership patterns
4. Farmers' production objectives
5. Beneficiary participation
6. Vertical integration
7. Monitoring results
8. Presenting results

Details of how to conduct such an audit using each factor are given in the following paragraphs.

Factor 1: Perceptual bias

The first step is to be aware of, and avoid, a number of common biases that influence what is selected as the topic for research. The following lists (developed from Chambers 1983) indicates some important biases:

Positive:	*Negative:*
near-urban locality	rural locality
dry season situations	monsoon mud
roadside farms	remote, roadless farms
rural elite	powerless people
males	females
modern practices	traditional approaches
progressive target people	nonadopters

Factor 2: Local socioecological environments

Subtle, but important, social and physical differences among localities affect the suitability of contemplated technical adjustments. It is therefore important to understand the social and biophysical parameters in detail before designing a study of potential technical innovation. For example, the rural elite are likely to oppose effective implementation of any project they view as being threatening to their position. While it is thus necessary to ensure that the better-off are not totally excluded from the benefits of an intervention, it usually is possible to design in a bias that ensures that the main flow of benefits is to the poor and disadvantaged.

ICRAF's diagnosis and design methodology (Raintree 1989) has been found useful in identifying agroforestry potentials and setting priorities for research. It is a way of avoiding lengthy, expensive, and slow questionnaire surveys through what has been termed *rapid rural appraisal* (Chambers 1983). This approach at least should answer the following questions:

What are the major social groups?
What subjects or issues generate conflict among these groups?
How is the land distributed among groups?
If a "commons" exists, who uses it for what?
What are the biophysical constraints on productivity?
Which social group experiences them most severely?
What institutional avenues of support are available?
Who, after all, really are the rural poor?

Factor 3: Traditional and statutory ownership patterns: Tree tenure as an example

The rights to use trees and tree products differ sharply from place to place, influencing the adoptability of agroforestry technologies (Fortmann 1987). Such rights may be traditional, or they may be statutory. In both cases, they can strongly influence who will plant trees, and where. In many parts of Africa, for example, individuals who plant trees retain the ownership of the trees and their produce; this ownership can be used to establish certain rights to the land under them, even where such rights did not exist before. By contrast, in much of South Asia, forests and certain kinds of valuable trees (such as sandalwood) are by law the property of the government, even if they are on private land. In such cases, the existence of trees on traditionally communal or private land makes the land, by legal definition, "forest" land and the property of the forest department.

Thus, the structure of ownership and use rights should be clearly understood in order to determine the likely distribution of benefits from an agroforestry technology. For such an analysis, a table like table 2.1 may be useful.

Table 2.1. Structure for analysis of ownership and use rights

	Categories of use of trees or tree products							
Category of user	Twigs, leaves	Fuel wood	Fruit nuts	Live-stock	Tree in situ	Timber (own)	Cash sale	Gain or loss
By gender								
Men								
Women								
By land ownership								
Landless								
Temporary								
Permanent								
By farm size								
Small								
Large								
By special users								

Factor 4: Farmers' production objectives

Research objectives should be established only after the production objectives for the client group or area are clearly determined. To do this, a detailed understanding of the farming system is required. Systems are, by nature, closed: most elements are already linked in complex ways to other elements, and no assumptions can be made about breaking these linkages.

What, really, are the shortages? Are there channels through which they can be met other than by attempting to produce them on location? How can the *details* of the research objectives be specifically tailored to primarily aid the *real* poor while not generating friction with the powerful rural elite?

As the details of research objectives evolve and become more fixed, a joint brainstorming session of all the research participants is sometimes helpful to speculate on what unexpected social outcomes may result from a successful achievement of each particular objective. The aim is to reveal ways in which marginal adjustments of the details of objectives or methodology can help ensure a final outcome that is in accord with explicit broad objectives.

In general, then, each agroforestry research objective should be accompanied by an initial hypothesis about who is likely to benefit. Then alternative hypotheses should be generated about likely distribution of benefits, and these (together with the initial hypothesis) should be tested against available information about the client group and area. If insufficient data is available for an adequate analysis, more should be sought. Failure to do so and to be fully confident of likely distributional effects could mean that initial objectives are frustrated by unexpected outcomes.

Factor 5: Beneficiary participation

A social audit of research proposals designed with the explicit objective of helping the rural poor needs to include an assessment of how the research will or can be carried out.

This means asking such questions as:

Does the client group understand the objective?
Do they accept that a successful result will really be to their benefit?
Are the research methods clear? Is the experimental design not just statistically valid, but also understandable by the client group?
Can the research be carried out actually on the farmer's own fields?
If not, what are the constraints? Will these constraints also be felt by farmers when they try to practice a successful result themselves?

Wherever possible, the research should be conducted with farmers' participation on their own fields, including due safeguards against private risk. This greatly enhances understanding and acceptability of the findings, and reduces the need for extension work later on. If the intervention works, it is already there in the field, and neighbors have been watching all along!

Factor 6: Vertical integration requirements

If an innovation resulting from the research requires additional external support or input, it should be sufficiently beneficial to make pursuing these outside requirements worthwhile. It should be apparent what steps are necessary to ensure that the supply of those requirements can be adequately institutionalized and their availability secured into the foreseeable future. Failure to do this creates opportunities for later exploitation as dependency on the innovations develops. It may even endanger initial successes, causing later failure to be blamed on factors not associated with the initial innovative technology.

For example, promotion of agroforestry for specialized cottage industries, such as tassar silk production or apiculture, can succeed only if the end product is of sufficiently high quality to satisfy accessible markets (Avery, personal communication) and if machinery and tools are available and can be repaired locally.

Factor 7: Monitoring results

Monitoring the *actual* outcome of a useful research result depends largely on what kind of innovation or project component is involved.

Small-scale micro-surveys, with few subjects and small sample size, are given to in-depth understanding and careful interpretation. They are usually more cost-effective and yield quicker results than large, complex surveys that generate overwhelming quantities of data.

The main components of such a small-scale survey could include the following:

- Input and output data from a project's activities.
- An analysis of the distribution of effort among component elements. (Are the strongly socially oriented components being neglected?)
- An assessment of the use being made of the data, innovation or inputs by the client group and by other groups in the area.

Such a survey requires regular (but not so frequent as to be a nuisance) visits to the field (not ignoring the biases listed in factor 1 above). These visits should continue for enough time to see the agroforestry system pass through a complete growth cycle, usually a period of several years, so that certain questions can be answered:

- What is the *net* increase in biomass production? What is its component distribution? Who is getting it?
- What are the side effects? Are there new sources of social tension?
- What adaptations or modifications have the farmers themselves introduced? Why? With what outcome?

Factor 8: Presenting results: Positive expression

Agroforestry is still a young discipline, competing for credibility, staff, and research funds with much older, better-established disciplines like agriculture and forestry. Sometimes agroforestry reports, papers, and talks tend to be apologetic about the idea of introducing crops onto forest land or trees onto cropland. Such a tendency is counterproductive, but it must first be acknowledged before it can be opposed. Two examples will help to illustrate the problem.

Use Positive Ways to Express Results

The way results of experiments and projects are expressed subtly, but significantly, biases their acceptance, even their meaning. Even the design of reporting forms tends to lock in how results are compiled and, therefore, how they are perceived, so attention must be paid to this aspect from the beginning.

For example, compare the following statements:

> Alley cropping of sorghum with *Leucaena* reduced sorghum yields by an average of 38% over all treatments. The *Leucaena* loppings yielded 1.6 t/ha.
>
> Hedgerows of *Leucaena* yielded an average of 1.6 t/ha of green fodder, while the grain yield of alley-cropped sorghum in the same system was maintained at 62% of pure crop yields, and the stover yield was 85%.

Tell All the Partners What Happened

No research should be considered complete until it has been communicated. Different reports should be developed for different audiences. An agroforestry researcher must not forget that various interest groups are following the research: sponsors, user groups, scientific colleagues, and scientific supervisors. Different forms of presentation of results are appropriate for each; and separate reporting guidelines need to be developed pertinent to each group's interest.

Scientists generally are experienced in presenting their results so that they can be understood by their peers: other scientists, journal editors, supervising scientific administrators. This is essential for the scientist's own career prospects. But sponsors and other institutional audiences also need to be convinced that *their* objectives are being met. Above all, for research to be useful (and to meet its original explicit objectives), it should be known and used by small farmers or other groups of rural poor (see Fortmann 1982). The best way of doing this is for the research to be conducted in front of their eyes so the results are prominent and visible.

The above social audit framework should help researchers understand what social results may be expected from their research, but it will not give the whole picture. Understanding the social objectives of agencies and individuals in research and implementation is equally important. That is the subject of the next section.

SOCIAL OBJECTIVES: EXPLICIT AND IMPLICIT

A key contention of this chapter is that researchers, as well as policy-makers, program implementors, and their institutions, have social objectives. This section examines the evidence behind this contention and proposes ways of reviewing research plans in order to ensure that intended social outcomes are, in fact, likely to occur. Furthermore, it is assumed that such social objectives generally are oriented explicitly toward greater social equity, that is, toward increasing the resources and assets of the poor (Chambers and Leach 1987).

However, a problem arises in that agroforestry research and application falls between the mandates of the two most closely involved government agencies—the departments of agriculture and forestry. Forest departments are responsible for growing trees; agriculture departments are responsible for maximizing production of arable crops. Sometimes neither institution is willing to recognize the benefits of mixing the two. The result is that the considerations that set departmental priorities seldom include the best interests of small farmers, who have little political clout (Mahiti Scheme Papers 1982, 1983).

What then are the social objectives of these individuals and organizations? How might they be expressed? And what effect can they have on research and on implementing the results of research?

The social objectives of any individual organization or program contain two elements: (1) those that are explicit in the printed and published documents about the program and (2) those that are not explicit but are present in the attitudes and behavior of the agencies or individuals concerned. This distinction is of critical importance, for while many program components may be designed and intended to strongly favor the rural poor, the executing agencies cannot be relied upon to have the same concerns despite what the explicit project document may say (Mahiti Team 1982, 1983).

Look, for example, at the explicit social benefits of agroforestry systems (and therefore of agroforestry research), which are generally perceived, or argued, to include the following:

Intensified land use resulting in increased production
Generation of employment for landless labor through increased demand for labor
Better seasonal distribution of labor needs and products
More equitable distribution of benefits
Better environmental quality and ecological stability (equally beneficial to all)
Improved community stability

These social objectives usually mean a transfer of resources and an increased flow of investment and benefits to the (usually rural) poor, including the creation of owned productive assets where none were before.

But this relatively straightforward set of objectives does not operate by itself. Potentially, there are also sets of *implicit* objectives to agroforestry research and social forestry programs that are obstacles to achieving these explicit social objectives. The most obvious is the maintenance and growth of institutional and individual resources, prestige, and influence. This type of implicit objective has been well described in the case of flooding in the plains as caused by alleged deforestation in the Himalayas (Thompson et al. 1986).

There are, of course, always practical obstacles to achieving these explicit objectives, the most difficult of them being the desire and ability of the rich and influential to capture (or retain) access to the major flows of public investment or resources (Leonard 1979, Hobley 1987). Also, the improved standard of living of those directly employed by social forestry programs may actually increase social tensions between them and the nonbeneficiaries. For example, as part of Perum Perhutani's Prosperity Approach, base camps on Java were constructed to improve the living conditions for the forest workers (*magerserman* community). One unexpected outcome of this activity was resentment of the neighboring villagers (Avery, personal communication 1988).

When broad, explicit objectives are translated and elaborated into specific activities, it becomes possible to more accurately predict (during the planning stage) what the likely outcome will be and to perceive (during monitoring and evaluation) the actual outcome of each component. When the implicit objectives are recognized at the planning stage, it becomes easier to design for the explicit ones. It is usually assumed by planning and funding agencies that only the explicit objectives are operative; consequently the implicit ones are often, even usually, overlooked or ignored. The explicit objectives are built in and tend to guide the design of reporting formats for monitoring progress. The result can be the neglect of important aspects of the project and misleading conclusions about overall successes or failures. Thus, it is clearly important to consider in advance the implicit as well as explicit objectives.

Which components are included in the project design and financed out of public funds (whether generated internally or through an aid grant or loan) is a political decision with explicit social consequences. But which components receive emphasis and absorb the lion's share of resources during implementation tends to be determined by the implicit objectives of the implementing agency.

CASE STUDIES

The reader has now been given (1) examples of how research itself shapes the social outcome of results, (2) a method to assess the social impact of proposed research programs, and (3) an overview of how social objectives influence the direction of research and the application of research results.

It is now time to look at two major programs of research and application in agroforestry, keeping in mind the above framework.

The Case of *Leucaena*

Even the most casual contact with agroforestry or social forestry is certain to include references to *Leucaena leucocephala*. Promotion of the biophysical *and* social benefits of *Leucaena* is one of the greatest success stories in developing countries in recent years. Its fast growth, ease of maintenance, wide adaptability, lack of pests and diseases, and multiple uses were promoted by prestigious institutions (University of Hawaii, National Academy of Sciences, and International Development Research Center). It was indeed the "miracle tree."

What has been the outcome of this vigorous promotion? *Leucaena* is now widespread in developing countries and is backed by a vast and institutionalized research network involving numerous scientists. But its advantages have not lived up to the promotion. On the one hand, the tree has serious site limitations, performing outstandingly well only on a restricted range of soils and climatic regimes; on the other, it has become a weed under some conditions (the Philippines). It is brittle and suffers in high winds (Bangladesh); in some localities it suffers from Fusarium dieback (India), and in most countries it is subject to an insect problem (*Psyllid*) that can reach devastating proportions.

An enhanced (and better-funded) research program to deal with these emerging problems is promoted on the grounds that the tree is now widespread and much basic information is already available.

Who has benefited, then, from the promotion of *Leucaena*? Certainly many farmers and rural poor in developing countries have benefited, and continue to do so. But it is not the universal miracle tree that was once portrayed. Some small farmers have invested only to be losers in the end. And it is not clear that a similar investment in screening and development of a wider range of potentially useful species, at a wider range of adaptive sites, would not have had a more stable result.

The other big beneficiary has been the research establishment (institutions and individual scientists) involved in testing and developing the group of species. Does this initial broad promotion demonstrate an implicit objective (even an unconscious one) on the part of the research establishment? Are there perceivable scenarios where the concentrated resources might have been better applied if hidden objectives had been recognized and if greater emphasis had been given to explicit objectives at an earlier stage?

The Case of *Eucalyptus*

The genus *Eucalyptus* includes several species that have been outstandingly successful as introduced exotics for intensive management for production forestry in several tropical developing countries. In India, the most common species include *E. camaldulensis* and *E. tereticornis*. Widespread promotion by forest departments was based on ease of propagation, broad site adaptability (including exceptional drought tolerance), rapid growth, good form, and above all, resistance to browsing by livestock.

In India, many sites recently classified as forest land had actually been used for generations as common pasture by local residents. Intensified population pressure, however, had led to some of these common pastures being exploited for domestic fuel and overgrazed to the point of denudation, making them prime candidates for reforestation.

In the presence of grazing pressure, the tree of choice was *Eucalyptus* for its browse resistance and other undoubted benefits. But as high-density plantations started to mature and the canopies closed, local herders found that grazing areas they had previously used (even though at a very low level of productivity) were no longer available. Similar situations also developed in Zambia (Ng´wandwe 1976). A groundswell of resentment against *Eucalyptus* started to emerge.

Eucalypts were seen as the cause of several problems: (1) decreasing groundwater levels, (2) absence of groundcover, and (3) slower recharge and reduced runoff. In addition, the main beneficiaries were seen to be the forest department and the major timber contractors.

These trees thus acquired a reputation of being "antisocial" in some, but by no means all, areas of India and Africa. Vigorous, even virulent, popular campaigns of opposition to them were organized, irrespective of site, species, or suitability (Shiva et al. 1984, Hoskins 1980).

Objective consideration of the scientific evidence, however, demonstrates that the tree is not inherently antisocial (Davidson 1983). Its vigorous growth no doubt causes it to consume water more quickly than slower-growing trees, but its *efficiency* of water use is higher: it produces *more* biomass per unit of water consumed than other tree species (Chaturvedi 1984). Moreover, the rooting depth of eucalypts (and indeed most trees) is too shallow to reach groundwater, at even the previous, higher levels. Increased irrigation by mechanized pumpsets is more likely the cause of groundwater depletion.

The marked absence of groundcover under some *Eucalyptus* plantations is attributable to high planting density (circa 1 m × 1 m) and, where wider spacing was practiced, to continued access by local people and livestock. The ground was regularly and literally swept clean of twigs and litter for use as fuel, and any green shoots appearing above ground were consumed.

Arguments about afforestation with *Eucalyptus* causing slower recharge and reduced runoff from upper catchments are more complex and have some foundation, but again, they are not directly relevant to the actual sociopolitical opposition actions, which were directed against plantations at lower elevations.

The fast growth certainly contributed greatly to the overall availability of wood fiber in the localities of plantations, of which some components (such as branch loppings and leaf litter) were available to the rural poor. The argument about beneficiaries is thus based less on the tree itself than on the way it is used.

What, then, were the explicit and implicit objectives of the anti-*Eucalyptus* campaigns?

The implicit objective of the campaign was to change *land-use* practices and distributional policies (Guha 1990). The tree was the explicit target because it was the effective instrument through which forest departments reasserted their claim to parcels of land over which they had lost control to local herders, who did not benefit from its plantation. This is a sociopolitical issue, not a technical one over the properties of a tree (see Raintree in press).

In fact, *Eucalyptus* has important advantages that can also benefit the rural poor. Its site adaptability, fast growth, and browse-resistant qualities are available to all. When it is planted at wider spacing, grass and even crops (on suitable sites) grow satisfactorily underneath. Competition for scarce rainfall during the cropping season can be controlled by timely canopy pruning. The sale of a single stick of wood is profitable to the small farmer as well as to the big. In fact, where land holdings are so small that they cannot produce enough for subsistene and the owners have no choice but to take off-farm jobs, planting low-maintenance *Eucalyptus* can help them generate some income (Campbell, personal communication 1988, Chambers, personal communication 1988, Shepherd 1988).

Who, then, were the winners and losers in the anti-*Eucalyptus* campaign?

Certain official institutions and nongovernmental organizations and individuals became associated with such campaigns and gained national prominence and influence through them. Small farmers in social forestry and other programs are the losers because they had reduced opportunities to take advantage of the tree's many benefits in appropriate agroforestry systems and also because of a studied neglect of *Eucalyptus* in agroforestry research programs.

Researchers in agroforestry have an important role to play in this controversy. By developing systems of management (spacing, lopping, pruning) in which the advantages of *Eucalyptus* are combined with crops and grazing in systems suited to the land quality available to small farmers, research can enable small farmers also to benefit from this valuable tree. Though *Eucalyptus* has no value as livestock fodder, its fuel value and marketability on maturity are complemented by the potential for annual returns from gum, honey, and essential oils from foliage.

Eucalyptus, therefore, should not be automatically excluded as a fit subject tree for research by the responsible scientist. Rather, *what* research is done on *Eucalyptus* will have a vital influence on who can benefit from it.

CONCLUSIONS

In this chapter, it has been argued that most social forestry and agroforestry research programs can agree on the broad objective of reducing social inequity, which is commonly expressed in the explicit objectives of research programs as well as of social forestry projects.

This central objective can be lost, however, anywhere during the design, implementation, execution, and monitoring stages due to competing or even conflicting interests (often unstated, though foreseeably implicit) of participating individuals, groups, and agencies. Researchers cannot assume that their work is neutral in terms of social impact. The shape and subject of research itself does influence who benefits from it.

The cases chosen as illustrations show how the outcomes of various initiatives have been at variance with the explicit objectives. This has commonly been the result of insufficient consideration at the planning stages of what factors may affect the outcome. Hidden, and therefore often neglected, factors include the unstated, implicit objectives of associated agencies (sponsors, planners, and executors) and even the nature of the interventions themselves.

Since there would be no large-scale agroforestry research and application projects without associated agencies and interventions, these factors require careful attention in order to ensure the highest probability of achieving the stated objectives. A social audit should help reveal how careful selection of details and marginal adjustments after testing of alternative hypothetical outcomes can best ensure adherence to objectives and the achievement of explicitly desired results. Overall, a greater transparency of process as well as practice is advocated. In the course of considering the factors involved in such an analysis, associated costs must also be kept in mind. Institutional reforms to reinforce a more equitable distribution of benefits entail additional costs in monitoring and enforcement. These costs need to be factored into the appraisal of the probable consequences of such reforms.

ACKNOWLEDGEMENTS

The ideas presented in this chapter evolved over many years through stimulating discussions with numerous colleagues, and it is difficult at this stage to remember the exact source of particular elements. I apologize to those whose contributions I fail to acknowledge, and hope they do not take offense.

Valuable criticisms and literature references were provided by several readers of early drafts this chapter, including Martha Avery, Diane Rocheleau, John Raintree, and Louise Fortmann. I gratefully acknowledge their help, but remain solely responsible for the final result.

LITERATURE CITED

Chambers, R. 1983. *Rural development: putting the last first*. London: Longman.

Chambers, R., and M. Leach. 1987. *Tree to meet contingencies: savings and security for the rural poor*. Social Forestry Network Paper 5a. London: ODI.

Chaturvedi, A.N. 1984. *Water absorption, transpiration, and biomass ratios of* Eucalyptus *species and other fast-growing trees*. Paper for National Seminar on Application of Science and Technology for Afforestation, ACT, Jaipur, Rajasthan, India.

Davidson, J. 1983. *Setting aside the idea that Eucalypts are always bad*. Working Paper, Project BGD/79/017, UNDP/FAO, Dhaka and Rome.

Fortmann, L. 1982. Taking the data back to the village. *Rural Development Participation Review*, Cornell University, 2(Winter): 13-16.

Fortmann, L. 1987. Tree tenure: an analytical framework for agroforestry projects. In: *Land, trees and tenure* (John Raintree, ed.). Proceedings of international workshop, 27-31 May 1985, Nairobi. ICRAF, Nairobi, and Land Tenure Center, Madison, Wisconsin.

Guha, R. 1990. *The unquiet woods*: ecological and peasant resistance in the Hiamalaya. Berkeley, California: University of California Press.

Hobley, M. 1987. *Involving the poor in forest management: can it be done? The Nepal-Australia project experience*. Social Forestry Network Paper 5c. London: ODI.

Hoskins, M. 1980. Community forestry depends on women. *Unasylva* 32(130): 27–32.

Kunstadter, P. 1980. Implications of socio-economic, demographic and cultural changes for regional development in northern Thailand. In: *Conservation and development in northern Thailand*. (J. D. Ives, S. Sabharsi, and P. Vorawrik, eds.). Tokyo: United Nations University.

Leonard, D.K. 1979. *Reaching the peasant farmer: organization, theory and practice in Kenya*. Chicago, Illinois: University of Chicago. 298 p.

Mahiti Project. 1982. *Field guide to social afforestation*. Mahiti Project, Dhandhuka, Gujarat, India. Mimeo paper.

Mahiti Team. 1982. *Focusing on the real issue: some glimpses from Dhandhuka Tahuka*. Regional workshop on block level planning, South Gujarat University, 4–6 October 1982, Surat, Gujarat, India. Mimeo paper.

Mahiti Team. 1983. *Why is social forestry not "social"?* Ford Foundation workshop on social forestry and voluntary agencies, 13–15 April 1983, Badkhal Lake, Haryana, India. Mimeo paper.

Ng'wandwe, C.O.M. 1976. African traditional land tenure and agricultural development: case study of the Kundu people in Jumbe. *African Social Research* 21:51–67.

Raintree, J.B. 1989. *Agroforestry diagnosis and design: overview and update*. Paper presented at the international symposium on planning for agroforestry, 24–27 April 1989, Washington State University, Pullman, Washington.

Raintree, J. B. In press. *The socioeconomic attributes of trees*. ICRAF/FAO collaborative project.

Shepherd. J. 1988. Social forestry and the poor in Karnataka: prospects and problems. *Appropriate Technology* 15(1): 17–19.

Shiva, V., S. T. Somasekhara Reddy, and J. Bandyopadhyay. 1984. The ecology of eucalyptus and farm forest policy in rainfed areas. In: *Eucalypts in India: past, present and future*. New Delhi: IBH.

Thompson, M., M. Warburton, and T. Hatley. 1986. *Uncertainty on a Himalayan scale*. London: Ethnographica.

APPENDIX : GLOSSARY OF TERMS

Some of the common terms used in this chapter and other papers that deal with the broad relations among agriculture, forestry, and land use are given below. These are broad understandings of the terms, not definitions, which tend to become restrictive.

amenity forestry Forestry for the purpose of recreation, pleasure, or general beautification of an area or a settlement.

beneficiary The intended person or people who should receive goods, services, or cash subsidies of a social or political program, usually of a social forestry program in this context. In community forestry, beneficiaries are more properly referred to as participants. Beneficiaries may by defined by any or several of a range of criteria including, for example, size of landholding (landless, small, medium, or large), type of land ownership (private, cooperative, corporate, leasehold from government), ethnic group or tribe, or class (poor, medium, or rich).

block plantation A plantation of a single or a few species of trees, usually in orderly, closely spaced rows. The term may include or overlap with woodlot or farm forestry.

communityforestry The practice of forestry by a community, where a community controls and is responsible for management of a forest for communal benefit, possibly including realization of financial returns through commercial sales as well as direct use of forest products. The forestland need not be owned by the community, but it is usually essential that the community have reliable long-term tenure. It may include elements of the cultivation of trees on private lands as a secondary component. The term is not really interchangeable with social forestry, although it often is used in that way.

community woodlot A block of trees managed by and for a community.

farm forestry Usually monoculture blocks of trees on private forest plantations or individual farms, but can also include narrow strips, or individual trees, or even mixtures with arable crops. The last is more commonly called agroforestry. The terms need not be mutually exclusive, and they can overlap.

farm woodlot A block of trees on a private farm, usually intended for fuelwood.

forest In India and in much of South Asia, this means land that has been "notified," or published, as being under the ownership or legal control of the forest department, whether or not it has trees on it. The term *forestland* has the same meaning. Elsewhere, of course, a *forest* is land dominated by stands of *trees*.

forestry The general theory and practice of management of forests and forestland; often extended to include any aspects of knowledge about trees, usually including bamboo in Asia, but excluding fruit trees.

industrial forestry Plantation forestry that has the supply of industrial feedstocks (for example, for matchwood, paper, or rayon) as its main objective.

plantation forest Generally uniform monoculture stands of exotic species planted by forest departments.

production forestry Management of forests (either natural or plantation) for the purpose of harvesting forest products (usually confined to those of commercial importance, like pulp or timber). This term is widely used by professional foresters to distinguish forests from which revenue can easily be obtained from those that are to be left largely untouched for reasons of watershed protection, erosion control, amenity, or other noncommercial use.

protection forest Forests from which people may exercise certain long-held rights to some kinds of forest products, by permission of the forest department that retains and exercises control of access and management.

protection forestry Forests intended to be protected from most or all kinds of extractive felling (in contrast to the above). The primary land use is often watershed protection (hence the term).

reserve forest Forest for which the right to manage and extract any forest products is reserved for the department of forestry.

wasteland Land that is currently producing useful biomass grossly below its potential. The reasons for underproduction may be many and varied, from technical (salinity, acidity or alkalinity, waterlogging) to social (disputed ownership or rights) or political (forest department or community ownership).

PART TWO

Designing Agroforestry Systems

CHAPTER 3

Tree Selection and Improvement for Agroforestry

Lert Chuntanaparb
K. G. MacDicken

Agroforestry is a land-use system that is increasingly regarded as an effective, low-cost means for minimizing the degradation of cultivated land and for maintaining or even increasing the productive capacity of agricultural ecosystems. Yet agroforestry systems have existed for centuries, and new, imaginative innovations will be required to improve existing practices. These innovations will necessarily include improvement of trees, just as improvement of cereal crops has been a critical part of the increases in agricultural production of the twentieth century. Examples of the potential role of tree improvement in agroforestry are the 20% to 30% gains in wood volume or value obtained through simple mass selection techniques for industrial tree improvement (Namkoong et al. 1980) and the dramatic improvement in *Leucaena leucocephala* yields through simple selection.

Rural communities in tropical countries have integrated forestry and agricultural practices for centuries. One of the most outstanding is the paulownia-based agroforestry system that has had an immense social impact on rural living (FAO 1978, Zhu 1981, 1988). In the 1970s, there was increasing recognition in tropical countries of the role forests and trees can play in increasing agricultural productivity, improving human welfare, alleviating energy problems, and conserving the environment (Turnbull 1984). According to an appraisal by the World Bank and the Food and Agricultural Organisation (1981), the main thrust of forestry in the 1990s will be reforestation with fast-growing trees, with special emphasis on multipurpose trees around homesteads, along boundaries, in village woodlots and in upland watersheds.

The goal of tree improvement for agroforestry is to increase the effectiveness of land and forest management for productivity, stability, and sustainability of land use for rural communities. Methods of establishing, managing, and harvesting trees as well as market considerations such as processing require-ments and profitability all lead to quantification of the economic and service values of trees. Invariably, increasing the value of a tree requires that more than one trait be improved. A fast growth rate is generally one of the most important traits, but for agroforestry, crown shape and root morphology can be equally important, while wood quality is less important. Because marginal sites are often considered for agroforestry, the ability to survive may be more important than the growth rate. In some cases, disease resistance is of paramount importance and may take precedence over all other traits.

This chapter describes several constraints to tree improvement in agroforestry within this framework and suggests refinements of these approaches.

STATE OF TREE IMPROVEMENT IN ASIA

Attempts to initiate or improve breeding programs for agroforestry tree species depends on information from existing tree improvement programs. In Asian countries, such information is generally scattered, diverse, and lacking in detail; little effort is made to record and compile data on tree improvement at local and national levels.

There are about 30 important tree species used in active plantation programs in tropical Asia. However, tree improvement programs were conducted only for the most important industrial species (for example, *Tectona grandis*, *Pinus caribeae*, and *Pinus merkusii*) and important high-yielding hardwoods such as *Paraserianthes falcataria*, *Eucalyptus* spp., *Gmelina arborea*, and *Leucaena leucocephala*.

Only *L. leucocephala* and *P. falcataria* have extensive agroforestry uses and have been improved for use in agroforestry systems. *Leucaena* use in agroforestry has been greatly enhanced through the selection and multiplication of arboreal Salvador-type varieties (Brewbaker 1987). Varieties such as K8 have yielded more wood than common varieties by 40% to 900% (Hu et al. 1980, Mendoza and Javier 1980) and more forage by 100% to 300% (Brewbaker et al. 1972, Oakes and Skov 1967, Instituto de Ciencia Animal 1979, Shih and Hu 1981). While these increases have been based on substantial research, the Salvador-type varieties were simple selections of superior phenotypes that propagate by seed.

There are likely to be few success stories as dramatic as that of the self-pollinated *Leucaena*, but there certainly is scope for improvement in many currently underexploited species. Research to produce such results is sorely lacking. There is much to be done to fill gaps in our knowledge about species requirements, criteria for selection, and beneficial combinations of trees and crops for agroforestry systems as well as how to manage and harvest them.

Although the economic feasibility of investing in breeding programs for agroforestry purposes may be questioned, preliminary cost-benefit ratios have been favorable for breeding pine species in the United States (van Buijtenen and Saitta 1972). The pan-tropical spread and use of the Salvador-type *Leucaena* varieties further demonstrates the economic impact of a simple tree improvement approach.

APPROACHES TO TREE IMPROVEMENT FOR AGROFORESTRY

The principles and practice of plant breeding for trees are well established (Wright 1976, Zobel and Talbert 1984), and they apply equally to industrial plantations and to smallholder agroforestry and community plantings. Superior phenotypes are selected from the best population and their breeding potential evaluated in clonal or progeny tests on typical sites with typical managements. Superior genotypes of cross-pollinated species are then propagated clonally or by seed and planted in special seed production areas or orchards where open or controlled pollination provides seed for plantations and, if needed, for further selection, testing, and breeding (Burley 1980). Zobel and Talbert (1984) identify five steps of which all beginning tree improvement programs consist:

1) Determine the species, or geographic sources within a species, that should be used in a given area.

2) Determine the amount, kind, and causes of variability within the species.

3) Package desired qualities into improved individuals to develop trees with combinations of desired characteristics.

4) Mass-produce improved individuals for reforestation purposes.

5) Develop and maintain a genetic base population broad enough for the needs of future generations.

These stages remain the same regardless of the objectives of the tree improvement program. Programs for agroforestry, however, do have several special characteristics and requirements.

Unique Requirements of Tree Improvement for Agroforestry

Tree improvement for agroforestry differs in two important ways from traditional practices. First, species selection should be made under agroforestry conditions, not plantation conditions. If trees are to interact positively with crops or livestock, they should be tested and selected under those conditions. For example, if an objective of tree improvement is to produce faster-growing trees that will grow with paddy rice, then selection should be made using paddy rice conditions. While it is both difficult and expensive to include intercropped annuals in experiments for selection, such trials should also be considered if increases in total productivity are desired.

Second, there are tree species potentially available for testing that have already been under some selective pressure on farms. Improvement programs for plantation tree species are generally conducted only by government, quasi-government, or private sector organizations, but many agroforestry species already have been selected and tested by farmers themselves. This is particularly true for trees with food or feed uses since the criteria for selection are easy to establish! Where this is the case, the tree improvement worker has access to materials that have already gone through a farmer-initiated improvement program. An example is the giant *Leucaena* varieties of Central America, which were collected almost entirely from parent trees that had been selected by local residents (J. L. Brewbaker, personal communication).

In addition, breeding trees for agroforestry has three problems not encountered in tree breeding for traditional industrial use (Burley and von Carlowitz 1984). First, the genetic base and genetic information are restricted; many existing trees and plantations have been derived from parents of unknown origin or ancestry. Second, the breeding of any crop for multiple characteristics is difficult because (1) at a given selection intensity, the rate of improvement of any one trait declines as additional characteristics are included; (2) the derivation of selection indices is complex mathematically, with limited techniques available for rapid assessment and large-scale screening of some traits; and (3) the material must be bred for a range of environments and management conditions. Third, inadequate professional and technical staff for tree improvement in agroforestry remains a problem.

Given these unique characteristics and constraints, old methods must be carefully reviewed and new approaches developed for tree improvement for agroforestry.

Accelerating Tree Improvement Research for Agroforestry

Traditional tree improvement involves literally decades of sustained effort, involving a series of species elimination and provenance and progeny experiments, which are outlined below. The urgent need for improved agroforestry species requires acceleration of this traditional process.

Traditional strategy for species and provenance testing

Ideally, the selection of species and provenance follows a systematic pattern of experiments (Burley and Wood 1976). Although this pattern is usually abbreviated in tropical countries due to time and resource constraints, the general sequence is as follow:

Stage 1: Elimination Trail (ET)

The objectives of the ET stage are (1) to test adaptability of species to the trial area quickly and at low cost and (2) to identify the most promising candidates and eliminate the least suited. In general, small plots of 16 to 25 trees are used. They are unreplicated on any one site but could be repeated on several different sites. The trial stage lasts for about 10 years. Assessments are only approximate—survival, height, and diameter (or girth) of one or two dominants are measured. In some cases, average height and diameter are evaluated through the whole trial, but survival is the most important measure of success at this stage. The ET stage drastically reduces the number of species for testing in later stages.

Stage 2: Growth Trial (GT)

Sometimes referred to as species performance trials (Cooling 1962, cited in Iyamabo 1969), the objective of GTs is to obtain information on such features as comparative performance, growth rates, and form of the most promising species of the ET stage on given sites. Growth trials are normally scattered to cover a range of areas; plots are usually larger (for example, 10 × 10 to 20 × 20 trees) and are replicated in randomized blocks. Spacing is the same as for normal plantations, and survival, vigor, form, height, and diameter (or girth) are periodically assessed for 10 to 15 years. These two stages are expensive and time consuming, and are often unjustified given resource constraints.

Stage 3: Provenance Trial (PT)

The importance of provenance in species performance makes it an integral part of species trial programs, and consequently, provenance trials feature prominently in tropical countries. For example, in Thailand. Indonesia, Malaysia, and the Philippines such trials have covered teak, *Eucalyptus*, pine, and *Leucaena* (Burley and Nikles 1973).

Although some species are of interest to several countries simultaneously, most of these provenance trials were planned and executed on a national basis. Recently, however, internationally organized trials have been established through the cooperation and direction of FAO, ACIAR, and USAID. Some of these trials have sought to reduce the amount of time required for screening and evaluating by relying on collaborative, multilocation experiments. An example is the Humid Zone Network Trials of Multipurpose Tree Species that is supported by the Forestry/Fuelwood Research and Development Project of USAID. These trials utilize a 3 × 2 × 3 factorial treatment design of two provenances of each of three species and three cutting treatments.

Available provenances of the most promising species may be included in the GT stage, but the more usual and better practice is to develop separate provenance trials concurrent with growth trials. The best provenances of the best species are then passed into plantation programs for subsequent selection, testing, and breeding.

The situation in agroforestry now almost exactly parallels that of industrial forestry 25 years ago (FAO 1986). There are problems in the choice of species, provenances, field trial methodology, and evaluation and breeding strategies. Little research has been conducted on species and provenance of multipurpose trees (Burley 1987a). These problems limit progress in the improvement of tree seed for agroforestry and require immediate action.

Modify screening practices for agroforestry

The ideal, three-stage screening process outlined above is both time consuming and expensive, and as a result, it is rarely used in its entirety in tropical countries. An example of the time frame for even an intermediate-level breeding program for tropical hardwood species, given in figure 3.1, would not produce seed commercially for over 15 years after the initial selections. That is too slow to meet the rapidly growing demand for quality seed for agroforestry programs (Turnbull 1984).

There are alternatives to this lengthy process of intense selection and breeding for tropical plantation species (Chuntanaparb 1975, Namkoong et al. 1980). Figure 3.2 provides an example of a program for reducing the time required for screening and selecting a multipurpose tree species (MPTS) with a short seed-to-seed cycle. How the time frame for breeding programs is reduced must be determined individually for each species due to differences in reproductive biology among species and the resources available for breeding. However, the following techniques may be considered:

Shortening Rotations at which Evaluations are Conducted

This may include the use of juvenile-mature relationships that allow early prediction of performance at maturity. This is more useful for such traits as fodder quality that are not likely to vary much between trees of different ages.

Nursery Screening for Desirable Traits

Tolerance of seedlings to acid soils in the nursery is one example of such a screening technique.

Flower Induction Techniques for Late-flowering Species

Flowering may be accelerated by using gibberellins in gymnosperms and dwarfing rootstocks with fruit trees or by using crown release, irrigation, fertilizer, girdling, or root-pruning treatments.

A large number of improved tree species for agroforestry is desirable; therefore it is appropriate to consider only those approaches that require minimal or intermediate levels of investment.

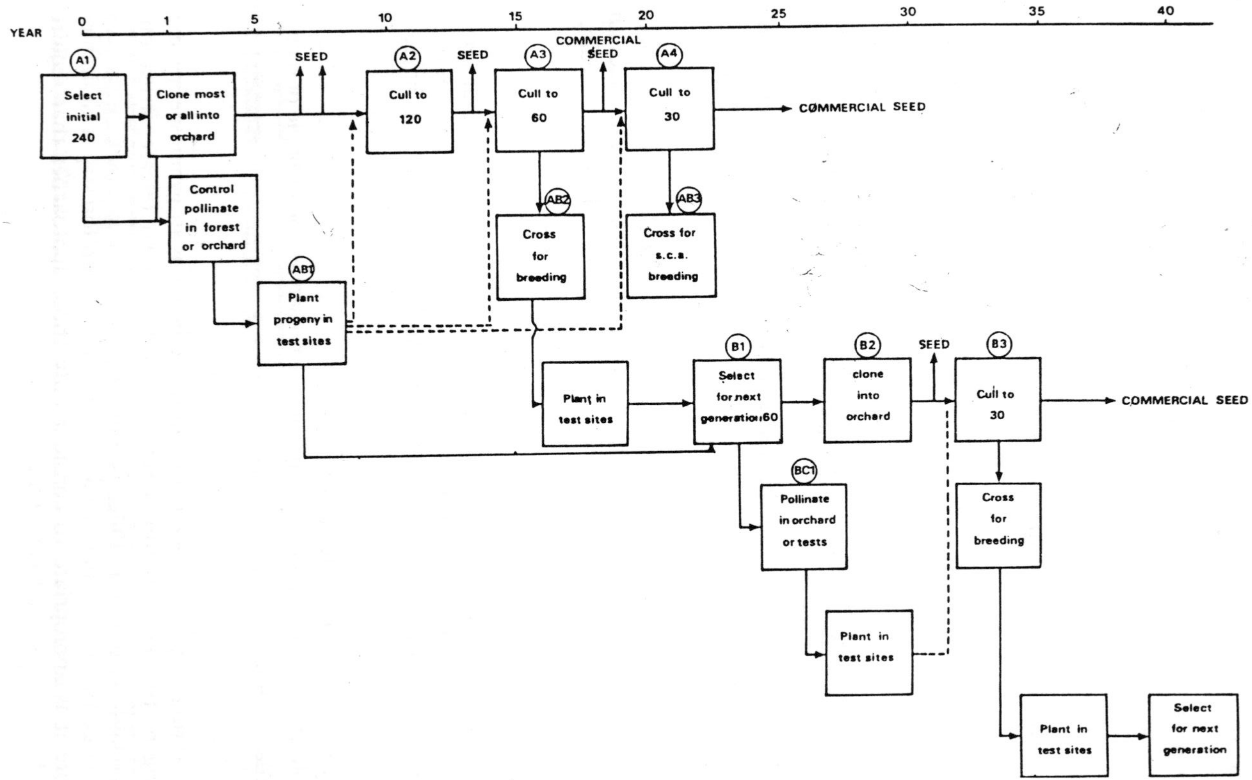

Figure 3.1. Stylized timetable for a program of intermediate intensity for breeding a cross-pollinated tropical tree species. Source: Namkoong et al. 1980.

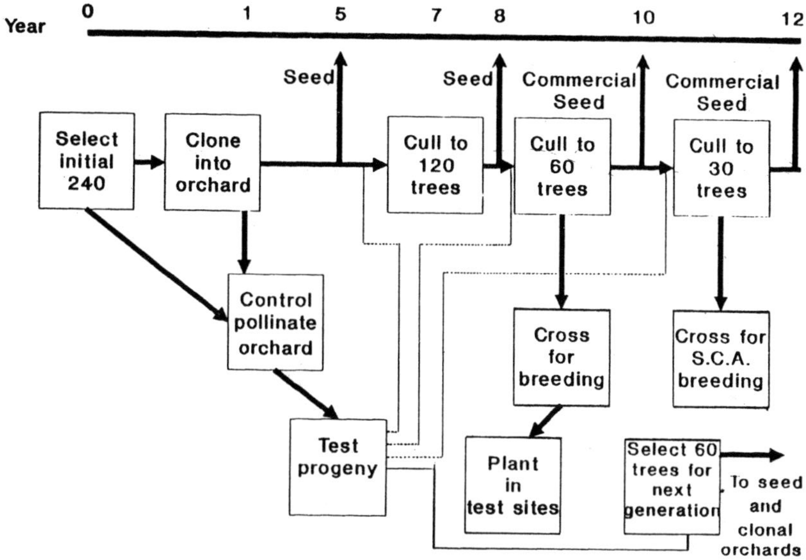

Figure 3.2. Accelerated screening schedule for MPTS with short seed-to-seed cycle. Adapted from: Namkoong et al. 1980.

Other Tools for Agroforestry Tree Improvement

A number of other tools are available to tailor tree improvement programs to the unique demands of agroforestry. These include the exploitation of relevant genetic resources, first stage selection with a broad range of species, and the creation of an effective cadre of so-called barefoot tree breeders.

Exploiting relevant genetic resources

Germ plasm resources have not been well explored or exploited, even for some tropical species that have been the focus of long-term tree improvement programs. Three types of sources exist for improving agroforestry tree species: natural populations, existing collections made for industrial purposes, and domesticated sources. Natural populations have been the primary focus of seed collection expeditions for many species. This invaluable yet diminishing resource, from which promising agroforestry species originate, must be conserved and utilized in breeding populations. Figure 3.3 describes a general process for evaluating germ plasm with an emphasis on those steps that may differ for agroforestry.

Provenance studies for plantation species have rejected some populations because of unsuitable stem morphology, but these existing collections should be reevaluated using new selection criteria to see if they include useful agroforestry materials. Unexploited genetic resources also exist in farm woodlots, which in some cases have been the focus of farmer selection for generations. For some species such as *Dalbergia sissoo*, these plantation and small farm plantings may be the most important remaining germplasm resource.

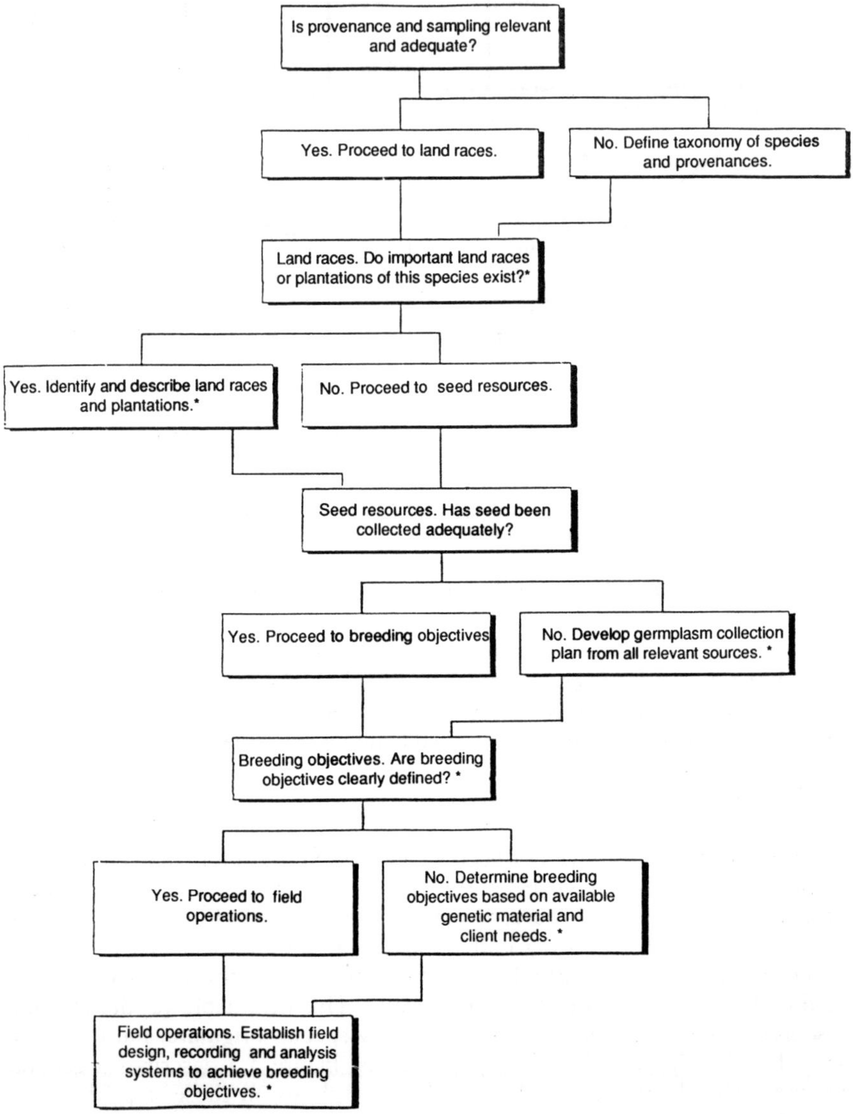

Figure 3.3. General steps required in improvement research for agroforestry.
*May be unique to agroforestry.

Increasing the number of multipurpose tree species in the first stages of selection

Substantial increases in production and value can be expected from many unimproved MPTS in just a few generations. Since the seed-to-seed cycle for most fast-growing tropical hardwood species is less than five years, significant gains of at least 20% to 30% should be expected in less than 10 years of selection. As previously noted, simple selection of several accessions of *Leucaena leucocephala* collected in the native range resulted in increases of wood yield of at least 40%. Research since the first release of the giant *Leucaena* varieties has failed to result in comparable genetic gains in either wood or fodder yield.

Similar potential exists for a wide range of MPTS for agroforestry use. The strategy of making collections and mass selections for a wide range of MPTS promises to provide not only the highest returns to labor for tree breeders but also much larger arsenal of improved species for use in agroforestry.

The rapid spread of a serious insect pest, the jumping plant lice *Heteropsylla cubana*, has proved that reliance on just one species of the genus, in this case *Leucaena leucocephala*, is economically risky. Had parallel selection and development programs been conducted for other promising species, such as *Sesbania sesban* or *Calliandra calothyrsus*, the economic impact of this insect infestation in regions such as Timor in Indonesia or the Central Visayas in the Philippines would have been greatly reduced. Early germ plasm collections of *Leucaena* by researchers in Australia, Indonesia, and the United States resulted in major advances in the selection and breeding of new *Leucaena* varieties and hybrids for resistance to the psyllid. With a few exceptions, such as the development of *Leucaena leucocephala* var. Cunningham, most of these collections had not until recent years been used for serious breeding programs. However, they have proven extremely valuable in selection and breeding for psyllid resistance. Collections of other MPTS for mass selection can also provide breeding populations that will have great value for future yield gains, environmental adaptability, and pest resistance.

Barefoot tree breeders

Farmers have long been in the business of domestication of both plants and animals. They have consciously maintained diversity, planted mixed fields to achieve natural crosses (that is, indigenous mating designs), practiced selection, and set up individual gene banks (Rhoades 1989). While much of the knowledge farmers have about conservation and use of plant germ plasm is based on crop plants and domesticated fruit tree species, it is highly likely that it extends to other favored tree species as well.

An important challenge for agroforestry tree improvement specialists is to build on their own ability to recognize and select plus trees for the farmers' economic benefit. The Chinese have successfully evolved the concept of community-based "barefoot doctors" who bring basic health care to rural populations and collect information on the use of herbal medicines. Social foresters have proposed a similar class of "barefoot foresters" who would live in local communities and work with farmers on selecting, propagating, and managing local forestry resources.

The tree breeders in this group would need a very basic knowledge of dendrology, genetics, experimental design, and seed handling, a strong dose of common sense, and access to well-trained forest geneticists. They would emphasize indigenous tree and shrub species that have been, and likely will always be, neglected by traditional forest tree improvement programs.

Cooperation between trained tree breeders and this new class of barefoot tree breeders would be essential. Table 3.1 provides an example of the division of responsibilities between traditional and community-based tree breeders. The training and encouragement of these farmer/tree breeders to select and propagate plus trees for their own benefit will greatly enhance the ability of tree improvement researchers to extend their knowledge to farmers in a productive manner.

Table 3.1 Skills required of trained tree breeders and barefoot tree breeders for agroforestry species

Trained tree breeder	Barefoot tree breeder
Study breeding systems for selected species	Identify suitable agroforestry tree species for improvement
Assist with taxonomy of selected species	Select and clone plus trees into clonal orchards
Know hand-pollination techniques	Train farmers in selection of ideotypes and seed collection
Maintain collections from a broad geographic area	Supervise early field testing of progeny from improvement programs

TREE SELECTION CRITERIA FOR AGROFORESTRY

Existing tree improvement programs generally use industrial plantation requirements to guide selection and breeding. For example, a major plantation breeding objective for *Acacia auriculiformis* might be to produce single-stemmed trees with clear boles to 16 m. In a fuelwood deficit area, farmers using *A. auriculiformis* in an agroforestry system might be more interested in trees with a low-branching habit, multiple stems, and high-coppicing ability. The selection for such characteristics is a critical step in the improvement of agroforestry species.

Choosing and improving species for agroforestry is complicated because the tree is expected to fulfill many poorly defined functions and because the desired traits for agroforestry species are seldom evaluated. The methods of securing, improving, conserving, and managing MPTS require considerable long-term effort and input. It is therefore crucial to carefully define selection criteria for appropriate species and clearly identify breeding objectives before agroforestry improvement programs begin.

Most traits for production functions such as yield are quantitative, inherited by the action of genes at many loci, each contributing minutely to the inheritance of the traits. Because many genes are involved, inheritance of these traits is immensely complex (Franklin and Stonecypher 1977). It is simpler, and therefore advisable, to work with as few traits as possible. In some cases, several traits can be combined mathematically into a composite trait of equal usefulness. For example, height and diameter can be expressed as volume; shading and wind effects can be combined as crown volume.

Some attributes are not simple characteristics, easily evaluated in terms of phenotypic value. The lack of criteria for service functions makes it impossible to use direct comparative measurement, but objective evaluation (that is, scoring) may serve instead. Indirect approaches have been proposed by Wood (1989) using the interrelationship between the tree attributes and the service and production functions that are required in actual agroforestry practices (table 3.2).

Table 3.2. Multipurpose tree characteristics for agroforestry systems

Attributes of the tree	Relationship of attribute with performance in agroforestry systems
Height	Ease of harvesting leaf, fruit, seed, branchwood; shading/wind effects
Stem form	Suitability for timber, posts, poles; shading effects
Crown size, shape and density	Quantity of leaf, mulch and fruit production; shading/wind effects
Multistemmed habit	Fuelwood and pole production, shading/wind effects
Rooting pattern (deep or shallow; spreading or geotropic)	Competitiveness with other components, particularly resource sharing with crops; suitability for soil conservation
Physical and chemical composition of leaves and pods	Fodder and mulch quality. Soil nutritional aspects
Thorniness	Suitability for barriers or alley planting
Wood quality	Acceptability for fuel and various wood products
Phenology: leaf flush, flowering and fruiting cycle: seasonality	Timing and labor demand for fruit, fodder, seed harvest; season of fodder availability; barrier function and windbreak effects
Di-/monoeciousness	Sexual composition of individual species community; important for seed production and pollen flow
Pest and disease resistance	Important regardless of function
Vigor	Biomass productivity, early establishment
Site adaptability and ecological range	Suitability for extreme sites or reclamation uses
Phenotypic or ecomorphological variability	Potential for genetic improvement; need for culling unwanted phenotypes
Response to pruning and cutting management practices	Use in alley farming, or for lopping or coppicing
Possibility of nitrogen fixation	Use in alley farming, planted fallows, rotational systems

Adapted from: von Carlowitz 1986.

No single species can grow on all sites, tolerate all types of management, or yield all types of products and services. Some species have received attention from tree breeders primarily because more is known about them, not necessarily because they are well suited to agroforestry. Many valuable species used extensively by farmers are neglected because they have escaped the attention of researchers.

Choosing a species for agroforestry requires careful review of these basic questions:

What are the intended uses of the planting, both immediate and future?
Are potentially promising species available?
What are the quality and condition of the planting sites?
Will the trees be used in combination with crop and livestock?
How will the trees most likely be managed?

Species, Provenance, and Ideotype Selection

Priority lists of potentially available tree species, either indigenous or carefully chosen exotics, have been developed in many regions of the world for specific uses and sites. The rankings of individual species are somewhat subjective and, no doubt, include personal biases that may make the list only indicative of general interest. The improvement of MPTS, which is costly as well as time consuming, calls for a process of short-listing the hundreds of known, potentially useful species to a manageable number.

Criteria and a methodology that focus on the evaluation of uses in the context of different ecological zones (von Carlowitz 1984) have been developed for rapid appraisal of MPTS. A simplified, systematic methodology for species selection was later developed and recommended by Glover and MacDicken (1987), an important part of which is an index that ranks end uses to compare each species' degree of multipurpose use (table 3.3). Such an approach is most useful if the values are weighted using local farmers' preferences (table 3.4).

Table 3.3. Multiple-use assessment of species

Species	Fuelwood	Fodder	Soil erosion	Food	Windbreak	Green manure	Total score
Gliricidia sepium	2	2	2	1	2	3	12
Erythrina poeppigiana	0	2	2	1	2	3	10
Acacia mangium	2	0	1	0	1	0	4

Comparing two fuelwood species for the humid lowlands, this example shows clearly that *Gliricidia sepium* has a much wider variety of uses than *Acacia mangium*

Source: Glover and MacDicken 1987.

Table 3.4. Weighted multiple-use assessment of species including farmer preferences

Species	Fuelwood	Fodder	Soil erosion	Food	Windbreak	Green manure	Total score
Gliricidia sepium	2 (6)	2 (4)	2 (2)	1 (0)	2 (4)	3 (9)	25
Erythrina poeppigiana	0 (0)	2 (4)	2 (2)	1 (0)	2 (4)	3 (9)	19
Acacia mangium	2 (6)	0 (0)	1 (1)	0 (0)	1 (2)	0 (0)	9
Weight co-efficient	3	2	1	0	2	3	

Using weighted values of: 3 = strongly desirable trait, 2 = desirable trait, 1 = useful, but not strongly desired, 0 = not useful. The values in parentheses are the species rating for each species × the weight coefficient.

Source: Glover and MacDicken 1987.

The example used in tables 3.3 and 3.4 shows that based only on biological characteristics, *Gliricidia sepium* has a much wider variety of uses than *Acacia mangium* but is essentially the same as *Erythrina poeppigiana*. However, when farmers' preferences are included, G. *sepium* is clearly more suitable for agroforestry use given the relative value of products needed by the end user. By understanding the relative value to the end user of products from MPTS, the breeder can more precisely select for traits to meet these expressed needs.

The value of selected seed sources can be best determined through provenance trials. The principles of species selection and provenance testing are well established, including the selection of plus trees for industrial purposes (Burley and Wood 1976, Namkoong 1979, Wright 1976).

Farmers using trees in an agroforestry situation are generally more interested in specific or multiple uses of an individual tree than in volume or biomass production (Chuntanaparb et al. 1988). They grow trees in homelots and fields for a wide range of uses (windbreak, fuelwood, shade, food, green manure, and fodder). These intended uses help define the concept of the ideal tree or ideotype for agroforestry. Ideotype has been defined as "a biological model which is expected to perform predictably, leading to greater quantities and qualities of crop yield in defined environment conditions" (Donald 1968, cited in Wood 1989). To fulfill the designed agroforestry system's objectives and to maximize land use in time and space, the associative or noncompetitive ideotypes are most desirable.

In principle, the selection of the best ideotype for agroforestry parallels the selection of plus trees for industrial tree improvement progams. The differences in practice are given in detail in the following section.

Realistic Genetic Selection Criteria

Genetic improvement for agroforestry consists of (1) locating and using the correct species, (2) using the best geographic sources within the best species, and (3) selecting and breeding the best individuals within the best sources of the selected species. The normal way to make maximum gains through seed production is to combine all three of these methods. A fourth method, when vegetative propagation is feasible, is mass-producing the very best trees to obtain maximum gain.

Selection is the process by which individuals with certain traits are favored in reproduction. It is a powerful tool in the arsenal of modern quantitative geneticists, but it can only act effectively on traits that vary due to differences in the genetic constitution of individuals. Selection cannot create variation, but rather operates on variations already in existence. If effective, it causes a change in gene frequencies that in turn alters genotypic frequencies.

If genes affecting a trait act in an additive manner, both on the allele and locus level, selection will tend to increase the frequency of the alleles producing the desired genotype. If gene action is mainly non-additive, selection can be used to choose individuals that, when crossed in specific combinations, produce offspring with desired traits. Unless selection is of the latter type, it operates on that portion of the genetic variance known as additive variance or variance of breeding values. Traits in which additive genetic variance makes up a substantial portion of the total variation will respond readily to selection (Kellison and Sprague 1971).

Many traits of forest tree species vary considerably among trees within populations. Selection will be effective in improving such traits, provided a large part of the variation is due to differences in genetic makeup and is of the additive type. Studies in a number of species have generated information on the magnitude of the additive genetic variance for important economic traits. For example, experiments for loblolly pine indicate that in this species additive genetic variance makes up a large portion of the variability in wood-specific gravity and a moderate portion of the variability in stem straightness and volume growth (Chuntanaparb 1973).

Various methods of selection can be used. The most common practice is selection on the basis of individual performance (that is, mass selection). This selection technique is the one most often used by tree breeders in the initial phases of tree breeding programs and is the simplest to undertake. It involves selection of individuals on the basis of their phenotypic values. It is frequently termed mass selection since the chosen individuals are put together en masse, in a seed orchard for example, for mating.

Individual selection is based on the principle that the average genotypic value of the selected individuals is more desirable than the average genotypic value of all individuals in the population. By mating only the selected individuals, progeny will be produced with a higher genotypic value than by mating individuals in the population at random.

In initiations of individual phenotypic selection, one of the first considerations is how much and what types of information are available upon which to base decisions. Inadequate and unreliable data in selection of criteria for plus ideotype and breeding lead to wrong decisions early in the planning stage and can cause even a highly sophisticated program to yield very little genetic gain. Frequently, research to discover this information must be undertaken during the planning and implementation phases of tree improvement programs.

According to the principles given above, individual tree selection uses selection criteria of both traditional forest and multipurpose tree species. These criteria are discussed below.

Traditional forest tree selection criteria

Selections are made by phenotypic evaluation of parent trees. The age of selected trees must fall within limits that assure maintenance of superior qualities of rotation age. Growth and form characteristics vary for each species, so the minimum number of acceptable phenotypes must be fitted to the species concerned.

In general, individuals are selected for traditional forest plantations that exhibit the fastest rates of growth while also possessing desirable form—straight boles, good crown features, good self-pruning, and lack of epicormic sprouts. Selections must be pest-free or exhibit the capacity for lower susceptibility to given insects and diseases than the average for the species. Specific gravity and fiber qualities may be determined, although it will not be possible to determine what limits to set on wood properties of a species until a fund of knowledge of wood variation and manufacturing requirements is obtained.

Selections are graded by subjective assessment of the individual tree either in uneven age natural stands or in plantations for commercially important species. In essence, phenotypic selection of an individual tree in traditional tree improvement programs generally focuses on production traits such as wood or timber and industrial uses.

Multipurpose tree species selection criteria

The selection criteria of the traditional tree improvement program are relatively simple, particularly when only a single trait is involved (for example, stem form). The process of general or specific adaptation to the site is complicated in agroforestry by the need for the size, form, phenological, and physiological characteristics of the tree component of specific agroforestry systems to be compatible with companion crops (Burley 1987b, Wood 1989).

Table 3.5 is an example of a desirable ideotype of *Gliricidia sepium* for agroforestry use in the tropics. This type of ideotype identification is recommended for the selection of breeding populations for field trials and also for the initial steps in a genetic improvement program for agroforestry.

Table 3.5. Example of an ideotype specification for *Gliricidia sepium* for agroforestry use in the humid tropics

Design Specifications

Products and service required:

In order of importance—green manure, windbreak, fodder, fuelwood, soil erosion, food

General selection criteria:

Vigor, freedom from pests and disease

Ancillary information:

Known—Nitrogen-fixing tree; tolerates poor soils, dry season, and shade; easy vegetative production; lopping and pruning possible

Required information—chemical composition (fodder value) of leaves and flowers

Ideotype Description

Crown—Preferably round, large diameters (crown/bole ratio 30/1 or more) with many branches positioned high up the stem

Stem—As straight as can be found in a population. Multistem phenotypes acceptable, but long boles important

Roots—Geotropic angled rather than horizontally extending lateral roots

Response to management—Prolific regrowth after pollarding and individual branch pruning; reliable coppicing response

Deciduousness—Low period of dry season leaflessness in comparison with the average tree of a population

Discussion

With green manure, windbreak, and fodder as major priorities, strong, wide, and dense crowns are most important. The modeling and selecting of an appropriate ideotype and subsequent phenotypic selection should concentrate on these tree attributes.

Adapted from: Wood 1989.

CONCLUSIONS

Tree improvement for agroforestry builds on the principles of traditional tree breeding, but differences in the objectives of agroforestry improvement programs require a modified approach. Such differences include the need to evaluate populations of trees for agroforestry under agroforestry conditions and the potential availability of populations already under some selection pressure for use in agroforestry. When trees are grown in conjunction with crops in intercropping systems, recording the yields of the intercrops as well as tree performance is necessary for accurate evaluation.

Several tools are available to the agroforestry tree breeder to accelerate improvement of agroforestry species, such as the modification of screening practices and exploitation of unique on-farm germ plasm resources for multipurpose tree species. The parallel selection and improvement of a number of potentially useful species promises to provide both increased genetic diversity and greater return to labor for the breeder. The training and encouragement of barefoot tree breeders is also a potentially powerful means of accelerating tree selection and improvement.

Selection criteria need to be carefully defined to meet the needs of the end user and the agroforestry system into which the trees will be placed. Appropriate ideotypes need to be identified and used to ensure that tree improvement will result in trees that meet the needs of both grower and consumer.

LITERATURE CITED

Brewbaker, J.L. 1987. Leucaena: a multipurpose tree genus for tropical agroforestry. In: *Agroforestry: a decade of development* (H. A. Steppler and P. K. R. Nair, eds.). Nairobi: ICRAF.

Brewbaker, J.L., D. L. Plucknett, and V. Gonzalez. 1972. *Varietal variation and yield trials of* Leucaena leucocephala *(koahaole) in Hawaii.* Hawaii Agriculture Experiment Station Research Bulletin 166. Hawaii.

Burley, J. 1980. Choice of tree species and possibility of genetic improvement for smallholder and community forests. *Commonwealth Forestry Review* 59(3): 311–325.

Burley, J. 1987a. Exploitation of the potential of multipurpose trees and shrubs in agroforestry. In: *Agroforestry: a decade of development* (H. A. Steppler and P. K. R. Nair, eds.). Nairobi: ICRAF.

Burley, J. 1987b. Strategies for genetic improvement of agroforestry trees. In: *Agroforestry for rural needs* (R.K. Khosla and D. K. Khurana, eds.). Solan, Himachal Pradesh, India: Indian Society of Tree Scientists.

Burley, J., and P. von Carlowitz (eds.). 1984. Multipurpose tree germplasm. *Proceedings of a planning workshop on international cooperation*. Nairobi: ICRAF. 298 p.

Burley, J., and G. Nikles (eds.). 1973. *Tropical provenance and progeny research and international cooperation.* Proceedings of IUFRO sysmposium, Nairobi, Kenya. Commonwealth Forestry Institute, Oxford University, UK. 597 p.

Burley, J., and P. J. Wood. 1976. *A manual on species and provenance research with special reference to the tropics.* Tropical Forestry Paper 10. Commonwealth Forestry Institute, Oxford University, UK. 263 p.

Chuntanaparb, L. 1973. *Inheritance of wood and growth characteristics and their relationships in loblolly pine (Pinus taeda* L.). Ph D Thesis, North Carolina State University, Raleigh, NC. 123 p.

Chuntanaparb, L. 1975. Planning breeding programme for tropical hardwood. *Natural History Bulletin Siam Society* 26: 179–197.

Chuntanaparb, L., W. Hoamuangkaew, and P. Sri-Aran. 1988. *Agroforestry in Dong Mun forest communities, Northeast Thailand.* Northeast Thailand Upland Social Forestry Project, Working Document 4. Faculty of Forestry, Kasetsart University, Bangkok. 81 p.

Cooling, E.N. 1962. *Procedures for the trial of exotic species in northern Rhodesia*. Lusaka, Rhodesia: Government Printer.

Donald, C. M. 1968. The breeding of crop ideotypes. *Euphytica* 17.

FAO. 1978. *China: forestry support for agriculture*. Forestry Paper 12. Rome: FAO.

FAO. 1981. *Genetic resources of tree species in arid and semi-arid areas. A survey for improvement of rural living in Latin America, Africa, India and Southeast Asia*. Rome: FAO/IBPGR.

FAO. 1986. *Tree improvement/seed improvement of multipurpose species*. GCP/RAS/IOB/JPN, Field Document 9. Bangkok: FAO/RAPA. 53 p.

Franklin, E.C., and R.W. Stonecypher. 1977. *Use of quantitative genetics in planning multi-trait breeding*. Third World consultation on Forest Tree Breeding, Canberra. FO-FTB-77-6/3.

Glover, N., and K.G. MacDicken. 1987. Soil and climatic requirements of NFT species. In: *Nitrogen fixing trees—a training guide*. FAO/RAPA 1987/15. pp 53–70.

Hu, T.W., T. Kiang, and W.C. Shih. 1980. The growth of planted *Leucaena leucocephala* in Taiwan. *Leucaena Newsletter* 1: 29–30.

Instituto de Ciencia Animal. 1979. *Los Pastos en Cuba Tomo 1: Produccion*. La Habana, Cuba: ICA.

Iyamabo, D.E. 1969. *Species introduction and growth in African savanna*. Second World Consultation of Forest Tree Breeding. Washington, DC. FO-FTB-69-2/6.

Kellison, R.C., and J. Sprague. 1971. *The practice of selection in forest tree breeding*. Tree Improvement (Short Course). School of Forest Resources, North Carolina State University, Raleigh, NC.

Mendoza, R.C., and E.Q. Javier. 1980. Wood yield potential of five ipil-ipil (*Leucaena leucocephala*) cultivars as affected by different cutting ages and population densities. *Leucaena Newsletter* 1: 27.

Namkoong, G. 1979. *Introduction to quantitative genetics in forestry*. USDA Forest Service Bulletin 1588. 342 p.

Namkoong, G., R.D. Barnes, and J. Burley. 1980. A philosophy of breeding strategy for tropical forest trees. *Tropical Forestry Paper 16*. Commonwealth Forestry Institute, Oxford University, UK. 67 p.

Oakes, A.J., and O. Skov. 1967. Yield trials of leucaena in the U. S. Virgin Islands. *Journal of Agriculture of the University of Puerto Rico* 51: 176–181.

Rhoades, R. 1989. The role of farmers in the creation of agricultural technology. In: *Farmer first* (R. Chambers, A. Pacey, and L.A. Thrupp, eds.). London: Intermediate Technology Publications. 218 p.

Shih, W.C., and T.W. Hu. 1981. The yields of forage of *Leucaena leucocephala* in Taiwan. *Leucaena Research Reports* 2: 55–56.

Turnbull, J. 1984. Tree seed supply: a critical factor for the success of agroforestry projects. In: *Multipurpose tree germplasm* (J. Burley and P. von Carlowitz, eds.). Nairobi: ICRAF. 298 p.

van Buijtenen, J.P., and W.W. Saitta. 1972. Linear programming applied to the economic analysis of forest tree improvement. *Journal of Forestry* 70(3): 164–167.

von Carlowitz, P. 1984. Rapid appraisal methodology for selecting priority multipurpose tree species and criteria for determining status and research needs. In: *Multipurpose tree germplasm* (J. Burley and P. von Carlowitz, eds.). Nairobi: ICRAF. 298 p.

Wood, P.J. 1989. Species selection for agroforestry. In: *Agroforestry: classification and management* (K.G. MacDicken and N.T. Vergara, eds.). New York: John Wiley and Sons.

Wright, J.W. 1976. *Introduction to forest genetics*. New York: Academic Press. 463 p.

Zhu, Z.H. 1981. China's great green wall. *American Forests* 87(5): 24–25,58.

Zhu, Z.H. 1988. A new farming system—crop/paulownia intercropping. In: *Multipurpose tree species for small-farm use* (D. Withington, K.G. MacDicken, C.B. Sastry, and N. Adams, eds.). Ottawa: Winrock International and IDRC. 282 p.

Zobel, B.J., and J.T. Talbert. 1984. *Applied forest tree improvement*. New York: John Wiley and Sons. 505 p.

CHAPTER 4

Plant Management in Agroforestry

M.G.R. Cannell

Any research plan needs to begin with a clear statement of objectives. The management of the experiments, the treatments, design, and measurements all follow from these objectives. There must be clear general objectives—such as to improve the well-being of poor people in defined social and biophysical circumstances—and clear specific objectives—such as to sustain yield at a certain level.

In many agroforestry experiments, it is tacitly assumed that the objective is to increase yield, and yield is often the only output measured. In fact, there are usually other, often more important objectives that may prohibit realizing maximum yields. Conserving the soil and its fertility may necessitate having a minimum number of N_2-fixing trees, perhaps deeply rooted ones, and a continuous ground cover. Minimizing the risk of food, fodder, or fuelwood scarcity may require growing species that are low yielding, or that are stress tolerant, or that yield in the dry season. Controlling weeds and easing the workload may depend on growing crops that cover the ground quickly, staggering planting and harvesting, growing low-yielding but problem-free crops, minimizing transport distances, and providing shade for people and animals. Finally, controlling pests and pathogens may require planting mixtures of crops that differ in height, smell, color, or susceptibility, or that provide shade to control such problems as banana leaf spot or cacao capsid damage. And to these rational, non-yield objectives may be added the "irrational" ones of tradition, culture, religion, and division of labor. Clearly, when defining the objectives of a research program, much needs to be known about the context in which the research will be done.

Having set the objectives, measurements need to be made that are capable of evaluating whether the objectives were met. If the plants are managed to yield biomass, then this can be easily measured, but the important factor may be the nutritive value of that biomass to people or animals, the time of year that it is produced, how much effort it took to produce it, or its cash value. It is rarely sufficient simply to weigh the product.

MANIPULATING TREES

Why is mainpulating trees important? The trees in agroforestry systems usually demand, or benefit from, individual management because of their persistence, size, and dominance, because they may become too large or unfruitful if unpruned, and because various products may be harvested from them over many years. The principles and practices of individual tree management depend on whether the product is a vegetative part—wood, leaves, pith, bark, or resin—or a reproductive part—fruits or seeds.

Manipulation for Vegetative Yield

Pruning is the principal method for managing the vegetative yield of trees. Its practices were divided by Huxley (1985a) into heading back and thinning out operations, and by Cannell (1983) into spreading, lopping, and branch pruning. Roots may also be pruned. Pruning of any sort will alter tree shape, total dry matter production, and the distribution of growth within the tree.

Following pruning, distal buds near the cut ends are stimulated to grow more than basal buds, and more shoots grow than would normally do so. Pruning near ground level produces longer shoots than pruning higher up, and the new shoots are often less periodic in growth, have fewer short shoots, and have greater apical dominance than shoots on unpruned trees. In positions where branches tend to grow horizontally, new shoots tend to grow vertically, and the highest, upwardly directed shoots usually attain dominance (Maggs and Alexander 1967, Wareing 1968 and 1977, Hallé 1978).

The potential dry matter production by any vegetation is closely related to the amount of solar radiation intercepted by the foliage (Monteith 1977, Charles-Edwards 1982, Cannell et al. 1988). Any pruning that removes foliage and decreases light interception will reduce the rate of dry matter production by the trees. However, the decrease in productivity may be less than expected (1) because the rate of photosynthesis of the remaining foliage may increase due to the increased sink/source ratios, better illumination, and increased share of root-originated metabolites and (2) because the leaves removed may be the older, more shaded ones, which have lower rates of photosynthesis than young leaves.

Pruning also alters the way new growth is distributed. By manipulating the height and branchiness of trees, the proportion of dry matter allocated to leafy shoots as opposed to large-diametered fuelwood or timber can be altered. Also, shoot pruning always temporarily checks root growth in proportion to the amount of foliage removed and for as long as it takes the tree to restore its original root-shoot functional balance (Alexander and Maggs 1971).

Although root pruning checks shoot growth, it can be beneficial when it enhances flower bud production in fruit trees or stimulates the regrowth of a more fibrous root system that may have greater access to soil phosphates.

How and when to prune

Research questions on pruning involve how to prune (for example, whether to coppice, pollard, or basal prune), how much to remove (what coppicing height, how severely to pollard), and when and how often to prune. These questions need to be decided with reference to the tree product required (green mulch, fodder, fuelwood, or timber) and the tradeoff between the productivity of the trees and of the crops beneath.

On the question of how to prune, hedgerows of *Leucaena leucocephala* can be coppiced severely for green mulch, less severely coppiced for fodder, or only branch pruned to produce poles. Trees in silvopastoral systems can be lopped or pollarded: *Prosopis cineraria* can yield more fodder and less fuelwood when completely lopped than when only partly lopped (Tejwani 1979).

On the question of how much to remove, it must be remembered that severe pruning can kill trees if followed by drought and that recovery will depend upon the amount of foliage and storage reserves (of carbohydrates and minerals) left on the trees. Pruned tea bushes are often left with a foliage-bearing "lung" branch. Coppicing small trees is often more life threatening than coppicing large trees with thick roots and storage reserves; pollarding is usually less life threatening than coppicing.

Several workers in Asia who examined the effect of cutting height on the amount of fodder and green mulch produced by *Leucaena* had conflicting results, probably because of differences in climate and soil. Pathak and colleagues (1980) found that the best cutting height was 30 cm (over the range 10 to 30 cm cutting every 40 to 120 days), while Krishna Murthy and Mune Gowda (1982) obtained the greatest fodder yield by cutting at 150 cm (over the range 15 to 150 cm, cutting every 40 to 70 days). Variable results have been obtained in other parts of the world (Robinson 1985). When *Leucaena* is grown in alleys in the semiarid tropics, the cutting height needs to be low (10 to 30 cm) to minimize competition with crops, whereas blocks of pure *Leucaena* might give more fodder if cut at 75 to 150 cm (Singh 1987). Also, the cutting height may be less critical as the interval between harvests increases. Clearly, few rules can be given without specifying the conditions and management objectives. If the objective is to provide off-season fodder, then the cutting regime is also constrained by the time of cutting.

On the question of the frequency of cutting, at least two good experiments have been done on lopping *P. cineraria* in India. Bhimaya and colleagues (1964) lopped trees growing near Jodhpur, Rajasthan (approximately 380 mm rainfall), either annually or once every four years, and Srivastava (1978) lopped trees growing near Mohindergarh, Haryana (approximately 450 mm rainfall), annually or once every two or three years. In both cases, annual lopping gave the greatest yield on a yearly basis, especially from the large, mature trees (figure 4.1). Traditionally, lopping of *P. cineraria* starts when the trees are about ten years old, is done in winter (leaf emergence is in summer), and is often complete, except for a few branches at the top (Bohra and Ghosh 1980). The trees evidently have a remarkable capacity to recover, and complete lopping annually yields more than partial or infrequent lopping (Bhimaya et al. 1964).

Hypothesis

The key question is how quickly new foliage is produced following pruning. This rate will depend upon the photosynthetic surface left (the leaf area), the amount of storage carbohydrate left (a function of the woody volume), and the supply of water, nutrients, and light during the recovery period. A testable hypothesis is:

> The rate of recovery from pruning increases as the area of foliage and woody volume left after pruning increases.

An experiment to test this hypothesis would use individual tree plots, would estimate the amount of foliage and stem + branch volume after pruning at

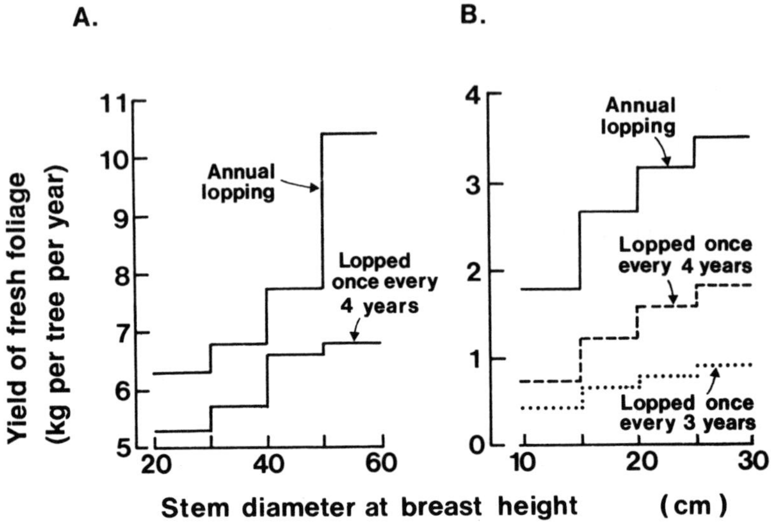

Figure 4.1. Mean annual fresh weights of leafy fodder harvested from trees of *Prosopis cineraria* in two studies in India with differet harvest intervals:
(a) Jodhpur, Rajasthan, derived from Bhimaya et al. 1964.
(b) Mohindergarh, Haryana, derived from Srivastava 1978.
Source: Robinson 1985.

different intensities, and would measure the amount of new growth after a fixed period. Clearly, there are numerous possible variations on this theme.

Manipulation for Reproductive Yield

The principles of managing trees for reproductive yield are contained in the science of the horticulture of fruit and nut-bearing trees. Management is needed, on the one hand, to encourage the trees to bear fruit, and on the other, to limit the fruit load in order to maintain a satisfactory balance between fruit and vegetative growth, thereby safeguarding future yield.

Cannell (1983) identified five specific problems requiring management. First, the juvenile period may need to be shortened by promoting early growth (by such measures as applying fertilizers or removing competition), by grafting onto flower-promoting rootstocks, or if all else fails, by partial girdling, root pruning, or bending the stems horizontally. Second, the trees may need to be pruned or handled so that high light levels are received during the periods of floral initiation at the points in the trees where floral initials form. The number of floral initials produced within the canopies of citrus, cashew, grape vine, oil palm, and coconut have all been related to some measure of solar radiation receipt. Trees that fruit on young shoots on the outside of the canopy, like mango, may need less pruning than trees like coffee that fruit on old wood, and some species may need to be well illuminated only at particular times. Third, to ensure cross-pollination, such measures may need to be taken as growing trees close together, interspersing pollinator trees, keeping bees, and timely, selective pruning to phase flowering with favorable weather conditions. The flowers of many fruit trees are receptive to pollen for only two to four days, and there must be sufficient pollen, vectors, and favorable conditions for pollen transfer. Fourth, a good fruit set does not ensure a good harvest: a large percentage of the fruit can be shed before it matures. At one stage or more (often during the period of rapid fruit swelling), young fruit seem to be weak sinks for mineral nutrients and carbohydrates compared with the shoot tips, especially when there are simultaneous leaf flushes or droughts. Preventive measures for fruit drop are alleviating water stress, applying fertilizers, removing shade, pruning to synchronize the sensitive stages of fruit growth with favorable climatic conditions, or removing shoot tips at the stage when fruit shedding is anticipated. Fifth, trees often bear biennially or irregularly. This is a serious problem in some cultivars of apple, citrus, and mango, and it is characteristic of nut tree crops. It is especially common in trees with prolonged periods of fruit development where fruiting overlaps with both vegetative growth and floral initiation for the next year. The best solution is to select regular bearing cultivars, or to prune so that a proportion of the trees are in their on and off years. To these five problems could be added many more concerning viruses, nutrition, replant problems, pests, and pathogens.

Case studies

Although fruit trees can be problem prone, there is enormous scope for rapid yield improvement by clonal selection within the large pools of genetic variation. Most species are readily propagated by cuttings or grafting, parthenocarpic genotypes can be selected, and a variety of pruning and training techniques can be employed. The potential for improvement is shown by the recent history of the kiwi fruit in New Zealand, pecan nuts in the United States, and macadamias in Hawaii. In India, the Central Arid Zone Research Institute has selected varieties of the desert apple, *Zizyphus mauritiana*, that are now popular in regions where rainfall is 250 to 450 mm because they can be propagated by budding onto the rootstocks of other *Zizyphus* species. A budded plant can start fruiting in 18 months and become fully bearing (producing over 50 kg fruit per year) in about five years (Shankarnarayan et al. 1987). Similar work has been done on most of the better known tropical fruit trees such as the mango, guava, avocado, and arecanut.

MANIPULATING TREE/CROP MIXTURES

The performance of trees and crops in a mixture will depend upon their relative ability to tap the resource pools of light, water, and mineral nutrients and their responses to suboptimal levels of these resources (Connor 1983). If one component captures more light and water, how does the other component respond to shade and water stress? In order to achieve desired objectives, the mixtures can be arranged differently in the vertical, horizontal (spatial), and temporal dimensions.

The Vertical Dimension

If water and mineral nutrients are plentiful and light is the main environmental resource limiting plant growth, then the dominant crop will be the one that grows tallest. One of the most effective ways of increasing total yield is to plant crops so that each becomes tallest in turn, or so that a multilayered canopy is established in which most of the light is intercepted by product-yielding plants all the time. Crops can be mixed that attain similar heights at different times; crops can be planted at different times; and understory crops can be given extra light at certain times of the year by depending on the leaf fall of deciduous trees or by tree pruning. If we regard the soil as part of the vertical dimension, then we can include here the full exploitation of nutrients and water by deep-rooting trees mixed with shallow-rooting crops.

Practices and case studies in Asia

The vertical dimension is most important in high-rainfall areas with fertile soils, such as in Kerala, India. Research at the Central Plantation Crops Research Institute, Kasaragod, Kerala, has focused on multistrata cropping, canopy shape and size, and rooting habits in order to increase total light interception and soil exploitation (Bavappa 1982, Ghosh et al. 1987). Such research is particularly important because at conventional spacings many plantation crops intercept only a fraction (often less than 50%) of the light, and some, such as coconut, have a restricted rooting zone. At 7.5 m × 7.5 m spacings, the roots of coconut in Kerala spread over only 23% of the area (Nair 1983). Nair and colleagues (1975) showed that cacao could be mixed with coconut with essentially no negative competitive interaction. Similarly, oil palm, rubber, and coffee plantations can be interplanted with a wide range of semiperennial and annual crops, including pepper, yams, banana, pineapple, arrowroot, and ginger. The home gardens and village forest gardens of Indonesia can contain over 100 species, and be arranged in five canopy layers (Michon 1983). Experiments in Sri Lanka and Kerala have been conducted with three canopy layers and 10 to 15 species, including crops such as jackfruit, breadfruit, avocado, mango, coconut, nutmeg, clove, papaya, arecanut, lime, banana, pepper, coffee, and a range of understory crops (Bavappa 1986).

Hypotheses

When designing experiments with multistoried canopies in fertile, high-rainfall areas, the following hypotheses might be addressed:

1) Total dry matter (biomass) production is closely related to the amount of solar radiation intercepted. In western Europe, both agricultural crops and broad-leaved trees produce about 1.5 g of dry matter per megajoule of intercepted total solar radiation (about 3.0 g/MJ of photosynthetically active radiation; Monteith 1977, Cannell et al. 1988). Conversion efficiencies may be lower in tropical regions because of higher respiration rates, but the principle is equally valid.

2) There is a maximum total leaf area index (LAI) for any site; the more fertile and less droughty the site, the greater the maximum LAI. Any increase in LAI of the overstory implies a corresponding decrease in LAI of the understory (figure 4.2).

3) Species differ in their response to shade, that is, in the degree to which they can modify their morphology and physiology to make use of low-light flux densities. Species with C_3 photosynthesis are likely to perform better in shade than species with C_4 photosynthesis.

4) Economic yield will depend upon the ability of shaded plants to partition a high proportion of their dry matter to the harvested parts. In general,

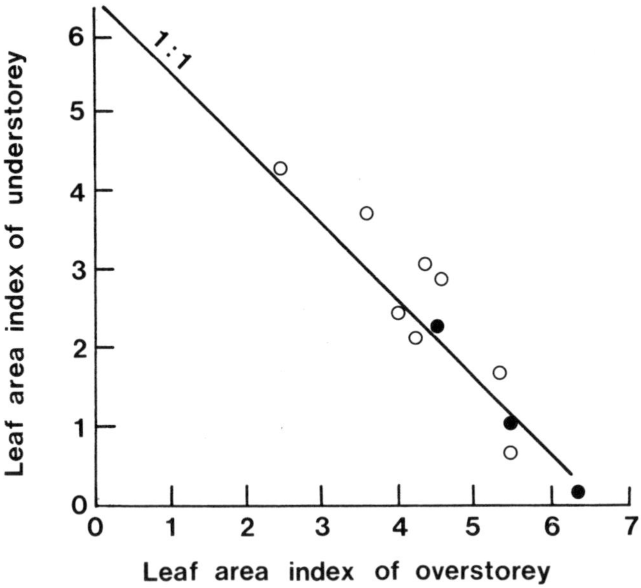

Figure 4.2. Relationship between the leaf area index of overstory trees and understory vegetation for a range of woodlands in Japan.
The line marks the 1:1 relationship.
Source: Satoo and Madgwick 1982. 0 = *Betula* spp. overstory; 0 = *Fagus crenata* overstory.

crops with vegetative yield, like yams and potatoes, have higher harvest indices in shade than crops that yield seeds or fruit (Cannell 1983).

5) The social benefits of the assemblage of crops will depend upon the timing of yield and labor demand.

The Horizontal Dimension

After it has been decided which crops are to be grown together, the decision about how to plant them must be made—should they be grown in blocks as separate monocultures, or should they be grown in intimate mixtures? These choices are especially important if nutrients and water are limiting; they become vital if any allelopathic toxins are produced by any of the crops or if particular crops require shady conditions.

The spatial arrangement of crops will determine the length of boundary or interface between them. The minimum areas of interface will occur when each crop is grown separately in blocks or zones. This arrangement is obviously best if the crops are mutually competitive so that their combined yield is depressed along each interface. The maximum areas of interface will occur when crops are grown in intimate mixtures. This arrangement is best if the crops are mutually noncompetitive or benefit each other. Intermediate degrees of mixing are given by row or strip planting.

In agroforestry, the particular interfaces that need to be investigated are those between trees and crops. In theory, the trees and crops can be mutually depressive or mutually promotive, or one of the two components can be depressed or promoted.

When the competition along a tree/crop interface has been determined (see below), it is possible to alter not only total yield, but also the balance of yield obtained from the two components by changing the spatial arrangement. The greater the area of interface, the more opportunity there will be for the dominant component to benefit and the weaker component to be suppressed. Consequently, the proportions of yield taken from the different components will diverge more and more from the proportions expected on the basis of the land area allocated to it (Huxley and Maingu 1978). As the area of interface increases, more land must be devoted to the weaker component in order to obtain the expected proportions of crop and tree yield.

Examining tree/crop interfaces

Huxley (1985b) considers experimental designs for studying tree/crop interfaces and points out that they can be studied wherever they occur simply by examining crops beneath trees. Alternatively, tree/crop interface experiments can be conducted by planting crops under existing trees or by planting trees within crops. All zonal or alley-cropping experiments automatically provide interfaces that should be examined. The observations required are basically transects of tree and crop yields across the interface, preferably with some measurements of the environmental conditions (light, temperature, and vapor pressure deficit).

The outcome of competition along tree/crop interfaces can be anticipated from current knowledge. Where forage legumes or grasses are grown beneath trees in fertile soils, there may be little competitive interaction, but where deep-rooted perennial grasses are grown beneath trees in infertile soils, the growth of both components may be depressed (figure 4.3a). On the other hand, the grain yield of sorghum or millet can be as great or greater beneath *Acacia albida* as in the open owing to improved nutrition, and the yield of pods from A *albida* may be greater on edge trees, irrespective of the presence of crops beneath (figure 4.3b).

Alley-cropping experiments have been widely planted in India and elsewhere using different alley widths. These experiments can be used to examine interfaces by taking transects or by measuring the yield of each row of crop and tree from the sole crop areas to the interfaces. This approach may be more cost-effective than trying to establish land equivalent ratios or relationships between alley width and crop yield as a percentage of the sole crop yield. The crucial information is whether yield is depressed along the interface. Then comparisons between crops can be made—is sorghum, for instance, more compatible with *Leucaena* than groundnuts?

Hypothesis

The null hypothesis for any examination of a tree/crop interface is that the yield of the trees and of the crop is the same at the interface as in the sole crop areas some distance from the interface. The expectation, however, is that the yield of the trees or crop will be depressed at the interface owing to competition for environmental resources. All interface studies should, therefore, include measurements of light, water, and nutrients in order to explain the outcome of this competition. Only then will it be possible to generalize the result to other situations. Where yield is depressed along the interface, it must be determined

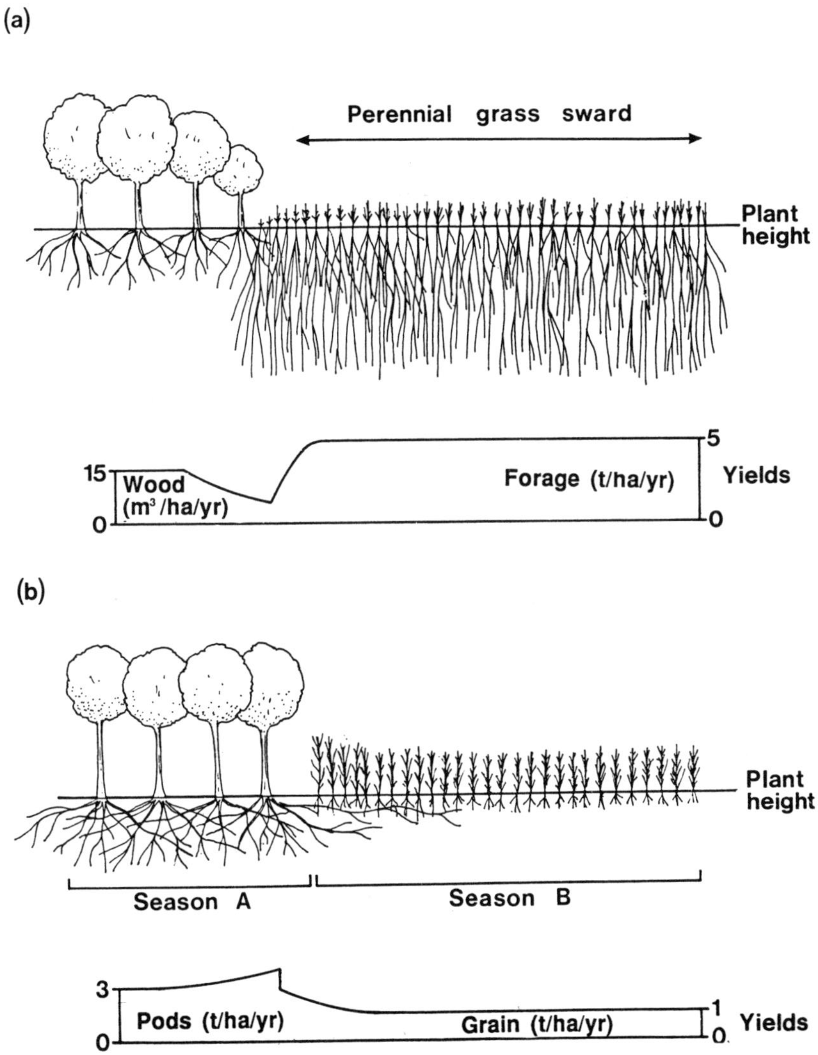

Figure 4.3. (a) Section across a tree/crop interface in which the growth of the tree species is inhibited by a deep-rooted perennial grass.
(b) Section across an *Acacia albida*/grain crop interface, in which the yield of both components is enhanced along the interface.
Source: Huxley 1985b.

whether this is due to shading or to below-ground interference for water or nutrients. Where crop yields benefit from the proximity of trees, it must be determined whether this is because of their response to shade, lower temperatures, higher humidity, nutrient inputs from the trees, or simply, erosion control.

The Time Dimension

If no one crop fully utilizes environmental resources (light, water, and nutrients) throughout the year, then yields will be increased by skillfully phasing the growth cycles of different crops. Skill is required (1) in choosing the species and cultivars according to their phenology, stature, habit, product, and other characteristics and (2) in relay planting or in staggering planting or sowing. If successful, this crop phasing will help ensure that soil water resources are fully utilized, the soil is protected from rain, weeds are suppressed, erosion is prevented, shade-intolerant crops (including most of those that produce seeds and fruit) are not overshadowed after the onset of floral initiation, and so on (Andrews 1974, Baker 1979, Fisher 1979). Subsistence farmers are, in fact, very skilled in these matters, and much could probably be learned from them (Bunting 1980, Okigbo 1980).

A number of variables need to be considered when choosing species that are phenologically compatible, including the length of the growing season, the time of sowing in relation to rainfall, and the natural phenophases and/or potential management treatments such as lopping of trees. Some tree species, such as A. *albida*, produce leaves prior to the onset of rains and shed their leaves during the early part of the rainy season. This behavior enables millet and groundnuts, for example, to be grown beneath the trees where they will be relatively unshaded and derive benefit from soil enrichment. There is, in this case, an almost ideal temporal separation in the phenology of the two component crops.

When studying the sequence of planting and cropping, two types of chronology may be considered—sequence and succession (Huxley 1983). *Sequence* is the time course of events among crops (including trees) utilizing the same unit of land. Cropping can be described as coincident, concomitant, overlapping, or interpolated, as shown in figure 4.4a. *Succession* refers to cycles of land occupancy. A single crop species or a sequence of different species may occupy the same unit of land either intermittently or continuously, and a second crop can be interpolated, grown concomitantly, or in an overlapping sequence, as shown in figure 4.4b.

So far we have considered chronological time. For plants, we should also consider thermal time and water time. If a unit of plant growth or development is equivalent to one day degree above a base temperature, then the time period to reach a given developmental stage will be defined by thermal time—a given number of day degrees or heat units. If a unit of plant growth is equivalent to the transpiration of a unit of water, then the time period to reach a given stage can be defined in terms of water use. In cool regions, it is important to grow crops that fully utilize all available thermal time; in semiarid regions, it is important to grow crops that fully utilize all available water time.

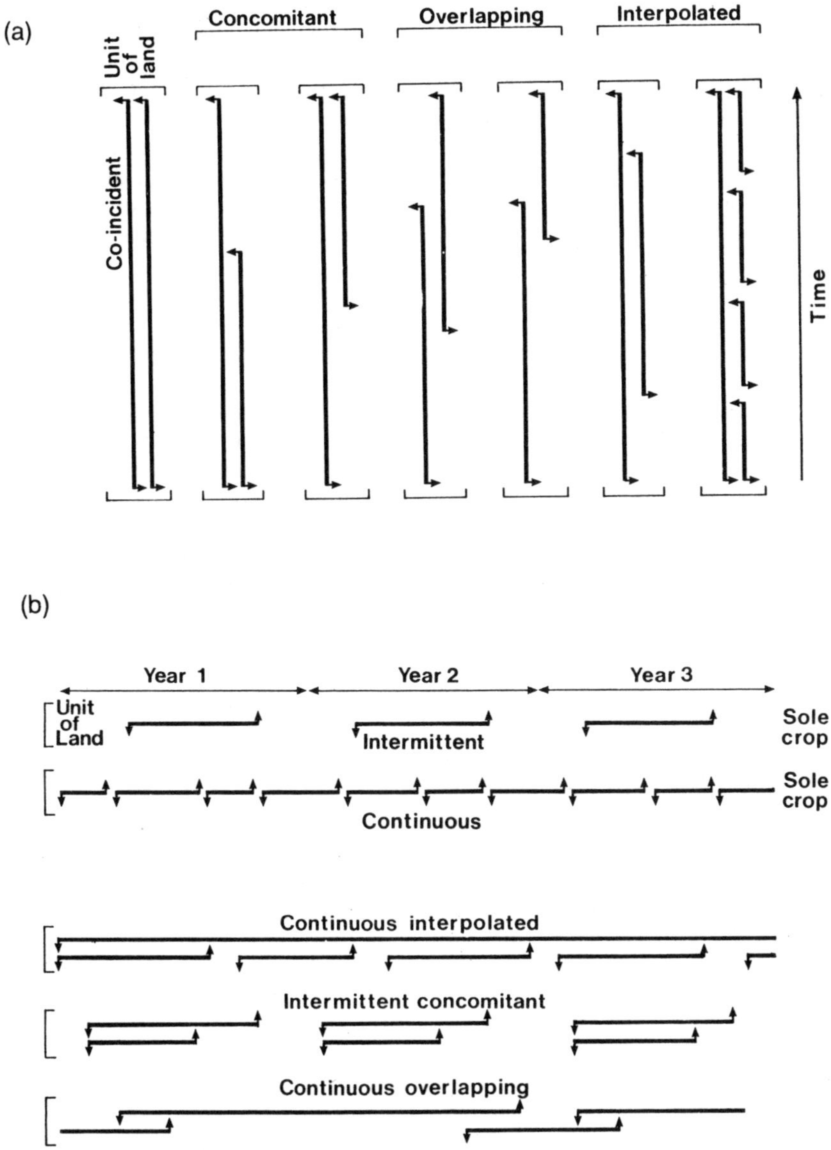

Figure 4.4. (a) Cropping sequences.
(b) Cropping succession. Downward arrows denote times of sowing or planting; upward arrows denote times of harvesting or clearing.
Source: Huxley 1983

Practices and case studies

All agroforestry practices and studies must consider the time dimension. If trees are grown to ameliorate soil conditions in rotational plots (tree fallows)—in sequence with field crops—then explicit consideration must be given to the length of the tree rotation relative to the crop rotation. The tree rotation must be managed so that the combined benefits of the improved soil conditions (that subsequently enhance crop yield) plus any tree products (fruit, fuel, timber) are greater than the crop yield foregone by growing the trees and the cost of planting and removing the trees. Critical parameters will be the rate of soil amelioration during the tree fallow period, and the rate of soil degradation under crop monoculture. There is surprisingly little information on these parameters, even for age-old traditional systems such as *Acacia* species fallows in the Sahel.

When trees are grown in mixtures with short-statured crops, consideration has to be given to the management of both components as the trees get larger and older and begin to bear fruit and seeds. In the simple case of *taungya*, the timber trees may be virtually unmanaged and crops need to be chosen that can withstand or avoid shade, or that are phenologically compatible with deciduous trees. The best companion crops are often forage crops or vegetative crops such as cassava, climbers such as yams and peppers, or tall reproductive crops such as bananas. The point is that the future development of the trees has to be anticipated when choosing the understory crop and the time of planting.

Trees grown for fodder, fuelwood, and fruit can be lopped or pruned, as discussed. Trees may be multipurpose or multimanagement: some trees may be pruned for fruit, others lopped for fodder, still others allowed to grow tall for shade and soil amelioration. Trees grown for different purposes might be planted on different patches of soil.

The timing of coppicing and lopping of *Leucaena*, for instance, must coincide with the interplanting of field crops, as well as with the period when fodder and fuel are needed. Sowing field crops depends on when future rainfall is anticipated and on day length and temperature conditions. There is considerable information on the optimum time for sowing rainfed crops (see Rijks 1967 for cotton) and on the thermal time to maturity (see Carr 1977 for maize). Future conditions influence not only the period to harvest but also the size of the harvest, which in turn usually depends on the completion of critical stages of development, such as tuber initiation or anthesis.

CONCLUSIONS

The purpose of this chapter has been to raise some of the main issues involved in managing trees and crops in agroforestry systems. The options are numerous; indeed, one of the advantages of agroforestry is the scope given to farmers to maximize yields and other benefits. Having defined the objectives, the trees can be managed as individuals, and crops that complement the trees in the vertical, spatial, and time dimensions can be planted with them. The failure and success of given mixtures and practices can perhaps be analyzed and best understood with reference to these three dimensions in which plants interact.

LITERATURE CITED

Alexander, D.M., and D. H. Maggs. 1971. Growth responses of sweet orange seedlings to shoot and root pruning. *Annals of Botany* 35: 109–115.

Andrews, D. J. 1974. Responses of sorghum varieties to intercropping. *Experimental Agriculture* 10: 57–63.

Baker, E.F. 1979. I. Mixed cropping in northern Nigeria. II. Cereals and cottons. III. Mixtures of cereals. *Exploratory Agriculture* 15: 33–40, 41–48.

Bavappa, K.V.A. 1982. High intensity multispecies cropping—a new approach to small scale farming in the tropics. *World Crops* March/April 1982, pp. 47–50.

Bavappa, K.V. A. 1986. *Mixed multi-storeyed cropping*. Paper presented at ICAR-ICRAF workshop on agroforestry, September 1986, CRIDA, Hyderabad. Nairobi: ICRAF.

Bhimaya, C.P., R.N. Kaul, and B. N. Ganguli. 1964. Studies on lopping intensities of *Prosopis spicigera*. *Indian Forester* 90: 19–23.

Bohra, H.C., and P.K. Ghosh. 1980. The nutritive value of loong (*P. cineraria*) leaves. In: *Khejri* (Prosopis cineraria) *in the Indian desert—its role in agroforestry* (H. S. Mann et al., eds.). Jodhpur, India: CAZRI. pp 45–50.

Bunting, A.H. 1980. The future of research on mixed cropping in tropical agriculture. In: *Opportunities for increasing crop yields* (R. G. Hurd, P. V. Biscoe, and C. Dennis, eds.). Boston: Pitman. pp 247–252.

Cannell, M.G.R. 1983. Plant management in agroforestry: manipulation of trees, population densities and mixtures of trees and herbaceous crops. In: *Plant research and agroforestry* (P. A. Huxley, ed.). Nairobi: ICRAF. pp 455–487.

Cannell, M.G.R., L. J. Sheppard, and R. Milne. 1988. Light use efficiency and woody biomass production of poplar and willow. *Forestry* 61: 125–136.

Carr, M.K.V. 1977. The influence of temperature on the development and yield of maize in Britain. *Annals of Applied Biology* 87: 261–266.

Charles-Edwards, D. A. 1982. *Physiological determinants of crop growth*. London: Academic Press.

Connor, D.J. 1983. Plant stress factors and their influence on production of agroforestry plant associations. In: *Plant research and agroforestry* (P. A. Huxley, ed.). Nairobi: ICRAF. pp 401–426.

Fisher, N.M. 1979. Studies in mixed cropping. III. Further results with maize-bean mixtures. *Experimental Agriculture* 15: 49–58.

Ghosh, S.P., G.M. Nair, N.G. Pillai, T. Ramanujam, B. Mohankumar, and K. R. Lakshmi. 1987. Growth, productivity and nutrient uptake by cassava in association with four perennial species. *Tropical Agriculture* 64: 233–235.

Hallé, F., R. A. A. Oldeman, and P. B. Tomlinson. 1978. *Tropical trees and forests, an architectural analysis*. New York: Springer-Verlag.

Huxley, P. A. 1983. Comments on agroforestry classifications: with special reference to plant species. In: *Plant research and agroforestry* (P. A. Huxley, ed.). Nairobi: ICRAF. pp 161–171.

Huxley, P. A. 1985a. The basis of selection, management and evaluation of multipurpose trees—an overview. In: *Trees as crop plants* (M.G.R. Cannell and J.E. Jackson, eds.). Monks Wood, Huntingdon, England: Institute of Terrestrial Ecology. pp 13–35.

Huxley, P. A. 1985b. The tree/crop interface—or simplifying the biological/ environmental study of mixed cropping agroforestry systems. *Agroforestry Systems* 3: 251–266.

Huxley, P. A., and A. Maingu. 1978. Use of a systematic spacing design as an aid to the study of intercropping: some general considerations. *Experimental Agriculture* 14: 49–56.

Krishna Murthy, K., and M. K. Mune Gowda. 1982. Effect of cutting and frequency regimes on the herbage yield of *Leucaena*. Leucaena *Research Reports* 3: 31–32.

Maggs, D. H., and D. M. Alexander. 1967. A topophysic relation between regrowth and pruning in *Eucalyptus cladocalyx* F. Muell. *Australian Journal of Botany* 15: 1–9.

Michon, G. 1983. Village-forest-gardens in West Java. In: *Plant research and agroforestry* (P. A. Huxley, ed.). Nairobi: ICRAF. pp 13–24.

Monteith, J. L. 1977. Climate and efficiency of crop production in Britain. *Philosophical Transactions of the Royal Society of London*/Series B 281: 277–294.

Nair, P. K. R. 1983. Agroforestry with coconuts and other tropical plantation crops. In: *Plant research and agroforestry* (P. A. Huxley, ed.). Nairobi: ICRAF pp 79-102.

Nair, P. K. R., Rama Varma, E. V. Nelliat, and K.V.A. Bavappa. 1975. Beneficial effects of crop combination of coconut and cacao. *Indian Journal of Agricultural Science* 45: 165–171.

Okigbo, B. N. 1980. The importance of mixed stands in tropical agricultural. In: *Opportunities for increasing crop yields* (R.G. Hurd, P.V. Biscoe, and C. Dennis, eds.). Boston: Pitman. pp 233–246.

Pathak, P.S., P. Rai, and R. Deb Roy. 1980. Forage production from Koo-Babool [*Leucaena leucocephala* (Lam.) de Wit.]. I. Effect of plant density, cutting intensity and interval. *Forage Research* 6: 83–90.

Rijks, D.A. 1967. Optimum sowing date for yield: a review of work in the BP52 cotton area of Uganda. *Cotton Growers Research* 44: 247–256.

Robinson, P.J. 1985. Trees as fodder crops. In: *Trees as crop plants* (M.G.R. Cannell and J.E. Jackson, eds.). Monks Wood, Huntingdon, England: Institute of Terrestial Ecology. pp 281–300.

Satoo, T., and H. A. I. Madgwick. 1982. *Forest biomass*. The Hague: Martinus Nijhoff.

Shankarnarayan, K. A., L. N. Harsh, and S. Kathju. 1987. Agroforestry in the arid zones of India. *Agroforestry Systems* 5: 69–88.

Singh, R. P. 1987. *Alley cropping in semi-arid tropics of India—the CRIDA experience*. Paper presented at an ICAR-ICRAF workshop on agroforestry, CRIDA, Hyderabad. Nairobi: ICRAF.

Srivastava, J. P. L. 1978. Lopping studies on *Prosopis cineraria*. *Indian Forester* 104: 269–274.

Tejwani, K. G. 1979. Soil fertility status, maintenance and conservation for agroforestry systems on wasted lands in India. In: *Soil research and agroforestry* (H. O. Mongi and P. A. Huxley, eds.). Nairobi: ICRAF. pp 141–174.

Wareing, P. F. 1968. The physiology of the whole tree. In: *Annals of Report of the E. Malling Research Station*. East Malling, Kent, UK.pp. 55–68.

Wareing, P. F. 1977. Growth substances and integration in the whole plant. In: *Integration of activity in the higher plant*. Symposia of the Society of Exploratory Biology 31. London: Cambridge University Press. pp 337–365.

CHAPTER 5

Biological Control of Insect Pests and Plant Diseases in Agroforestry Systems

Zhao Liping

The temporal and spatial diversity of species embodied in agroforestry systems is presumed to reduce the risk of damage by insect pests and pathogens. Currently, there is little direct evidence specifically from agroforestry systems to support this premise. Research involving agricultural and horticultural cropping systems, however, suggests that vegetational diversity frequently results in significant reduction of insect pest problems (Altieri and Liebman 1986), thereby laying the foundation, in terms of both ecological hypotheses and methods, for investigating biological control in agroforestry systems.

Trees add a new structural and genetic dimension to multiple cropping systems. Not only is this vegetational change reflected in the magnitude of their effect on pest populations (Perrin 1980), but the complexity of the systems also make it difficult to predict cause and effect relationships. The effects of or interactions between specific plant mixtures and specific insect pests or diseases constitute the bulk of current biological control literature; results cannot necessarily be extrapolated to agroforestry systems.

Insect pests and plant diseases can be biologically controlled in two ways: (1) by altering the genetic characteristics of plants and insects (for example, to repel insects in the case of plants and to produce sterile matings in the case of insects), and (2) by manipulating the elements of the ecosystem that support or are subject to insect and disease attack. While the first method is outside the scope of this chapter, the technologies and practices of the second are particularly suited to agroforestry research and application.

The critical research question is to what extent can agroforestry systems be designed to control or reduce insect and disease damage? This chapter briefly describes both the current theories of biological control mechanisms in polycultures and the techniques available for identifying and evaluating plant/insect/disease interactions, and it explores how agroforestry systems might be designed to achieve successful biocontrol.

EFFECTS OF VEGETATIONAL DIVERSITY

There is a large body of anecdotal literature describing various means of biocontrol from many traditional systems of agriculture. But it was not until the late 1970s that researchers began to test and evaluate polyculture systems, especially their pest reduction capacities. Although evidence has come in on both sides, the results predominantly support the theory that polycultures reduce insect pest and disease damage. An excellent review is provided by Altieri and Liebman (1986) in reference to multiple cropping systems.

The variety of responses reported from different polycultures illustrates the difficulty in predicting response due to the specificity of the interactions. A specific system may be effective in some areas and ineffective or counter-productive in others (Cromartie 1981). If all or most crops in a mixed system are palatable to a polyphagous insect, for example, it is likely that this pest will stay longer and become more numerous (Speight 1983). In addition, generally successful polyculture systems do not reduce equally all important pests (Risch et al. 1983). Management practices are by necessity based on incomplete and conflicting evidence.

Ecological Hypotheses of Biological Control

To interpret the results from research describing significant decreases in insect pests in polyculture systems, Root (1973) offered two ecological hypotheses: (1) natural enemy and (2) resource concentration. These hypotheses are briefly described below as a framework for investigating and understanding biological control interactions. Readers are referred to work by Altieri and Liebman (1986), Risch (1981), and Root (1973) for more detailed discussions.

The *natural enemy hypothesis* contends that polycultures support a larger and more diverse population of the natural enemies of insect pests than monocultures. Predators tend to be polyphagous and have broad habitat requirements; thus predators would be expected to encounter alternative prey and microhabitats in complex environments (Altieri and Liebman 1986). Polycultures indirectly control insect pests by offering improved habitats for their predators.

The *resource concentration hypothesis* contends that insect herbivore populations can be controlled directly by the concentration or spatial distribution of their food plants. Monocultures provide concentrated food sources and homogenous physical conditions that attract many herbivores. In polycultures, the resource concentration of any one plant species is lower, thus making it more difficult for insect herbivores to locate the host species. In addition to fewer host plants per unit area in polycultures, other factors, such as shading (Risch 1981), color (Cromartie 1981), and odor (Shahjahan and Streams 1973), appear to influence insect pest habitation, as discussed below.

Mechanisms of Biological Control

The numerous studies indicating reduced insect pests in various polycultures prompted researchers to try to identify the causes or mechanisms underlying biological control. The mechanisms identified thus far (for example, by Hasse and Litsinger 1981) can be divided into physical and chemical interactions.

Physical interactions

Crop Mixtures

Insects may find it difficult to locate their hosts because of the presence of nonhost crops. Companion crops may provide camouflage (Altieri and Liebman 1986), act as a mechanical barrier to the dispersal of the herbivores (Root 1973, Kennedy et al. 1959), or physically repel the pests because of unpleasant morphological features such as hairy leaves (Levin 1973).

Color and Texture

Some pests prefer a crop of a particular color or texture (Cromartie 1981). In the Philippines, after fewer corn borers (*Ostrinia furnacalis* Guenee) were found in maize intercropped with peanuts than in pure maize, the peanut crop was replaced by rows of green burlap sacks. The corn borer still avoided the maize, suggesting that the color green is repulsive to the pest (IRRI 1974). It is also commonly noticed that aphids colonize plants more readily when they stand out against a background of bare soil (Kring 1972).

Microclimatic Conditions

Intercropping systems can be manipulated to provide light, temperature, humidity, and wind conditions that are unfavorable for pests but favorable to pest predators. Shade from trees can interfere with the host-seeking and/or reproducing behavior of some insects (Yang et al. 1988, Risch 1981); increased humidity can favor the parasitization of pest eggs by *Trichogramma* species (Pu 1978) or infection of pests by entomophagous fungi (Jaques 1983).

Pollen and Nectar Sources

Compared with monocultures, polycultures can provide more pollen and nectar sources attractive to, and sustaining for, pest predators. For example, *Ageratum conyzoides* Linn bears flowers throughout the year that provide pollen for predacious mites and favor their colonization and buildup in citrus orchards (Mai et al. 1979).

Ground Cover

Increased ground cover may favor certain predators, particularly those that feed by night and hide by day. Carabid beetles (*Harpalus rufipes* DeGeer) were much more destructive to early instar *Pieris rapae* caterpillars in weedy than in weeded brussels sprout plots (Dempster 1969).

Alternate Food Sources

The presence of different insect herbivores in polyculture systems can encourage predators to remain when their principal insect pests are rare.

Chemical interactions

Repellents

Plants that produce repellents, oviposition and feeding deterrents, and toxicants when interplanted with a host crop can deter host recognition (Dethier et al. 1960, Schoonhoven 1968).

Attractants

Plants that produce attractants, arrestants, excitants, and stimulants can promote host recognition, oviposition, and feeding (Dethier et al. 1960, Schoonhoven 1968). When interplanted with crops, they may reduce feeding damage to the main crop by diverting the pest.

Mixtures

When a mixture of various volatile chemicals are produced by different plants, insects are confused and find it difficult to locate hosts, to feed, and to reproduce (Altieri 1986).

Enhancement of Predator Efficiency

Some plants produce chemicals that enhance the efficiency of predators. Parasitization of corn earworm eggs by wild *Trichogramma* wasps was promoted by applying extracts of the weed *Amaranthus* species (Altieri et al. 1983).

Odor

When a pest predator or parasite uses the odor of the plants on which the host/prey feeds to orient itself to its host or prey, intercropping with other crops may mask the odor and inhibit the host- or prey-seeking behavior of the pest predators (Shahjahan and Streams 1973, Monteith 1960). Such situations should be avoided in crop manipulation and system design.

EVALUATING POLYCULTURE SYSTEMS FOR PEST AND PATHOGEN REDUCTION

Traditional cropping systems, which often incorporate biological pest control (Mateson et al. 1984), have evolved through centuries of systematic experimentation by generations of smallholder farmers. These systems have been adapted exceptionally well to local conditions and are based on countless distinct crop varieties and mixtures. Many plant protectionists are now examining effective traditional pest management systems to develop new technical packages for smallholder farmers. How plants interact with each other and the environment to affect insect populations and insect behavior and the spread of disease is not yet clearly understood.

Client Problem

Polycultures require increased labor input (intensified management) and complicate alternative management methods (such as chemical pest control). Land managers lack the information necessary to predict plant/insect/disease interactions and thus cannot effectively design agroforestry systems to specifically apply biological control strategies.

General Methods

When designing experiments (see also Mead, this volume), the most convincing results are those obtained from field experiments that are set out in randomized block or split plot design and have several replications. The plots should be large enough (usually 100 m^2) to permit them to have pest effects comparable to those of local farming systems. There should be sufficient space between plots to minimize interference. Plant density, plant size, and plant quality may all differ in polycultures from those in monocultures. Thus, it is important to control these differences in experiments to make it clear that pest reduction is the specific result of plant diversity rather than of interrelated and confounding effects of plant density, host plant size, and quality.

A single plant of one species may not necessarily be equivalent to a single plant of another species. Thus, when interplanting two species, it is necessary to determine what number of substitute plants exert the same ecological pressure as the replaced plants from the same area. This plant equivalence can be calculated according to the ratio of the component crop in pure stands. For example, in cowpea/maize interplanting, three cowpea plants are equivalent to one maize plant. With this technique, the plant population pressure in mixed and pure stands can be kept constant (Karel et al. 1982).

The methods used to identify the underlying mechanisms and evaluate the effect of different polyculture systems on insect pests are outlined below.

Identifying the Mechanisms

Plant density and crop arrangement patterns

The effects of plant density on insects can be tested by setting several population levels in the experiment (Bach 1980). To test for the effect of different spatial arrangements of crops, several patterns of mixed cropping, row intercropping, and intra-row mixed cropping should be tried. All these variations can be tested in experimental plots to determine which ones are best for pest regulation.

Physical replacement technique

The physical replacement technique, which demonstrates the physical effects a component crop may have on pest regulation, replaces the crop with abiotic substitutes that exert the same or similar physical effects on the pests. The above mentioned replacement of peanut plants with green burlap sacks in a maize/peanut intercrop system to test for color aversion (IRRI 1974) is one example of the physical replacement technique. Another was devised by Risch (1981). He observed that there were far fewer beetles on beans planted with maize than in bean monocultures. The beans grown with maize were more shaded than beans in monocultures. To test whether shade influenced the feeding behavior of the beetles, causing them to emigrate from shaded area to light area, Risch constructed two large shade screens and suspended them 80 cm above the ground. One screen allowed 65% light transmission, the other only 25%. Squash and bean plants were grown in the greenhouse and placed under these screens. The result of counting beetles on the plants over a series of days showed that there were always significantly more beetles under the light shade screen than under the dark screen.

Risch (1981) tested the shade hypothesis further by placing maize stalks among potted bean plants and erecting a light screen over them. Potted bean plants without maize stalks were also placed in a nearby area with the same kind of light screen over them, allowing the same amount of light to reach the plants.

Risch consistently found many more beetles on the beans without maize stalks, indicating that maize physically inhibited colonization by these beetles in ways other than by just increasing the overall shade.

Diffusion experiments

Movement of insects in relation to food sources may play an important role in regulating insect populations in polyculture plots. In- and out-migration patterns can be studied by diffusion experiments or mark and recapture techniques (Risch 1981, Bach 1980). Insects collected from natural habitats or raised artificially in the laboratory can be marked with a small dot of testers brand paint just outside the plots or in the centers of the plots. After a period of time, samples are taken and marked insects are captured and recorded. Sometimes, it is necessary to aspirate the naturally occurring individuals from plants just before the release of marked individuals to ensure clean fields for each experiment.

Another way to release marked insects inside the plots is to place dormant (chilled) individuals directly on plants. It is then possible to determine both the movement pattern within the plots and the tenure time per plant (the average time an insect spends on a plant of each component crop). Wetzler and Risch (1984) grew maize, squash, and bean plants in pots until all the plants were in flower. Approximately half the maize plants had large numbers of corn aphid (*Rhopalosiphum maidis*). The authors placed 50 beetles (*Coleomegilla maculata*) on five aphid-infested maize plants, 50 on five bean plants, and 50 on five squash plants (10 beetles per plant). The beetles were chilled to approximately 6°C before being placed on the plants. The number of beetles remaining on the plants was counted approximately every 10 minutes for a period of 100 minutes to determine the tenure time on each of the component crop plants. Naturally occurring individuals in plots can also be marked in situ and released and sampled later to determine their movement patterns in monoculture versus polyculture systems.

Microclimatic factors

The first step in identifying one or more microclimatic factors as contributing to a system's pest reduction capacity is through direct measurement. Efficient measurement apparatus are readily available. Microclimatic conditions such as light intensity or the extent of shade, temperature at the soil surface or in the air, wind velocity or air circulation, and relative humidity are suggested to affect pest populations. These conditions can differ significantly between polycultures and monocultures.

Chemical effects

Just as in physical replacement techniques, plant extracts can replace living plants. Plants can be crushed or homogenized and chemicals extracted with hot water, buffer, or other organic solvents such as ethanol and acetone, then sprayed on companion crops to test their effects on pest populations. If the extract exerts the same or similar effect on pests as the intercropped plant itself, its role in pest regulation can be determined. Further tests can be conducted in the laboratory under more strictly controlled conditions. When the effects of the extracts have been demonstrated, the chemicals responsible for the effects can be purified and characterized with phytochemical techniques (Wang 1985, Huang and Guan 1986, Cole 1976).

Quantifying the Damage

Insect counts or damage estimates

The extent of insect problems can be expressed in terms of insect population density or plant damage. Techniques for insect population estimation include not only direct counting or visual inspection sampling, but also sampling with apparatus such as D-vac insect suction machines, Malaise traps, pitfall traps, sticky traps, and pan traps. When counting insects is very laborious, as in the case of aphids and leaf hoppers, it is preferable to estimate plant damage (Nordlund et al. 1984, Karel et al. 1982).

Parasitization and predation

One method to estimate parasitization is to sample host egg masses in the field and bring them back to the laboratory where they are incubated and the percentage of egg damage is checked (Nafas and Schreiner 1986). Another method is to place egg cards or sheets in the field for a period of time and then return them to the laboratory where the numbers of eggs parasitized and removed by predators can be counted at the same time (Altieri and Schmidt 1986). Trap sampling or direct counting of predacious arthropods is sometimes possible.

Discussion

Experiments focused on identifying the mechanisms and testing the ecological hypotheses of biological control have emerged mainly in the last decade.

In her work on the striped cucumber beetle (*Acalymma vittata* Fab) in cucumber (*Cucumis sativus* L.) monoculture and polyculture systems with maize (*Zea mays* L.) and broccoli (*Brassica oleracea* L.), Bach (1980) separated the effects of host plant density, total plant density, and taxonomic diversity through the design of appropriate field trials and analysis of results. She found that all measures of growth (leaf area, growth rate, and vine length) and reproduction (fruit production and number of flowers) of cucumber plants were most strongly affected by diversity, but also were affected by plant density. Both per plot and per plant values were greater in monocultures than in polycultures. The number of beetles was strongly correlated with the total amount of plant growth and reproduction in monocultures but was not in polycultures. For plots with equal amounts of leaf area, monocultures had a greater number of beetles by an order of magnitude. Her work convincingly showed that differences in host plant quantity did not explain the differences in beetle abundances between monocultures and polycultures.

Not all research is so carefully designed, however. A number of researchers have found higher numbers of herbivores in monocrops than in polycrops, but few of them discuss the possible role of the host plant or total plant density effects as opposed to those of taxonomic diversity itself. Their data is usually reported only in terms of the numbers of herbivores per unit area, or per unit sampling effort, not in terms of the numbers of herbivores per host plant per unit of plant biomass (for example, the number of beetles per plant divided by the mean size or dry weight of the plants in that plot, Risch 1980). Altieri and colleagues (1978) found fewer *Diabrotica balteata* on beans interplanted with maize, but they provide no data showing that bean plants were not smaller in the interplanted treatments than in the monocrops. It is not possible, therefore, to determine to what extent differences in the number of herbivores might be due to differences in plant diversity rather than to host plant size and/or quality.

What data is collected, how it is interpreted, and how it is used to support conclusions must also be carefully considered. For instance, Wang (1985) studied several mechanisms by which one insect pest can be suppressed in a polyculture system. He found that intercropping sugarcane with oilseed rape could significantly reduce the damage caused by sugarcane stem borer (*Chilo influscatellus*) because rape seedlings planted earlier in the spring grew quickly and blossomed when sugarcane seedlings emerged. The author found that (1) rape plants masked sugarcane seedlings and made them difficult to locate by stem borer adults during oviposition; (2) extracts from different parts of rape plants had repelling and toxic effects on stem borer adults, the extract of stem and pod showing the strongest effect; and (3) the flowering rape plants attracted many more natural enemies to the interplanted plots. The population of natural enemies could be 5 to 10 times higher in diculture than in monoculture plots. The author did not provide data to assess the relative importance of these mechanisms for suppressing the population of stem borers. Though the population of natural enemies was higher in dicrop than in monocrop, there was no data to show that actual rates of parasitization or predation of the target pest were really higher in dicrop than in monocrop.

DISEASE REDUCTION IN MIXED CROPPING SYSTEMS

Biological control of diseases in agroforestry systems has received even less attention than insect pest interactions. Most research on the subject comes from agricultural multiple cropping systems, where one crop is a buffer against disease for a companion crop (Browning 1975, Thresh 1982).

Diseases can be transmitted by insects and soilborne pathogens. Insect-borne diseases are influenced by the mechanisms discussed above regarding plant/insect pest interactions. Soilborne pathogens also appear susceptable to plant manipulation. Mixed cropping systems are credited with buffering against disease losses by delaying the onset of the disease, reducing spore dissemination, and modifying microenvironmental conditions (Altieri and Liebman 1986).

Mechanisms of Disease Intervention

Several mechanisms have been suggested to explain disease reduction in diversified farming systems (Altieri and Liebman 1986, Liebman 1986, Keswani and Mreta 1982). Evidence to support the mechanisms listed below (with the exception of the interception or flypaper mechanism), are well documented (see, for example, Burdon and Whitbread 1979, Altieri 1987, Johnston et al. 1978, ICRISAT 1984).

Lower-density planting

Polycultures are suited to planting susceptible species at a lower density than in monocultures; this decreases the spread of diseases by reducing the amount of tissue per unit area that is infected and that can subsequently serve as a new source of inoculum (for example, see Burdon and Whitbread 1979).

Interception

Resistant plants interspersed among susceptible plants in polyculture can reduce the wind velocity or act as mechanical barriers. By acting as "flypaper," they intercept disease inoculum spread by wind or rain splash and prevent it from infecting susceptible companion plants. This has been suggested as the mechanism for reduction of Ascochyta disease on cowpea when grown with maize (Moreno 1979).

Interference with vector behavior

Insects transfer pathogens during pollination and feeding. In polycultures, nonhost plants interfere with host-seeking behavior of insects and make them less likely to fly directly from one susceptible plant to another. This reduction in the transfer of inoculum by insects may lead to a reduction in the severity of disease. This can be achieved by planting a taller nonhost crop either around the main crop or in a mixed intercropping pattern (Altieri 1987, Thresh 1982, Zitter and Simons 1980). Decoy crops, targeted primarily at nematode populations, offer another example. Decoy and trap crops (see table 5.1) are planted to make nematodes waste their infection potential on nonhost crops (Palti 1981).

Microclimate modification

Some intercropping systems have better air circulation than others, reducing humidity and changing the temperature of the microhabitats. The severity of several pea diseases was reduced when pea vines climbed up associated cereals, rather than lying matted on the ground (Johnston et al. 1978).

Suppression within the rhizosphere

Excretions from roots or microbes on the roots of one crop species may adversely affect the soilborne pathogens of companion crop species. This appears to be the mechanism responsible for the reduced incidence of *Fusarium* wilt of pigeon pea when it grows in association with sorghum (ICRISAT 1984).

Methods

Research involving the reduction of insect-borne diseases in polycultures may use many of the methods discussed under biocontrol of insects without significant modifications. Determining chemical interactions below-ground, however, is difficult and usually requires pot experiments and increased laboratory testing. These two methods are briefly described below, and the

Table 5.1. Crops for the reduction of pathogen populations

Crop	Pathogen	Decoy, trap or inhibitory crop
Brassicae	*Plasmodiophora brassicae*	Rye-grass, *Papaver rhoeas, Reseda odorata*
Potato	*Spongospora subterranea*	*Datura stramonium*
Olive	*Verticillium albo-atrum*	*Tagetes minuta*
Tomato, tobacco	*Orobanche* spp.	Sunflower, safflower, lucerne, flax, chickpeas and others
Various	*Striga asiatica*	Sudangrass
Eggplant	*Meloidogyne incognita, M. javanica*	*Tagetes patula, Sesamum oriental*
Tomato	*M. incognita, Pratylenchus alleni*	*T. patula*, castor bean, chrysanthemum
Tomato	*M. incognita*	*T. patula*, groundnuts
Narcissus, tomato, okra	*Meloidogyne* sp.	*T. patula*
Soybean	*Rotylenchulus* sp. *Pratylenchus* sp.	*T. minuta, Crotalaria spectabilis*
Various	*Pratylenchus penetrans*	*T. patula*, hybrids of *Gaillardia* and *Hellenium*
Various	*P. neglectus*	Oil-radish (*Raphanus oleiferus*)
Oats	*Heterodera avenae*	Maize
Susceptible potato varieties	*Trichodorus* sp. *Globodera Pallida*	Asparagus resistant varieties
Beet	Cyst nematodes	Crucifers
Pineapple	Root-knot nematodes	Tomatoes
Trees	*Armillaria mellea*	*Hyparrhenia* spp.
Potato	*Heterodera rostochiensis*	Mustard
Rubber	*Ganoderma pseudoferreum, Formes noxious, F. lignosis*	Creeping legumes
Cotton	*Fusarium* sp.	Peppermint

Source: Palti 1981.

reader is encouraged to refer to the methods section of Horsley's chapter (Allelopathy, this volume) for additional information.

Pot experiments

After inhibitory plants are grown in the field for enough time to allow their inhibitory effects to be expressed, soil from the rhizosphere can be used in pot tests. The soil is inoculated with the target pathogen, and host plants are grown in it. If the soil has inhibitory effects on the pathogen, it means some chemical agent(s) has been released into the soil and is responsible for the disease reduction observed in both field and pot tests. By growing inhibitory plants in pots with sand, peat, or silt soils, the toxic chemicals may be collected and concentrated and then purified and characterized with analytical techniques (Li 1988, Baker and Cook 1974).

Laboratory tests

The inhibitory effects of extracts from soil and plants/roots on pathogens can be observed directly in petri dishes. Soil or plant material is first extracted by adding water or a buffer and homogenizing at low temperatures. The solid contents are removed by centrifuging, and the supernatant obtained is filtered first with Whatman paper, then with a milipore filter. The sterilized filtrate is mixed with melted agar medium at 50°C, then poured onto plates. Fungal or bacterial pathogen is inoculated on the plates and incubated at optimum growth temperatures to allow colony formation. The diameter of the colonies or weight of mycelium or other biomass criteria is measured for comparison with controls where sterilized water was added to the medium instead of the soil or plant extract. If a toxic chemical is volatile, its effect can be tested by inoculating spores or cells of the pathogen on plates, inverting the petri dishes, and pipetting different concentrations of the toxic material into the lids of the dishes. The fumigation action of the extracts can be expressed in terms of biomass reduction of the pathogen relative to controls treated with water or buffer solution.

Discussion

The practical application of pot and laboratory tests is demonstrated in two studies on the various effects of peppermint (*Mentha haplocalyx* var. *piperascens*). Farmers in Jiangxi, China, found that after growing peppermint for only one year in a field heavily infested with Fusarium, they could grow cotton continuously for nine years without serious losses from *Fusarium* wilt. Li (1988) studied the effectiveness and mechanisms of the inhibitory effect of peppermint on *Fusarium* species. First, peppermint was grown in the field for one cropping season. Soil was taken from peppermint rhizosphere and used to culture cotton plants in pot tests. Soils from the rhizosphere of crucifers were used as controls. All the potted plants were inoculated either with the pathogen's mycelium or with heavily infested soils at the amount of 0.4% or 50% of the weight of the pot soil respectively. Thirty days after planting, wilt incidences were 23.8% and 41.2% for these two inoculation treatments, while the incidences in corresponding controls were 67.8% and 81.7%, respectively.

The results indicated that extract from peppermint soil could inhibit germination of spores by 98.2% and reduce the growth of mycelium by 62.5%. Peppermint shatter mixed with inoculated soil at the rates of 0.5% and 1% and peppermint oil either mixed with or used to fumigate infested soil at rates of 0.025% and 0.05% in pot tests both showed inhibitory effects to Fusarium wilt. A complete inhibition of spore germination was achieved when the spores in PDA plates were fumigated with a 2,000 ppm to 10,000 ppm diluted solution of peppermint oil (Li 1988).

DESIGNING AGROFORESTRY SYSTEMS FOR BIOCONTROL

The variety of responses recorded for a given polyculture makes it impossible to adequately predict changes in pest control capacity in response to specific polycultures. Thus "designing" agroforestry systems with biological control as an objective is premature. Nonetheless, in the process of understanding systems interactions and refining measurement techniques, it is important to give consideration to presumed modes of biological control within agroforestry systems.

It is first of all important to determine which of the two methods of cultural control of pests, resource concentration or natural enemies, should be emphasized in biocontrol strategies for agroforestry. Risch and colleagues (1983) assessed their relative importance by a combined empirical-theoretical approach. They discovered that, among the 150 monophagous species and 48 polyphagous species studied, more polyphagous species (43%) were present in diversified systems than monophagous species (10%) (table 5.2).

Table 5.2. Numbers of monophagous and polyphagous herbivore species in diversified agroecosystems compared with monocultures for annual and perennial cropping systems*

System	More abundant	No difference	Less abundant	Varied	Total
Annual					
Monophagous herbivore	3	15	58	23	99
Polyphagous herbivore	16	2	11	12	41
Total annual	19 (13.6)	17 (12.1)	69 (49.3)	35 (25.0)	140
Perennial					
Monophagous herbivore	12	1	34	4	51
Polyphagous herbivore	5	0	2	0	7
Total perennial	17 (29.3)	1 (1.7)	36 (62.1)	4 (6.9)	58
Total monophagous herbivore	15 (10.0)	16 (10.7)	92 (61.3)	27 (18.0)	150
Total polyphagous herbivore	21 (43.8)	2 (4.2)	13 (27.1)	12 (25.0)	48

* A perennial system is one in which at least one component crop is a perennial. Numbers in parentheses indicate percentage: 150 studies were reviewed.
Source: Risch et al. 1983.

Although it is generally accepted that natural enemies are more important and effective control agents in less disturbed, perennial systems, Risch and colleagues (1983) recorded greater (14%) herbivore populations in the 58 diverse perennial systems than in the 140 diverse annual systems. These results suggest that, in general, resource availability is more important to insect pest populations than is the presence of natural enemies. Therefore, crop composition and arrangement in time and space should be central to pest management in agroforestry systems.

Component species in a successful agroforestry system should be (1) botanically unrelated and therefore less likely to have a set of common pests, thus minimizing the number of polyphagous pests in the system, (2) ecologically compatible to reduce direct competition or adverse interactions between them, and (3) socioeconomically acceptable so that yields or profits are higher in polycultures than in monocultures to induce farmers to plant them.

Litsinger and Moody (1976) identify several factors that can be manipulated in the choice of component crops for pest management in multiple cropping systems (figure 5.1).

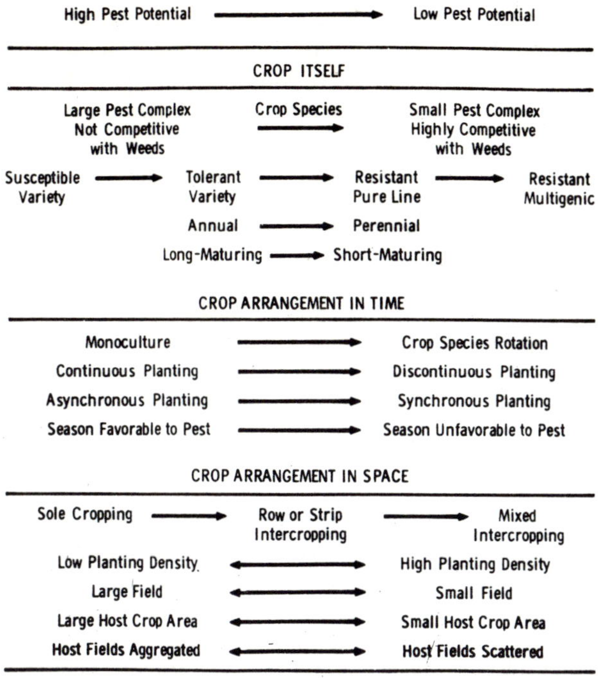

Figure 5.1. Kinds of crops and their arrangement in time and space evaluated as to the potential development of pest problems. Some effects are seen to be high in pest potential, some intermediate, and some low.

Source: Litsinger and Moody 1976.

Different crops have different pest complexes, some of which are very large, others very small. For instance, few pests have been reported on cassava (*Manihot esculenta*) (Lozano and Booth 1974), but solanaceous crops have large pest complexes that are even larger in the tropics than in temperate regions (Wellman 1968). Since system stability relies heavily on thriving trees, sowing or planting crops susceptible to polyphagous pests or multihost pathogens such as *Sclerotium rolfsii*, *Verticillium dahliae*, or root-knot nematodes, must be avoided at all costs if the tree crops can be affected (Palti 1981).

How each system is designed depends on answers to the following questions:

1) What are the important local pests?

2) Is more than one crop in the system under consideration host to a particular pest?

3) If so, is it possible to omit or replace one or more of those crops with nonhost or resistant crops?

4) Is it possible in this manner to make every important pest in the system monophagous?

5) Does a particular crop have an adverse effect on its pest or the pest of associated crops if properly arranged in the system?

6) If so, is the possible mechanism of the effect physical, chemical, or biotic?

7) What is the best way to exploit the effect(s)?

If new crop species are introduced into an area, care must be taken that exotic pests are not brought along with the new plant material (Rawat 1968, Lin 1984) and that indigenous insects do not become pests because the newly introduced plants provide them with a preferred food source or a more favorable eco-niche (Rao 1970). High-yielding, but pest-susceptible, varieties should not be used unless economical pest management technologies have been developed.

Strategies for Establishment

Agroforestry systems may be established in two ways: transformation and rehabilitation. Transformation is the process by which natural wild vegetation is gradually replaced with useful species that fill the same functional and structural niches as their wild predecessors (Oldman 1981). Pests can be kept in check if the transformation process does not drastically disturb populations of local pest predators. Rehabilitation begins with a simple system, such as mono- or diculture, and gradually changes it to a more diversified agroforestry system.

It is in this rehabilitation process that serious pest problems can occur if, for example, the crop complex is wrong, is improperly arranged in time and space, or is poorly managed. Irrigation of lucerne and vegetables intercropped with oil palm has been shown to accelerate the passage of the bayound pathogen, *Fusarium oxysporum* f. sp. *albedinis*, in the palms' stem vessels. Intercropping young peaches with melons and other vegetables has led to severe attacks of the root-knot nematode, *Meloidogyne arenaria*, on peaches (Palti 1981). Cocoa grown under the shade of *Leucaena* suffered more seriously from attacks by defoliating lepidoptera than cocoa grown under thinned forest canopy because the pests were able to use the *Leucaena* as an alternative food source (Room and Smith 1975).

Examples of Successful Systems

Crops can be grown in several temporal and spatial designs (such as strip-cropping, intercropping, mixed cropping, and as cover crops and living mulches) within each plot, thus optimizing the use of limited resources and enhancing the self-sustaining and resource-conserving attributes of the system. An important consideration in designing rotation is the stability of the cropping systems in terms of their pest regulation properties.

Peasants in southern China have developed many intensively managed agroforestry systems with pest suppression mechanisms. One of these consists of four components: pond cypress (*Taxodium ascendens*), rice, wheat, and fish. Pond cypress planted around the field reduces wind velocity, increases relative humidity, and acts as a buffer against temperature fluctuations. One crop each of wheat and rice are grown annually between the trees. In such a system, not only may the yield of rice increase by an average of 10%, but the crude protein content of the rice may increase by 0.62% compared to rice in plots without cypress windbreaks (Shi and Gao 1986). Because pond cypress are an alternate habitat for predacious spiders, they keep plant hoppers in check (Zhang, personal communication). Fish in the rice fields, a very important cash crop, can eat larvae of mosquitoes and fallen leaves and weeds, thus improving water quality and the physical condition of the soil. Rice-wheat rotation can also have a negative impact on pests in the system. All these factors combine to suppress pest populations, reduce the amount of chemical fertilizers and pesticides applied in the system, and improve growing conditions for crops. In one example, peasants annually produced 4.43 t/ha of wheat, 7.88 t/ha of rice, and 360 kg/ha of fish, while the cost of chemical pesticides was reduced from 84 yuan/ha to 28 yuan/ha following the adoption of the system (Zhang 1987).

Polyculture systems such as this that incorporate biocontrol techniques are compatible with the traditional farming experiences of small landholders; they do not need external resources, either chemical or financial, for their continued application; and because they are simple and efficient, they can be taught farmer to farmer.

In the end, however, what must characterize biocontrol techniques for small-scale agroforestry systems is adaptability, which is closely related to increasing productivity or reducing costs and risks. Small landholders are very practical people who will adopt a cropping system only when they can increase yields or profits by doing so. It is thus important to know whether a system of pest reduction increases yields or reduces pest control costs (FAO 1971).

CONCLUSIONS

Research in biocontrol of pests and pathogens in agroforestry systems must be directed to one principal question: what are the mechanisms for effective biological control? There is much anectodal and traditional evidence to support the claims of benefits from polyculture systems. The variation reported in research results, however, suggests that the interactions might be species specific, but the interactions are obviously not well understood. Carefully designed and conducted research to determine these mechanisms specifically within agroforestry systems is needed.

LITERATURE CITED

Altieri, M.A. 1986. *Agroforestry: the scientific basis of alternative agriculture*. Boulder, Colorado: Westview Press. 227 p.

Altieri, M.A., D. K. Letourneau, and J. R. Davis. 1983. Developing sustainable agroecosystems. *Bioscience* 33: 45–49.

Altieri, M.A., and M. Z. Liebman. 1986. Insect, weed and plant disease management in multiple cropping systems. In: *Multiple cropping systems* (C. A. Francis, ed.). New York: Macmillan. pp 183–218.

Altieri, M.A., and L. L. Schmidt. 1986. The dynamics of colonizing arthropod communities at the interface of abandoned, organic and commercial apple orchards and adjacent woodland habitats. *Agriculture, Ecosystem and Environment* 16: 29–43.

Altieri, M.A., A. van Schoonhoven, and J. D. Doll. 1978. A review of insect prevalence in maize (*Zea mays* L.) and bean (*Phaseolus vulgaris* L.) in polycultural systems. *Field Crops Research* 1: 33–39.

Bach, C.E. 1980. Effects of plant density and diversity on the population dynamics of a specialist herbivore, the striped cucumber beetle. *Acalymma vittata* (Fab.). *Ecology* 61: 1515–1530.

Baker, K.F., and R. J. Cook. 1974. *Biological control of plant pathogens*. San Francisco, California: W.H. Freeman and Company. 433 p.

Browning, J.A. 1975. Relevance of knowledge about natural ecosystems to development of pest management programs for agroecosystems. *Proceedings of American Phytopathology Society* 1: 191–194.

Burdon, J. J., and R. Whitbread. 1979. Rates of increase of barley mildew in mixed stands of barley and wheat. *Journal of Applied Ecology* 16: 253–258.

Cole, R.A. 1976. Isothicyanates, nitriles and thicyanates as products of autolysis of glucosinolates in cruciferae. *Phytochemistry* 15: 754–762.

Cromartie, W.J. 1981. The environmental control of insects using crop diversity. In: *CRC Handbook of pest management in agriculture* (D. Pimentel, ed.). Boca Raton, Florida: CRC Press. pp 223–250.

Dempster, J.P. 1969. Some effects of weed control on the numbers of the small cabbage white butterfly (*Pieris rapae* L.) on Brussels sprouts. *Journal of Applied Ecology* 6: 339–345.

Dethier, V.G., L. Barton-Browne, and C.N. Smith. 1960. The designation of chemicals in terms of the responses they elicit from insects. *Journal of Economic Entomology* 53: 134–136.

FAO. 1971. Crop loss assessment methods. In: *Manual on the evaluation and prevention of losses by pests, disease and weeds* (L. Chiarappa, ed.). Rome: Commonwealth Agricultural Bureaux and FAO.

Hasse, V., and J.A. Litsinger. 1981. *The influence of vegetational diversity on host finding and larval survivorship of the Asian corn borer,* Ostrinia furnacalis. IRRI Saturday seminar. Entomology Dept., IRRI, Los Baños, the Philippines.

Huang, X., and Z. Guan. 1986. The effects of alcoholic extracts and dusts on tomato-plant used as oviposition deterrent and antifeedant to the cabbage butterfly. *Plant Protection* 12(1): 17–19.

ICRISAT. 1984. Annual Report for 1983. Patancheru, India.

IRRI. 1974. IRRI Annual Report for 1972. Los Baños, the Philippines. 246 p.

Jaques, R.P. 1983. The potential of pathogens for pest control. *Agriculture, Ecosystem and Environment* 10(2): 101–126.

Johnston, H.W., J.B. Sanderson, and J.A. MacLeod. 1978. Cropping mixtures of field peas and cereals in Prince Edward Island. *Canadian Journal of Plant Science* 58: 421–426.

Karel, A.K., D.A. Lakhani, and B.J. Ndunguru. 1982. Intercropping of maize and cowpea: effect of plant populations on insect pests and seed yield. In: *Intercropping* (C.L. Keswani and B.J. Ndunguru, eds.). Proceedings of the second symposium on intercropping in semi-arid areas, 4–7 August 1980, Morogoro, Tanzania. Ottawa: IDRC. pp 102–109.

Kennedy, J.S., C.O. Booth, and W.J.S. Kershaw. 1959. Host finding by aphids in the field. I. Gynoparae of *Myzus persicae* (Sulzer). *Annals of Applied Biology* 47: 410–423.

Keswani, C.L., and R.A.D. Mreta. 1982. Effect of intercropping on the severity of powdery mildew on green-gram. In: *Intercropping* (C.L. Keswani and B.J. Ndunguru, eds.). Proceedings of the second symposium on intercropping in semi-arid areas, 4–7 August 1980, Morogoro, Tanzania. Ottawa: IDRC. pp 110–114.

Kring, J.B. 1972. Flight behavior of aphids. *Annual Review of Entomology* 17: 461–492.

Levin, D. A. 1973. The role of trichomes in plant defense. *Quarterly Review of Biology* 48: 3–21.

Li, Y. 1988. A preliminary study on use of the medical herb peppermint in control of cotton Fusarium wilt. *Scientia Agricultura Sinica* 21(3): 65–69.

Liebman, M. 1986. Polyculture cropping systems. In: *Agroforestry* (M.A. Altieri, ed.). Boulder, Colorado: Westview Press. pp 115–125.

Lin, X. 1984. Reform farmland's habitat, breed natural enemy and control diseases and pests. (Chinese) *Journal of Ecology* 1: 47.

Litsinger, J.A., and K. Moody. 1976. Integrated pest management in multiple cropping systems. In: *Multiple cropping* (P. A. Sanchez, ed.). American Society of Agronomy Special Publication 27. pp 293–316.

Lozano, J.C., and R.H. Booth. 1974. Diseases of cassava (*Manihot esculenta* Crantz.). *Pest Articles and News Summaries* 20(1): 30–54.

Mai, X., M. Huang, W. Wu, D. Ouyang, Z. Lin, Z. Li, R. Zhu, and Z. Li. 1979. Control of citrus red mites by augmentation of natural population of *Amblyseius* spp. in hillside orchards. *Natural Enemies of Insects* 1: 52–56.

Matteson, P.C., M.A. Altieri, and W.C. Gagne. 1984. Modification of small farmer practices for better pest management. *Annual Review of Entomology* 29: 383–402.

Monteith, L.G. 1960. Influence of plants other than the food plants of their host on host-finding by tachinid parasites. *Canadian Entomology* 92: 641–645.

Moreno, R.A. 1979. Crop protection implications of cassava intercropping. In: *Intercropping with cassava: proceedings of the international workshop, 27 November-1 December 1978, Trivandrum, India* (E. Weber, B. Nestal, and M. Campbell, eds.). Ottawa, Canada: IRDC.

Nafus, D., and I. Schreiner. 1986. Intercropping maize and sweet potatoes. Effects on parasitization of *Ostrinia furnacalis* eggs by Trichogramma chilonis. *Agriculture, Ecosystems and Environment* 15(2-3): 189–200.

Nordlund, D.A., R.B. Chalfant, and W.J. Lewis. 1984. Arthropod populations, yield and damage in monocultures and polycultures of corn, beans and tomatoes. *Agriculture, Ecosystems and Environment* 11: 353–367.

Oldeman, R.A.A. 1981. The design of ecologically sound agroforests. In: *Viewpoints on agroforestry* (K. F. Wiersum, ed.). Wageningen, The Netherlands: Agricultural University.

Palti, J. 1981. *Cultural practices and infectious crop diseases*. New York: Springer-Verlag.
Perrin, R. M. 1980. The role of environmental diversity in crop protection. *Protection Ecology* 2: 77–114.
Pu, Z. 1978. Principles and methods for biological control of insect pests. (Chinese) Beijing: Academic Press. 261 p.
Rao, B.S. 1970. Pest problems of intercropping in plantations. In: *Crop diversification in Malaysia* (E. K. and J.W. Blencowe, eds.). Kuala Lumpur: Yan Seng Press. pp 245–252.
Rawat, R.R. 1968. Pest problems in cropping systems. In: *Proceedings of symposium on cropping patterns in India*. New Delhi: ICAR. pp 565–571.
Risch, S.J. 1980. The population dynamics of several herbivorous beetles in a tropical agroecosystem: the effect of intercropping corn, beans and squash in Costa Rica. *Journal of Applied Ecology* 17: 593–612.
Risch, S.J. 1981. Insect herbivore abundance in tropical monocultures and polycultures: an experimental test of two hypotheses. *Ecology* 62: 1325–1340.
Risch, S.J., D. Andow, and M. A. Altieri. 1983. Agroecosystem diversity and pest control: data, tentative conclusions and new research directions. *Environmental Entomology* 12: 625–629.
Room, P.M., and E.S.C. Smith. 1975. Relative abundance and distribution of insect pests, ants and other components of the cocoa ecosystem in Papua New Guinea. *Journal of Applied Ecology* 12: 31–46.
Root, R.B. 1973. Organization of a plant–arthropod association in simple and diverse habitats: the fauna of collards (*Brassica oleracea*). *Ecology Monogram* 43: 94–125.
Schoonhoven, L. M. 1968. Chemosensory bases of host plant selection. *Annual Review of Entomology* 13: 115–136.
Shahjahan, M., and F.A. Streams. 1973. Plant effects on host finding by *Leiophron pseudopallipes* (Hymenoptera: Breconidae), a parasitoid of the tarnished plant bug. *Environmental Entomology* 2: 921–925.
Shi, Z., and Z. Gao. 1986. On the ecological efficiency of shelter-belt network and its yield-increasing effect in paddy fields. (Chinese) *Journal of Ecology* 5(2): 10–14.
Speight, M.R. 1983. The potential of ecosystem management for pest control. *Agriculture, Ecosystems and Environment* 10: 183–199.
Thresh, J.M. 1982. Cropping practices and virus spread. *Annual Review of Phytopathology* 10: 193–218.
Wang, Y. 1985. Influence of some substances from rape plant on sugarcane stem borer (*Chilo influscatellus*). *Natural Enemies of Insects* 7(1): 30–33.
Wellman, F. L. 1968. More diseases on crops in the tropics than in the temperate zones. *CEIBA* 1: 17–28.
Wetzler, R.E., and S.J. Risch. 1984. Experimental studies of beetle diffusion in simple and complex crop habitats. *Journal of Animal Ecology* 53: 1–19.
Yang, J., L. Jin, S. Qu, G. Luo, and W. Ceng. 1988. Effects of mixed forest of Chinese fir and Homana on population of insect pests. (Chinese) *Journal of Ecology* 7(1): 45–47.
Zhang, Z. 1987. A preliminary approach on the model of artificial compound ecosystem for agroforestry. *Rural Eco-Environment* (4): 44–46.
Zitter, T.A., and J.N. Simons. 1980. Management of viruses by attraction of vector efficiency and by cultural practices. *Annual Review of Phytopathology* 18: 269–310.

CHAPTER 6

Silvicultural Applications of Agroforestry

C.B. Briscoe

Agroforestry techniques are commonly applied by small farmers, and related research is rightly concentrated on small farm implementation. Nevertheless, agroforestry can be used as an important tool to meet the objectives of large landowners as well. The possibilities are just beginning to be realized for using agroforestry on large farms and industrial and public forest lands.

The silviculture requirements of landowners and users vary widely as a function of size of landholding and public or private ownership, although some requirements may be common to many. The need to increase profits and decrease operating expenses, for example, is an objective common to small and large landowners alike. However, the degree to which money dictates management practices is usually more evident in industrial and large farm operations; self-sufficiency and labor requirements often override the profit motive on small farms. That is not to say that self-sufficiency cannot also be an important factor for large operations, as will be illustrated below. Additionally, many large landholdings are government or public forest lands. Inherent in the management of these public lands is the responsibility to address social and political issues. Such diversity of requirements cannot be realized by any single silvicultural practice—only the diversity of agroforestry activities offers opportunities to realize all of them. Nonetheless, applying agroforestry to silviculture systems lags far behind small farm applications.

The purpose of this chapter is to identify some of the differences between industrial and small farm applications, provide examples of large-scale agroforestry operations, and identify future research activities. Agroforestry systems designed to meet management efficiency, industrial self-sufficiency, and social objectives are described below with reference to specific case studies.

LARGE OWNERSHIP

Large-scale applications of agroforestry distinguish themselves from smalholder applications mainly by their ability to influence markets and test and incorporate new technology, as well as their need for extensive (versus intensive) management. One of the most important contributions of industry and other large owners is the opening of new markets and expansion of old ones, largely due to having better funding and more facilities for all aspects of marketing. Maintaining large-scale operations requires ever-expanding markets, and their potential volume of sales justifies the costs and efforts necessary to open a new market.

A market for pulpwood was instantly created, for example, when the Paper Industry Corporation of the Philippines (PICOP) was established. Initially, most of the raw material came from company-managed lands. As this supply dwindled, PICOP looked to small farmers for additional pulpwood. As a result, PICOP developed an excellent program for promoting and supporting woodlots for pulpwood on small farms where the trees were combined with agricultural crops during the first several years to cushion the delay in economic return to the farmer. The magnitude of the market for raw material created by PICOP could not have been created by small farmers alone.

The second major contribution of industry is the development and application of new or improved technology. Coffee, tea, vanilla, black pepper, and cacao are customarily produced as understory species on small- to medium-sized holdings. Large farmers and industrial owners, however, not only developed the original markets, but typically had the funds or credit to be the first to test fertilization, new pruning systems, processing machinery or methods, or new cultivation techniques that might increase yields or lower costs. Industry has led, and continues to lead, the development and/or application of technology for increased yields and higher returns.

Industrial size is not, however, without problems. Marketing large quantities of products through extensive distribution networks and being dependent on infrastructure, government policies, and available labor, is qualitatively different from supplying one's own needs. For the industrial producer, efficiency is attained by producing a limited range of products in maximum quantities. Pulpwood is a paradigm of such products. It is available from a wide range of sites and is marketable by the thousands of tons from a single species with virtually no price differential by size, form, or season.

A farm, on the other hand, may be its own primary market for silviculture products. It is best supplied by limited quantities of a wide variety of products: posts, vegetable stakes, fuelwood, roof beams, fodder, fruit, lumber, and bedding. Economically, the most favorable market for the individual farmer is always his own needs. In a self-sufficient system, there are no freight costs to truckers or markups to brokers, wholesalers, or retailers. There is, however, an obvious limitation on demand for a given product.

Large-scale operations usually require extensive rather than intensive land management. Detailed knowledge of the microsite is difficult or impossible to obtain for these managers. A small farmer may know that "two bulls had a fight right over there. Their tearing up of the surface began the erosion that stripped off most of the topsoil." Such intimate knowledge is not possible over thousands of hectares, even if the land manager happened to be present when such happenings occurred. There is also only a small probability that a hired manager will remain on the same management block for long periods.

Four apparent reasons why industrial managers are reluctant to expand into agroforestry are the needs to diversify, establish new networks, absorb start-up costs, and manage greater complexity. First, agroforestry requires diversification, whereas industrial operations are predicated on some degree of specialization. This specialization offers industrial managers relative financial and organizational simplicity that they are reluctant to forgo. Second, expanding into agroforestry forces an organization to acquire new skills in fields not directly related to current practices. Existing channels and contacts are usually inefficient and often unreliable. Third, during preparation and start-up, costs will rise immediately, but increased income will be delayed. Adding forestry operations to an agribusiness, or agricultural operations to a forestry business, generally requires assuming a significant delay in economic returns. Fourth, incorporating agroforestry activities may increase the complexity of land tenure issues, especially on public lands.

Despite the reluctance of some land managers, agroforestry systems have been successfully used by large operations in both tropical and temperate environments. The joint production of trees with other crops and/or animals serves a variety of management objectives, as the cases below demonstrate.

AGROFORESTRY TO INCREASE MANAGEMENT EFFICIENCY

The long history of intermixing trees with forage (McQueen 1978) and trees with agricultural crops (such as *taungya*) has been and continues to be dominated by trial and error management. The assumption is common that sole production of either trees or crops is more efficient than joint production. The successful adoption of agroforestry systems, by profit-making enterprises has, however, demonstrated that the joint production of trees with other agricultural sectors can be technically feasible and economically attractive. Formal analyses (for example, dynamic programming, Riitters et al. 1982) has augmented this process. Such is the case of a tree/livestock combination that was implemented in response to an unforeseen management problem of an industrial forest enterprise.

Jari Florestal e Agropecuaria, a large company, installed a pulp mill in Brazil near the equator. Supplying the raw material for the pulp mill involved clearing native forest and planting tree species that would supply long fiber for the pulp mix. Part of the plantations were located at low elevation on infertile, sandy soils, where 225 cm of rainfall was distributed fairly evenly throughout the year. The company also owned a substantial herd of cattle, pastured on infertile, poorly drained savanna on the flood plain along the Amazon (Briscoe 1979).

Client Problem

Native grasses regenerated naturally in the pine plantations, and their control constituted a major cost in pulpwood production.

Materials and Methods

Company personnel set about solving this problem through a series of agroforestry field trials that combined pulpwood production with cattle production. The research and development used to solve the problem was applied, in the truest sense of the word, but varied in detail in response to particular aspects of the problem.

The initial species selection trials in replicated and randomized complete blocks identified Honduras pine (*Pinus caribaea* v. *hondurensis*) as the best species for pulpwood production in terms of survival and overall growth. Nelore cows, a race of zebu (*Bos indicus*), had been selected over previous years of herd management. This selection process was supported by intensively monitored comparisons of breeds and crossbreeding, an operation not feasible for a small farmer with a limited number of animals. Modifications of tree and grass spacing, thinning regimes, and cattle management schedules (that is, when to introduce and remove cattle from each pasture) were decided in conference between livestock and forestry personnel, then tried. As experience was gained, the discussions dropped sharply in length and frequency.

Tree/livestock trials

Two of these trials are briefly described below. A more detailed report of the most important problems and the corrective actions taken are included in the appendix to this chapter. Although no formal research design was prepared, inputs and outputs were monitored carefully. The results from the agroforestry trials were compared to the production from adjacent areas still managed under the old methods, which served as experimental controls.

Assumption 1

Cattle put into the pine plantations would eat the abundant native grasses, reducing cleaning costs and limiting competition to the pines while increasing the cattle weight.

Trial. Eighteen dry cows (from a herd of more than 1,500 animals) were put into a 100 ha plantation of two-year-old pines.

Results. Five animals died from eating poisonous plants; the other 13 survived and gained slightly more weight than comparable animals in the savanna.

Assumption 2

Sowing palatable and nutritious grasses and providing minerals would crowd out the inferior native grasses and greatly reduce the desire of cattle to sample a wide range of plants, including toxic ones.

Trial. Panicum maximum cv. Colonial was broadcast seeded on 25 ha (from an annual planting of 4,000 ha) of newly planted pines spaced at 3 m × 3 m. Salt enriched with phosphorus and minor elements was set out in covered stations. Twenty-five dry cows were put in the plantation when the grasses were fully grown.

Results. There was no cattle mortality, and weight gains were approximately double that of equivalent cattle in the savannas. Improving the quality and quantity of grass reduced the hazard from toxic plants. However, the tall grass completely dominated the pines, and the cattle accentuated the damage by trampling the trees. Pine mortality was nearly 90%.

Discussion

Incremental changes continued each year, on progressively larger areas, as the results became more positive. Sowing the grass in strips between the trees (versus broadcast seeding), for example, provided sufficient time for the pine to become established without competition from the grass, which thus significantly reduced the pine mortality noted above. To maintain grass production through the sixth year, the spacing between the pines was modified (to 2.25 m × 4 m) to keep the original 1,111 trees/ha, but to provide more light for a longer period between the rows. Repeated soil analyses and bioassays indicated no measurable soil chemical degradation while monitoring continued during the first 10 years. The sandy soils limited compaction to the bedding and holding grounds.

The agroforestry tree/livestock combination essentially doubled the overall net annual financial yields per hectare realized from pulpwood plus meat. An additional economic gain resulted from reducing the need to buy more expensive meat in distant markets and transport it to the project (Internal reports, Jari Florestal e Agropecuaria, 1979–1981).

Agroforestry did not increase the management cost appreciably. Research and development costs were an insignificant percentage of the management budget. The necessary skills were developed by both livestock and forestry personnel learning together and from each other. The cost reduction of controlling forbs and woody weeds more than paid for the costs of erecting and maintaining pasture fences. Plant and cattle wastes, including phosphorus from the mineral supplements, appeared to maintain soil fertility despite harvesting of meat and timber.

This case study illustrates that it is possible and profitable to add livestock production to a forestry operation, beginning on a very small scale and expanding as success warrants. The only essential organizational expansion is the inclusion of an individual with expertise in managing livestock (a common skill that may already be present) plus a nucleus of the desired animals. The problems were manageable for a large landowner, but could easily have bankrupted a small farmer. The solutions, once determined, are independent of size, just as useful to a small farmer as to an industrial one.

AGROFORESTRY FOR INDUSTRIAL SELF-SUFFICIENCY

Self-sufficiency is an objective frequently attributed to small landholders, but it may be a management objective of larger landholders as well. Forestry plantations, by their very nature, are often far removed from towns and markets, and operations with many employees living on-site are responsible for providing food and other necessities to their employees. The joint production of food and trees through agroforestry can alleviate the need to purchase supplies from distant markets.

The nearest source of adequate supplies of maize (*Zea mays*), cassava (*Manihot edulis*), and other food and feed crops for workers and livestock was more than 500 km by plane, and considerably more by riverboat. To see whether the costs of transportation could be reduced through self-sufficiency, Jari Florestal e Agropecuaria initiated several tree/crop combination trials. Lands were being cleared on infertile sands and fertile clay loams for both pine and melina (*Gmelina arborea*). Pine averaged 14 m^3/ha/yr on the sandy soil and only minimally more on the good clay loam; melina varied much more with microsite, but averaged about 28 m^3/ha/yr on the clay loam and virtually nothing on the sands. Maize produced about 3 t/ha on the sandy soil and 5 t/ha on the clay loam; cassava averaged 12 t/ha on the sandy soil and 18 t/ha on the clay. Food crops were tried with melina on both the clay loam and the sandy soils.

Client Problem

Supplies of food crops from local sources for company employees and their families were inadequate.

Materials and Methods

Intercropping annual crops with two types of tree plantations on two different soil types was examined. Tree species were selected on the basis of formal trials. Crop species were selected as needed from among those grown in the region, although maize varieties were also tested formally. Several short-season annuals (rice, beans, tomatoes, and squash, and virtually any non-climber) were also intercropped. Maize and cassava planted with melina on clay loam are used here for illustration.

Tree/crop trials

Assumption 1

Maize could be grown between the rows of melina just as grass was being grown in between pines.

Trial. Immediately after planting or sowing melina at 3 m × 3 m spacing, maize was sown. Maize was harvested after 120 days.

Results. The melina grew slightly faster the first year with the maize, and there was no long-term reduction of melina growth. The melina did not appear to affect the development of the maize during the first 120 days of intercropping. Maize yield was the same as in monoculture, approximately 5 t/ha.

Assumption 2

Cassava could be grown, just as was maize, between the rows of melina.

Trial. Cassava cuttings were planted immediately after the melina and harvested at 15 months.

Results. Melina growth was not noticeably affected by the associated cassava, but yield of cassava was only 56% of monoculture production on the same site (Internal reports, Jari Florestal e Agropecuaria 1979–1981). Cassava development was very good for about nine months, until the melina canopy began closing, when the reduced light reaching the cassava apparently greatly decreased production. (Cassava varieties exist that produce mature tubers within nine months, but none were locally available or tested.) At 15 months, cassava yields were 10 t/ha as compared to 18 t/ha in sole cropping. Tubers were noticeably smaller than from normal sun-grown crops. Although not measured, there was also obvious damage to the melina roots due to harvesting the cassava.

Discussion

The melina crowns were measured to project the planting density necessary to provide full sunlight to the cassava for 15 months. It was determined that a 40% decrease in the number of trees, corresponding to a 50% reduction in annual melina yield per hectare, would be required to provide full sunlight in asymmetrical spacings. A 45% decrease would be required for square spacings. One result of this trial was that both cassava and melina were found to yield more as monocultures, even on the clay soil.

In contrast, a simultaneous field trial of cassava intercropped with pine gave very good cassava results on both clay and sandy soils and the pine appeared to suffer neither short-term nor long-term effects from root damage. Sufficient pine plantation space was available to accommodate both pasture and cassava, so the cassava/melina trials were terminated. If further cassava/melina trials had been desirable, fast-maturing varieties of cassava would have received first attention.

The maturation period of the intercrop and the spacing and growth rate of the trees must be balanced to obtain the desired combined production. The difficulty in accomplishing this increases with longer maturation periods and lesser shade tolerance of crops and with site variability and sensitivity of the tree.

One benefit of intercropping maize or cassava with melina was that weeding costs were virtually eliminated. Also, although sowing and harvesting the maize was an additional production cost, these costs were about one-third the cost of purchasing and transporting maize from distant markets. Given that the company raised 60,000 chickens and 4,000 pigs, this resulted in large savings. The primary objective of the tree/livestock trials was to decrease management costs; an additional benefit was the company's increasing degree of self-sufficiency in meat and staple production for its employees.

With either large or small landholdings, any combination of trees or crops can be grown by adjusting culture techniques. The decision depends on desired products, costs and prices, and landowner preferences. In practice, most decisions normally depend on refinements based on multiple iterations rather than on basic or even applied, formal research. An understanding of plant physiology and a knowledge of species characteristics can greatly hasten determination of a satisfactory system of intercropping.

AGROFORESTRY FOR SOCIAL OBJECTIVES

Combining trees and agricultural crops during establishment of forest stands for one to three years is a practice of long standing in both temperate and tropical environments. This practice is known by several names, *taungya*, *tumpangsari* (Indonesia), *shamba* (East Africa), and is perhaps the most widely recognized type of agroforestry. The *taungya* method, which originated in Burma, was initially designed to provide an alternative to shifting cultivation. It is also a method to reduce tree planting and tending costs. As a reforestation method, *taungya* has produced mixed results. Champion and Seth (1968) describe it as a striking success, but more recent analysis by Evans (1982) and Wiersum (1981) found that additional research is desirable. The renewed interest in *taungya* on government lands is precipitated primarily by the steadily increasing pressures on forest land and, again, as an alternative to shifting cultivation.

Ninety-eight percent of the forest land in Indonesia is government owned. Java represents only 2% of the forest land area, yet supports over 63% of the country's population. This makes population density, averaging 570 people per square km, a very important feature of the forest area (Soekiman 1977).

The competition for food, fodder, and fuelwood on Java is intense. In 1973, the State Forestry Corporation of Java, Perum Perhutani, established a community development program called the Prosperity Approach, to provide enough resources or additional income for the people living near the forests to lessen their dependency on the forests. This dependency has resulted in overuse of the forest area, reduced standing stock, and increased soil compaction and soil erosion.

The constant and increasing pressures on forest land have promoted government officials to explore alternative land management options once again. In 1987, Perum Perhutani implemented a new social forestry program that included renewable contracts to farmers and promoted farmer participation in forest management. These contracts granted access to forest land for establishing and maintaining agroforestry systems. In contrast to the conventional two-year term of *tumpangsari*, these contracts can be renegotiated indefinitely (Winrock International 1988). The Java Social Forestry Project is described briefly as an example of agroforestry used primarily to meet social objectives. The reader is directed to the report by Stoney and Bratamidhardja (1989) for a more detailed description.

Client Problem

Encroachment, primarily by the rural poor, on state forests has undermined traditional forest management strategies and hampered reforestation campaigns. The customary *tumpangsari* system, in which farmers contract with Perum Perhutani to plant and tend trees for two years in exchange for the right to intercrop on forest lands, has not proven effective in offsetting the demands on forest lands (Winrock International 1988). Allowing increased use of forest land for food crops by rural farmers, however, increases the risk for Perum Perhutani of permanently losing the state-managed forest land base to agriculture and private stewardship.

Materials and Methods

Perum Perhutani initiated diagnostic studies at 13 research sites on Java to identify factors that contribute to the continuing encroachment on forest lands. Lack of available farm land and inadequate rural incomes were identified as primary factors contributing heavily to the pressure on forest land. Based on the results from these studies, Perum Perhutani began pilot activities designed to address some of the socioeconomic as well as technical considerations that contribute to forest degradation. Agroforestry systems were identified as the key to integrating the social and biological objectives.

Social forestry trial

Trial

In 1987, Perum Perhutani agreed to a pilot program in which the local people have annually renewable rights to designated reforestation areas for an estimated half timber rotation. Farmers can grow any crop from a list of approved crops. In return, the farmers must plant along contours for erosion control and not damage timber trees in any way. The tenure rights cannot be sold, but they can be cancelled for gross mismanagement.

Currently, the farmer is paid for planting and early tending of the trees. Hedgerow and fruit trees are available at no cost or at a heavily discounted price. The fertilizer and seed for approved crop species are subsidized. Technical assistance and marketing advice are available. If the program is successful in increasing the yields of food, tree crops, and other forest products, while reducing soil and water losses, there are tentative plans to grant successful farmers an additional parcel of land.

Results

As this is a relatively new project, measurable results have not yet been reported. But it is already apparent that the success of this program depends on overcoming some of the biological, social, and political conflicts inherent in managing natural resources jointly for social and biological objectives.

Discussion

The Java Social Forestry Project clearly illustrates the interaction between social, biological, and economic factors that underlie most forest management today. It also illustrates the need for an interdisciplinary research approach, more obvious in this example because of its stated social objective, but present in most forest management systems nonetheless. Four topics outlined below need immediate research attention.

Crop composition over the rotation

It is officially assumed that shade-tolerant species (for example, cardamom, ginger, taro) will replace the intolerant crops (such as rice or maize) as the trees mature. It is not evident that this is economically viable. Research investigating the crop/tree interactions and the economic feasibility of the various crop combinations is required.

Plant management of the tree crop

The relatively open-grown trees may require increased cleaning and pruning to preserve timber quality; intensified tree culture requires increased labor by the farmers. Research to identify and evaluate the different plant management alternatives (see Cannell, this volume) and, correspondingly, appropriate compensation for the workers is necessary.

Effectiveness of contour plantings

Thus far, the practice of planting hedgerows along the contours is not widespread. In addition, hedgerow density is often inadequate to reduce erosion. Additional erosion barriers, such as noninvasive grass and/or double hedgerows on the steeper slopes, may be required. Research to refine and improve the effectiveness of contour plantings would be useful. Sociocultural research directed at increasing the use of this practice may be needed.

Sociocultural implications

Agroforestry applied to forest lands requires increased biological, social and political management by the forest department. The current system assumes the land now belongs to, and will remain the property of, the government under the management of the forest department. Whether or not this is, or should be, the case is a concern to many. Sociocultural research directed at providing insight into farmers' and foresters' perceptions of land tenure issues and into the social processes involved in tenure issues would be particularly beneficial to the program.

FUTURE RESEARCH DIRECTIONS

Agroforestry research has focused primarily on small farm applications. The results from some of this research (for example, species interactions) can be incorporated into designing agroforestry systems for large farms and for industrial and public forest lands. The basic lack of biological information, however, limits all agroforestry design and implementation, regardless of scale.

An underlying assumption of agroforestry research is that this system permits sustainable cultivation of agricultural and timber crops, but this assumption must be translated into quantifiable, testable hypotheses if agroforestry is to become less dependent on trial and error and provide more predictable results. For example, a hypothesis might be, "Agroforestry systems can produce sufficient crop yields and tree growth to significantly decrease the abuse of forests and forest land." To make such a hypothesis testable, the species involved, the measures of crop yield (kg/ha) and tree growth (m^3/ha/yr), and the type of abuse (for example, soil degradation, unrestricted wood cutting) must be specified.

Experience, observation, and preliminary testing suggest a number of specific hypotheses; examples are given below. Definitive studies are needed to determine what guidelines are reliable, or to define their limitations.

1) In the humid and subhumid tropics, a light shade (10% to 15% ground covered vertically with tree crown) produced fodder and forage of maximum palatability, quantity, and nutritional value.

2) Except where temperatures are uniformly high, fruit and grain production is maximum under full sunlight.

3) In temperate regions and at high elevations in the tropics, shade with root competition reduces yield of all crops, vegetative and fruit.

4) Without root competition, all crops benefit from light shade during hot weather.

5) Windbreaks improve crop yield at distances of 2 to 15 times windbreak height if, and only if, wind is frequent, violent, hot, or dry.

6) In alley cropping, a double row of trees produces nearly double the biomass of a single row, but increases competition with the crops by less than 50%.

7) A high canopy reduces crop yields less than a low canopy.

8) A broken canopy with frequent sun flecks reduces crop yield less than does uniform shade.

Large-scale silvicultural applications of agroforestry are dominated by trial-and-error management. This is caused in part by the lack of existing information necessary to design agroforestry systems, but is also due to a general skepticism about the benefits to be gained from formal research. As the number of successful silvicultural applications has increased, more formal research methods have been applied, but the number of field trials is tiny in comparison to the potential applications.

Differences in diversification, networking, costs, and complexity are some of the reasons industrial managers have been reluctant to expand into agroforestry. Economic analysis and marketing research can address some of these concerns. Unraveling and evaluating the biological and management complexity inherent in agroforestry systems requires formal investigation focused first on species/site interactions and cultural operations.

LITERATURE CITED

Briscoe, C.B. 1979. Integracao silivi-agropastoril: Jari Florestal e Agropecuaria Ltda. In: *Proceedings of 13a Reun Anua, Soc Brasileira para o Progresso da Ciencia*, Fortaleza, Brasil.

Briscoe, C.B. 1983. Integrated forestry-agriculture-livestock land use at Jari Florestal e Agropecuaria. In: *Proceedings of ICRAF plant research and agroforestry, 1981*. Nairobi.

Champion, H.G., and S.K. Seth. 1968. *General silviculture for India*. Government of India. 511 p.

Evans, J. 1982. *Plantation forestry in the tropics*. Oxford: Clarendon Press. 472 p.

McQueen, I.P.M. 1978. Agroforestry in New Zealand. In: *Integrating agriculture and forestry* (K. M. W. Howes and R. A. Rummery, eds.). Townsville, Queensland, Australia: CSIRO. pp 71–79.

Riitters, K., J.D. Brodie, and D.W. Hann. 1982. Dynamic programming for optimization of timber production and grazing in ponderosa pine. *Forest Science* 28(3): 517–526.

Soekiman, A. 1977. Community development programmes in the forests of central and east Java. In: *Indonesia: an evaluation of its merits and prospects*. Misc/77/25. FAO, Rome. 23 p.

Stoney, C., and M. Bratamidhardja. 1989. Identifying appropriate agroforest technologies. In: *Farmers and forests: land management alternatives in Southeast Asia* (Mark Poffenberger, ed.). New Hartford, Connecticut: Kumarian Press.

Wiersum, K.F. 1981. Tree gardening and taungya on Java: Examples of agroforestry techniques in the humid tropics. In: *Viewpoints on agroforestry*. Wageningen, The Netherlands: Agricultural University. pp 165–171.

Winrock International. 1988. *Indonesia: Java Social Forestry*. Project profile. Morrilton, Arkansas: Winrock International. 4 p.

APPENDIX

The most important problems and corrective actions that were taken in the livestock/tree trials discussed in the chapter are briefly outlined here.

Toxic plants

The initial trial was simply for cattle to graze the native grasses, forbs, and woody perennials in the established pine plantations. To prevent repetition of the poisioning that killed nearly 30% of the first lot of cattle, abundant palatable and nutritious grass, which the animals preferred, was sown, and the toxic plants were killed when encountered.

Grass competition to the pine

Broadcasting grass seed gave excellent establishment and growth was vigorous, resulting in intense competition, physical overtopping by the grass, and trampling by grazing cattle. Nearly 90% of the pines were eliminated.

The competition to the pine by the grass was reduced to nominal amounts by seeding grass in two rows 1m apart, centered between tree rows 4m apart. This separated the grass from the nearest pine by 1.5 m. By the second year, the sown grass had matured and self-seeded the entire area, but the pines were then large enough to be virtually unaffected by grass competition and to escape casual trampling by the cattle.

Grass suppression by the pine

After six years, the grass was perceptibly less dense and growing less vigorously because of suppression by the overtopping pines.

To provide maximum sunlight to the grass, pine planting was changed from the original 3 × 3 m spacing to a spacing of 2.25 m apart in rows 4 m apart. The modified spacing kept the same original 1,111 trees/ha, but provided more light longer between tree rows.

To maintain the grass yield necessary for meat and milk production, it was necessary to thin the pines at about ages six and ten years. Thinning would not otherwise have been necessary for the pines at age six, but it did have the advantage of providing long-fiber wood to the pulp mill at an earlier date and, of course, increased diameter growth of the residual trees. To maintain rapid tree growth, the thinning at age 10 would have been necessary with or without cattle.

Control of cattle movement

Pasture management consisted primarily of control of cattle movement; this requires close herding or fences. Because these pine plantations were cleared by logging followed by broadcast burning, there was considerable debris in some areas. Close herding was very nearly impossible; fences were used, and cattle were shifted between pastures as forage condition indicated. Although the amount of debris on the ground decreased rapidly, the growing pines also interfered increasingly with cattle handling, so the fences remained essential.

Soil degradation

No measurable chemical degradation was descernible over the first 10 years. Because the sites selected were on sandy soils, compaction did not occur despite year-round use and 225 cm of rainfall. The exception was on bedding and holding grounds where the animals congregated often and at high density. The lost pine yield on such sites was nearly complete, and it is assumed that much of the 15% reduction of tree yield was on bedding grounds.

Bedding grounds were, therefore, limited to the minimum number necessary and, where possible, were located (1) on the sites of lowest inherent productivity and (2) far from high-quality sites. Fortunately, the cattle preferred bedding down on sandy ridgetops, which were generally the least productive sites and fairly common.

Use of resources

When the cattle were transferred to the pine plantations, the number of water buffalo (*Bubalus bubalis*) was expanded in the lower-grade natural pastures of the original pastures, flood plain, and savanna. They used such lands more efficiently than had the cattle, particularly the flood plain. A brief test of the buffalo in the young pine plantations was disastrous for the pine.

PART THREE

Investigating Biological Interactions

CHAPTER 7

Interactions of Light, Water, and Nutrients in Agroforestry Systems

C.K. Ong

Many people regard agroforestry as a logical solution to the severe problems of fuelwood shortage and degradation of marginal agricultural land in the tropics. A major assumption behind this view is that intercropping trees and arable crops substantially increases biomass production per unit area because the roots of trees can exploit water and nutrients below the shallow roots of crops, and a mixed canopy can intercept more solar energy. Although there is a vast amount of knowledge about single species stands in both agriculture and forestry (Huxley 1983), evidence verifying these assumptions is lacking even in alley cropping, the most common agroforestry system.

The current status of agroforestry research is similar to that of intercropping about 15 years ago: it is described as a complex system involving numerous interactions among species in relation to physical resources, pest levels, and environmental modifications (Baker 1974). A systematic approach to intercropping research at ICRISAT and elsewhere for a decade has revealed some of the major underlying principles that govern the substantial yield advantages of intercropping over sole cropping (Willey 1979). Such principles have important implications for the present effort to examine the interactions of physical resources in agroforestry systems.

This chapter deals with the biophysical principles of crop productivity, examines the factors that promote positive or negative interactions between trees and crops, and describes some examples of agroforestry experiments. To illustrate some of these concepts, examples are given from work on intercropping at ICRISAT that represent the most systematic information available on mixed species cropping.

BIOPHYSICAL CONCEPTS OF CROP PRODUCTIVITY

The four basic biophysical elements affecting crop productivity are light, water, nitrogen, and certain other nutrients, particularly phosphorus and potassium. How each of these contributes to crop yield is briefly discussed below.

Light

When water is not limiting, the dry matter (W in g/m^2) produced by vegetation is linearly related to total intercepted radiation. Squire and colleagues (1987) represent this relationship by

$$W = Sfe\,dt \qquad (1)$$

where

- S = is the total radiation (mean of daily totals) (MJ/m^2)
- f = the fraction of mean daily insolation intercepted by the canopy
- e = the amount of dry matter formed per unit radiation intercepted (conversion coefficient) (g/MJ)
- t = the duration of crop growth in days

The value of S varies from 12 to 30 MJ/m^2 in the tropics. The leaf area of the vegetation determines f at any time, and f can be related to the leaf area index by an extinction coefficient that depends mainly on the orientation and distribution of foliage.

Figure 7.1a illustrates the differences in the seasonal trend of f for pearl millet (t = 80 days), groundnut (105 days), and pigeon pea (180 days) at ICRISAT, where mean annual rainfall is 800 mm. Pearl millet intercepted a total of 600 MJ/m^2, groundnut 940 MJ/m^2, and pigeon pea 1,560 MJ/m^2 during the 1986 growing season (ICRISAT 1986). Intercropped groundnut and pigeon pea (figure 7.1b) intercepted 15% more solar radiation than did sole-cropped pigeon pea mainly because the rapidly developing groundnut canopy reached maximum f by 45 to 50 days, while the slower-growing pigeon pea took 90 to 100 days. The groundnut/pigeon pea combination is a good example of the temporal sharing of physical resources; each makes major demands on resources at different times of the season leading to improved light interception (Willey et al. 1986). This is probably the most common cause of higher productivity in intercropping. Another mechanism for increasing biomass production in intercropping is the improvement in the conversion coefficient (e) by a spatial sharing of solar radiation, typical of the millet/groundnut combination. For example, a combination of one row of millet and three rows of groundnut resulted in a 28% increase in biomass, largely due to a 27% improvement in the value of e (Marshall and Willey 1983).

These examples of positive interactions for light between crop species also provide a physiological basis for the management of mixed canopies for agroforestry interventions. The first strategy is to adopt the temporal sharing concept (for example, the groundnut/pigeon pea system) by pruning the tree so that a fast-growing crop like sorghum can intercept most of the solar radiation during the rainy season and allow the tree to regrow after the removal of sorghum during the dry season (figure 7.1c). This is basically the principle of hedgerow management in alley cropping, although light interception is seldom measured.

Another strategy is to allow the tree canopy to intercept radiation during the early part of the season when the water supply is favorable but the crop canopy is too open to intercept more than a small fraction of incident solar energy. Once the crop canopy becomes nearly closed, the trees should be pruned (figure 7.1c). Both strategies should result in more seasonal light interception than with sole cropping and, hence, more biomass production per unit area.

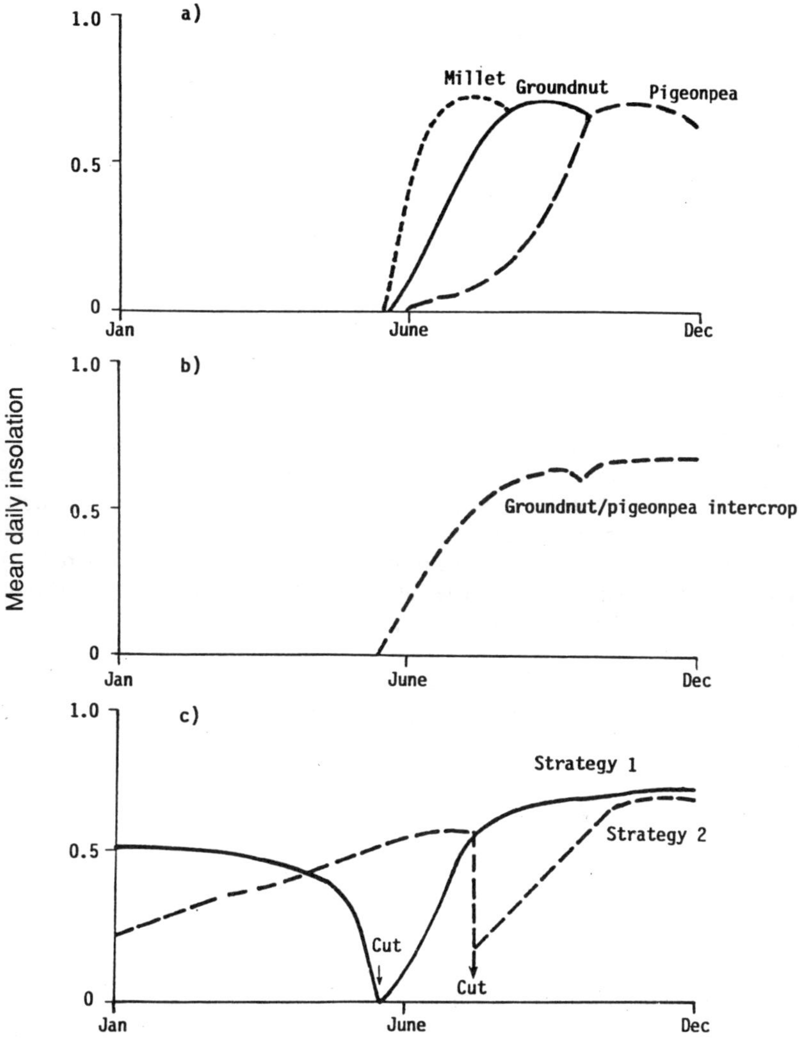

Figure 7.1. Diagramatic representations of cropping and agroforestry systems.
Source: ICRISAT 1986.

The improved grain yield of intercropping is often associated with increased partitioning of dry matter to the grains or pods. For example, in a sorghum/pigeon pea combination that has a 60% yield advantage, Natarajan and Willey (1980a) reported an increase in the harvest index of pigeon pea from 19% when sole cropped to 32% when intercropped. Thus, although the intercropped pigeon pea accumulated only 41% of the sole crop biomass, the seed yield was 70% of that of the sole crop yield. Such improvements in partitioning to reproductive parts have not yet been reported for crops in agroforestry systems.

Measurement of light interception

The fraction of incident light intercepted by a stand is usually measured by placing quantum sensors or tube solarimeters above and below the canopy. Tube solarimeters that integrate radiation over an area of 20 mm × 90 mm are commonly used with crops planted in narrow rows (Szeicz et al. 1964), but in plantation or forestry systems, a travelling quantum sensor is more practical (Norman and Jarvis 1974). Another method for estimating the interception of a canopy is through use of fisheye pictures. An 8 mm fisheye lens should be used with an ordinary camera placed on the ground or above the understory canopy in diffuse light. For details of calculations see Fuchs et al. (1984).

Water

Water shortage is a major restraint to plant productivity in the tropics, even in regions where rainfall is high (> 2000 mm), because distribution of rain is seldom uniform, and losses due to runoff deprive roots of moisture. Equation 1 can be used to examine how water shortage affects the components that contribute to crop productivity by reducing both f and t. Further details are provided by Squire and colleagues (1987). A more direct approach is to analyze productivity in terms of the limiting factor itself, namely, the supply of water. Therefore, dry matter (W) can be represented as the amount of transpired water (E) and the amount of dry matter produced per unit of water extracted (q), which is strongly dependent on the saturation deficit of the air (D) (Squire et al. 1987). This relationship may be expressed as

$$W = E(qD) / D \tag{2}$$

where qD is a conservative quantity with a characteristic value for each crop species. In pearl millet, Squire and colleagues (1987) reported a total biomass of 6 t/ha and q of 3.9 g/kg in Hyderabad, India, and 1.7 t/ha and q of 2.1 g/kg in Niamey, Niger, during the dry season. In that comparison, D was 2.4 kPa in Hyderabad and 4.0 kPa in Niamey.

In summary, the dry matter produced by a crop is determined both by the amount of water extracted and by the evaporative demand of the air (D). This has particular relevance to the management of agroforestry systems since a common objective is to promote the growth of trees beyond the cropping period when the saturation deficit is high.

Returning to the evidence from intercropping, the best example of a positive interaction for improved water use is the case of a pearl millet/groundnut intercrop described by Vorasoot (1982). In this study, however, total transpiration was similar for both sole and intercropped groundnut. The improvement in q of the intercrop was consistent with an increase of e, which is explained by the lower radiation level at the groundnut canopy. In a sorghum/pigeon pea system, the total amount of transpiration is similar to that of monocropped pigeon pea, although there is less evaporation from the soil surface in the first two months of the intercrop (Natarajan and Willey 1980b). In the above examples of intercropping, the rooting depth was similar to that of monocropped groundnut or pigeon pea; therefore it is not surprising that there was no appreciable difference in e. In contrast, with agroforestry combinations involving deep-rooting species, the total water uptake should increase substantially.

Measurements of water uptake

Methods of measuring total evapotranspiration from soil plus vegetation (Et) or transpiration (Ev) are variable, and their suitability depends on the objectives of the experiments. It is not possible to give a comprehensive discussion of all the methods within the scope of this chapter, but a review of the methods of measuring transpiration in agroforestry systems is available at ICRISAT (Black 1986). The methods included here are for determining how the total water of a specific agroforestry or water-use system is partitioned between the tree and crop components. Not considered here are methods that require complex measurements (such as micrometeorological approaches) or are unsuitable for drought-prone environments (for example, pan evaporimeters and mass transfer models).

The most common method is the soil water balance approach (Mcgowan and Williams 1980) that measures consumptive water use by crops on a wide range of soils. It is by far the best method for measuring stored soil moisture, but it is unsuitable for variable or stony soils, deep-rooted species, and areas where roots reach the water table. The second most widely used method in crop studies is the porometery approach (Black et al. 1985), but it is technically demanding and difficult for tall trees. Its main advantage is the inclusion of diffusive resistance (stomatal conductance) terms and its suitability for stressed and unstressed vegetation. It is useful for monitoring the transpiration of understory crops. The stem flow technique (Marshall 1958), particularly suited for the measurement of transpiration of trees, is the third method. Recent advances in electronic and data-logging devices have overcome the technical complexity involved, but they remain costly. At ICRISAT, a combination of all these methods has been used to partition the transpiration of a scattered tree system since there is no ideal method for both trees and crops. However, measuring transpiration by the methods listed above is not very useful unless the growth and development of both species is also regularly measured.

Nitrogen

Association between a nonlegume and a legume is generally considered beneficial because it is assumed that the nitrogen economy of the combination is improved (see Avery, this volume). Experimental evidence, however, is not conclusive because few measurements of nitrogen fixation have been made, and nitrogen effects are very often confounded with other interactions. After a decade of research on the nitrogen economy of intercropped legumes and cereals at ICRISAT, it became obvious that direct or current transfer of nitrogen to cereals is negligible, and cereals benefit only after the harvest of the legume (Willey et al. 1986). A similar conclusion is reached in a recent review of cereal/legume intercropping systems by Ofori and Stern (1987). An important consideration is the total amount of nitrogen that a legume might return to the soil. Unfortunately, even the residual benefit of intercropping with a legume is not well understood; much speculation is based on findings in sole legume cropping where an equivalent of 15 kg to 40 kg/ha of fertilizer N can be provided (ICRISAT 1987).

Current evidence suggests that residual nitrogen from an intercrop legume is the main benefit, but this benefit is not simply a proportion of the contribution observed in sole legume cropping. Several studies have compared the response of monocropped cereals to cereals intercropped with legumes, but have not considered monocropped legumes. For example, Nair and colleagues (1979) and Searle and colleagues (1981) reported substantial responses of wheat planted after intercropped maize/cowpea, maize/groundnut, and maize/soybean combinations but none after a sole-cropped legume. Recent evidence from ICRISAT suggests that the residual effect of pigeon pea/sorghum on subsequently planted maize is negligible, whereas sole-cropped pigeon pea increased grain yield of subsequently planted maize by 57% (Kumar Rao et al. 1983).

In one of the most careful estimates of the nitrogen budget of an intercrop, Ofori and Stern (1987) showed that a maize/cowpea intercrop resulted in a loss of 14 kg/ha of N from the soil, compared with a loss of 21 kg/ha of N for monocropped maize; but monocropped cowpea produced a gain of 36 kg/ha of N.

Several factors could have contributed to the low beneficial effect. First, the cereal in the intercrop is more competitive than the legume and greatly reduced the growth of, and presumably the amount of nitrogen fixed by, the legume. There is evidence that the nitrogen-fixing capacity of the legume is less in an intercrop system, although the dry matter per plant was virtually unaffected (Nambiar et al. 1983). Furthermore, most of the nitrogen of the intercropped legume is translocated to the grains, which are removed by harvesting. Unpublished data from ICRISAT indicates that there is virtually no difference in the economic value of yield from a millet/groundnut intercrop or a rotation of groundnut-millet-groundnut over a three-year period. This is probably due to the lower residual benefit of intercropped groundnut.

Other Nutrients

The relative uptake of phosphorus(P) and potassium(K) in intercropping and mono-cropping has received considerable attention. In general, legumes are poor competitors for P when intercropped with cereals because legumes have slower-growing root systems. For example, Lai and Lawton (1962) found that maize was more efficient than field beans in taking up ^{32}P-labeled fertilizer. Competition for P began early in a maize/pigeon pea system and continued until final harvest, when total P uptake by the maize was reduced by 25% and by the pigeon pea by 70% (Dalal 1974). Studies on nutrient interactions in intercropping are exclusively concerned with the competition effect on the component crops, and none has examined the long-term consequences of intercropping.

Reports on the competition for potassium in intercropping are similar to that observed for P in that cereals are more competitive than legumes. In a sorghum/pigeon pea system in India, for instance, the uptake of K by pigeon pea was reduced by 87.5% compared to a negligible change for the sorghum (Natarajan and Willey 1980b).

In sharp contrast, experience with alley cropping in Nigeria shows that the prunings of two-year-old *Cassia*, *Gliricidia*, and *Flemingia* are capable of providing huge inputs of N, P, and K into a maize system. Even without the addition of prunings, maize near the hedgerows performed better than those in the middle of the alleys, implying that there was a transfer of nutrients by root turnover or some other means of below-ground improvement (Yamoah et al. 1986). The author speculated that the benefit of alley cropping may be related to an improvement in both physical and chemical properties of the degraded Alfisol since the sole maize roots were restricted.

These observations on intercropping provide valuable guidelines for planning agroforestry systems. A relevant comparison is the common assumption that the presence of a leguminous tree species brings about direct and residual nitrogen benefits to associated crops. Such benefits are probably minimal in many agroforestry systems when the trees are pruned frequently and the fodder is taken away from the site. Nutrient enrichments are regularly reported only when large quantities of prunings are used as mulch (Kang et al. 1985).

ENVIRONMENTAL MODIFICATIONS

In semiarid India, changes in radiation level have some impact on leaf temperature and the water status of intercropped groundnut during the post-rainy season. Harris and Natarajan (1987) suggested that the amelioration of the microenvironment was particularly important during peg formation, allowing more pods to be formed and leading to a 12% to 68% improvement in the harvest index. In contrast, measurements of the microenvironment (wind speed, temperature, humidity) in an alley-cropping system involving *Leucaena* and pearl millet showed that the modification was slight and unlikely to have a major effect on crop growth because conditions during the rainy season were already favorable (Corlett et al. 1989). In this experiment, the hedgerow height was kept at 0.7 m. The environmental modifications, therefore, were small, and as the millet started to elongate, there was no appreciable difference in the height of the two canopies.

In another alley-cropping trial where *Leucaena* hedgerows were allowed to grow to a height of 3 to 4 m with alleys 10 m wide, the main modifications were reductions in wind speed and light. In intercropping, such changes in the microenvironment should result in the changed plant development and growth reported by Harris and Natarajan (1987), provided competition for soil moisture is not severe. Results from both alley-cropping experiments clearly showed that in semiarid India, environmental modification is relatively unimportant compared to the root competition between trees and crops (Singh et al. 1989).

LAND EQUIVALENT RATIO

The concept of land equivalent ratio (LER) is the most widely accepted index for evaluating the effectiveness of all forms of intercropping (Willey 1979). It is the ratio of the area under sole cropping to the area under intercropping, at the same level of management, that gives an equal amount of yield. It is the sum of the fractions of the yields of the intercrops relative to their sole crop yields. This is expressed as

$$\text{LER} = \frac{X_i}{X_s} + \frac{Y_i}{Y_s} \qquad (3)$$

where X and Y are the component crops in either an intercrop (i) or a sole crop (s) system. When LER=1, there is no advantage to intercropping over sole cropping. When LER > 1, more land is needed to produce a given yield under sole cropping than is required to produce that amount of yield by each component as an intercrop. An important criterion for such calculations is that the choice of sole treatment of each crop should be the optimal treatment for the site. The use of a suboptimum population, for example, for even one of the crops, would favor intercropping. The land equivalent ratio is generally used for comparing biomass or grain production but, due to lack of measurements, has rarely been used to analyze resource utilization.

The land equivalent ratio is a very effective tool to separate the light interaction in mixed canopies and is best illustrated by the work of Marshall and Willey (1983). They measured the light intercepted by each of the components in a millet/groundnut system during the rainy season in Hyderabad, India, and expressed the total intercepted radiation and energy conversion coefficient (*e*) in terms of LER (table 7.1). The intercropped millet intercepted twice as much light as the sole-cropped millet per row; the intercropped groundnut intercepted only 73% of the light that sole-cropped groundnut did, but produced the same biomass.

Table 7.1. Total intercepted radiation (MJ/m^2) and conversion coefficient (g/MJ) of a millet/groundnut canopy in terms of land equivalent ratio (LER), all values in ratios

	Millet	Groundnut
Fraction of total population	0.25	0.75
Total intercepted radiation	2.12	0.73
Conversion coefficient	0.97	1.46
Component LER	0.51	0.80
Combined LER	1.31	

Source: Marshall and Willey 1983.

In contrast, while the conversion coefficient (*e*) of intercropped millet was almost unchanged, the value of *e* increased by 46% for the intercropped groundnut. Multiplying the fraction of plant population by the total intercepted radiation and conversion coefficient gave a component LER that showed both component crops exceeded their relative performance as sole crops, resulting in a combined LER of 1.31. This is one of the best examples of a positive interaction for resource utilization in intercropping. The land equivalent ratio is further used to demonstrate the above and below-ground interactions in the same millet/groundnut system (Willey and Reddy 1981). Polythene sheet partitions were installed to a depth of one meter to separate the root system, and irrigation was provided to bring soil to field moisture capacity at sowing. Total dry matter at final harvest showed that root partitioning had a negligible effect on LER, suggesting that yield advantages were largely due to the better use of light and not to the transfer of nitrogen from the legumes or more uptake of water.

AGROFORESTRY EXAMPLES

Agroforestry research in the humid and semiarid tropics is confined largely to *Leucaena* in various alley-cropping arrangements (Kang et al. 1985, Singh et al. 1988). A common conclusion from such studies is that by incorporating trees with arable crops the biomass production per unit area is increased substantially. This conclusion, however, is based on a comparison of the productivity of sole crops and mixed alley crops without an appropriate sole *Leucaena* treatment. Even when available, a sole *Leucaena* treatment is rarely managed optimally for biomass production. The most common practice is to prune the *Leucaena* in the same way as for an agroforestry treatment.

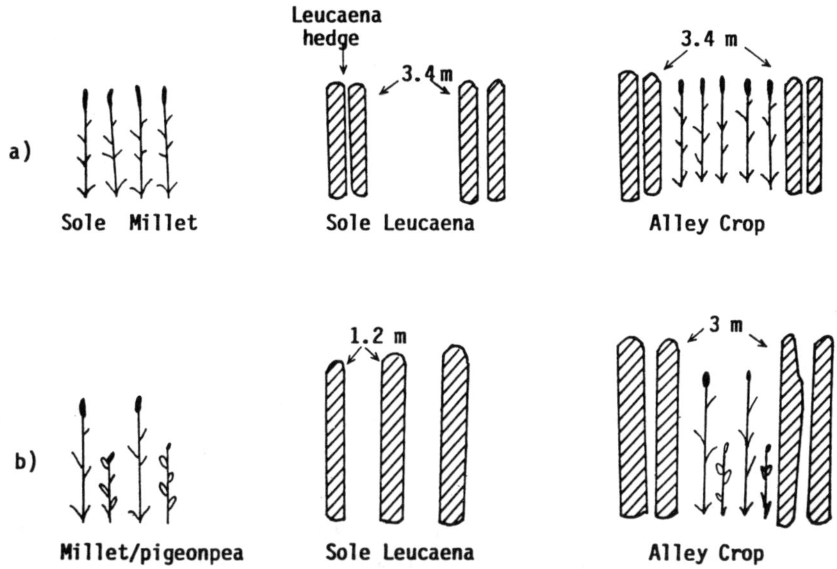

Figure 7.2. Examples of agroforestry trials with *Leucaena* and millet on Alfisols.
Source: ICRISAT 1986.

Example A

A typical example of this kind of trial was described by Corlett and associates (1989), where the sole *Leucaena* treatment consisted of hedges of double rows of *Leucaena*, spaced 60 cm between the two rows of each hedge and 25 cm between trees within each row. These hedges were spaced at 3.4 m in both sole-cropped *Leucaena* and alley-cropped plots incorporating pearl millet. The millet row spacing was 47 cm, both within the alleys (five rows in each alley) and in the plot of monocropped millet (figure 7.2a). The hedgerows were cut to 70 cm prior to sowing the millet, at 30 days after sowing, and again at millet harvest (another 40 to 45 days).

During the rainy season, both sole-cropped millet and the alley system produced about the same biomass (6.5 t/ha), while the sole *Leucaena* produced 4 t/ha (table 7.2). The high productivity of the millet was due to the greater value of light conversion efficiency, although the amount of energy intercepted was less than for the alley crop. The value of *e* for the alley crop was remarkably similar to that of the sole-cropped *Leucaena*, perhaps indicating that the *Leucaena* canopy was more dominant than the millet canopy. Because the *Leucaena* was pruned 30 days after the millet was sown and because the hedges were widely spaced, the sole-cropped *Leucaena* intercepted the least radiation. Therefore, this treatment did not represent the optimal management of *Leuceana*. In the following year, when the *Leucaena* was uncut during the whole rainy season, the biomass produced by the *Leucaena* reached 7.1 t/ha compared to 7.3 t/ha for the alley crop. The values of *e* for both years were quite consistent for *Leuceana* but were slightly lower for millet in 1987. However, the amount of radiation intercepted by the sole *Leucaena* was almost double that of 1986. Therefore, any calculation of LER in 1986 would grossly overestimate the advantage of alley cropping. Results from both years indicated that there was no

Table 7.2. Biomass, total light interceptions, and light conversion efficiency of crops and alley crops during rainy seasons at ICRISAT, Hyderabad, India

	Total biomass (t/ha)	Total intercepted radiation (MJ/m^2)	Light conversion efficiency (g/MJ)	Crop yield (t/ha)
1986				
Millet	6.56	575	1.14	1.98
Leucaena	4.01	450	0.89	—
Alley crop	6.53	725	0.90	1.60
1987				
Millet	4.97	512	0.98	2.01
Leucaena	7.14	870	0.82	—
Alley crop	7.35	919	0.80	0.94

Source: Corlett et al. 1989.

positive interaction for light and that, when sole-cropped *Leucaena* canopy was managed optimally, there was no advantage in alley cropping.

Example B

In the previous example, the alley crop was compared in the second and third year of the trial, and the sole-cropped stand of *Leucaena* was suboptimal in plant population. The productivity of the three systems should be compared over a longer period to arrive at a more realistic assessment. This consideration is important because *Leucaena* grows very slowly in its first year. In an experiment conducted near example A, again on shallow Alfisols, *Leucaena* cv. Cunningham was established in June 1984 in a randomized block design. Again, *Leucaena* was grown in paired rows (figure 7.2) but at two different alley spacings, 3.0 m and 5.4 m, and hedges were pruned as in example A. However, the sole *Leucaena* treatments were kept at a spacing of 1.2 m × 0.25 m, which was optimal for canopy development and dry matter production.

During the year of establishment, the *Leucaena* grew slowly and was not pruned. Dry matter production in 1984 and 1985 was low in all three treatments, probably because of exceptionally low rainfall (table 7.3). The

Table 7.3. Biomass production of crops, *Leucaena*, and alley crops on a shallow Alfisol, ICRISAT, India

	1984	1985	1986	1987	1988*	Total
Total biomass (t/ha)						
Annual crop	3.82	2.53	7.16	8.00	–	21.51
Sole *Leucaena*	–	5.04	10.33	11.15	6.17	32.70
Alley crops — 3.0 m	2.66	3.60	11.34	9.30	5.13	32.03
Alley crops — 5.4 m	3.17	2.18	10.25	7.62	4.20	27.42
Off-season biomass (t/ha)						
Sole *Leucaena*	–	0.40	3.30	5.13	6.17	15.00
Alley crops — 3.0 m	–	0.22	2.70	3.74	5.13	11.80
Alley crops — 5.4 m	–	0.09	2.02	3.10	4.20	9.41
Annual rainfall (mm)	655	577	713	879	115	
Annual crops	M/Pp	C	M/Pp	G	–	

*Off-season values January to June.
M = Millet; Pp = Pigeon pea; C = Castor; G = Groundnut.

biomass production of all treatments responded markedly to more rain during 1986 and 1987. Over the four-year period, sole-cropped *Leucaena* still produced the greatest amount of biomass followed closely by the yields in alley cropping. The annual crop treatment produced 66% of the biomass of sole-cropped *Leucaena* over the same period. From 1985 onwards, sole-cropped *Leucaena* was generally the most productive system, even though it was pruned twice during the cropping season. The superiority of the sole-cropped *Leucaena* stand over the alley crop was clearer during the off-season from January to early June. The greater production must indicate that sole-cropped *Leuceana* exploited more water than the alley crop since the value of *e* was similar for sole-cropped *Leucaena* and the alley crop (table 7.2). Even in this trial, the sole-cropped *Leucaena* could have been managed better, however: a calculation of LER would still favor alley cropping.

Example C

The above examples highlight the main deficiencies of agroforestry trials that emphasize the comparison of alley crops with sole crops without exploring the best ways of managing sole-cropped *Leucaena*. Without such a control, it is wrong to assume that alley cropping is the most productive means of introducing trees into a farmer's field. At present, there is no complete data to test the hypothesis that *Leucaena* yield is better under alley cropping than under sole cropping. However, there is sufficient data to estimate the productivity of sole-cropped *Leucaena* by using equations 1 and 2.

This trial was designed to examine the effect of alley width and the effect of cropping intensity on the productivity of agroforestry systems on shalow Vertic Inceptisols (figure 7.3). Five hedgerow spacings of *Leucaena* were increased by 0.9 m, from 1.95 m to 5.55 m. Within each spacing, a 1/1 row arrangement of sorghum and pigeon pea was sown at the start of the rainy season. The crop row width was constant (0.45 m) so that there were two rows in the 1.95 m alleys and 10 rows in the 5.55 m alleys. Cropping intensity was altered by reducing the two crop rows nearest the trees at 8 m intervals until a single *Leucaena* row remained. Agronomic results were reported in the 1986 ICRISAT annual report.

In 1986 and 1987, the biomass production was examined by measuring the total amount of solar radiation intercepted by *Leucaena* hedgerows and by three cropping systems: sole-cropped sorghum, sole-cropped pigeon pea, and intercropped sorghum/pigeon pea. Solar radiation interception by *Leucaena* in the sole and alley crop systems was between 850 and 900 MJ/m^2 during the rainy season (170 to 300 Julian days) and between 1,000 and 1,700 MJ/m^2 during the dry season (figure 7.4). Despite the greater interception of radiation in the dry season, the value of *e* was very low (0.21 ± 0.06 g/MJ), and the dry matter production ranged from 1.5 to 3.7 t/ha. In contrast with the dry season, *e* in the rainy season was three times higher (0.61 ± 0.03 g/MJ), and biomass production ranged from 5.2 to 5.5 t/ha. This value for *Leucaena* was lower than that reported in example A, but the value for sorghum was similar to that for millet (1 g/MJ).

When the value of *e* for sole-cropped *Leucaena* and the alley crop was expressed for the whole year, it ranged from 0.32 to 0.45 g/MJ (table 7.4). In this trial, the *Leucaena* hedges were pruned twice during the crop season at 36 and 90 days after the crops were sown. Calculations indicate that, by allowing

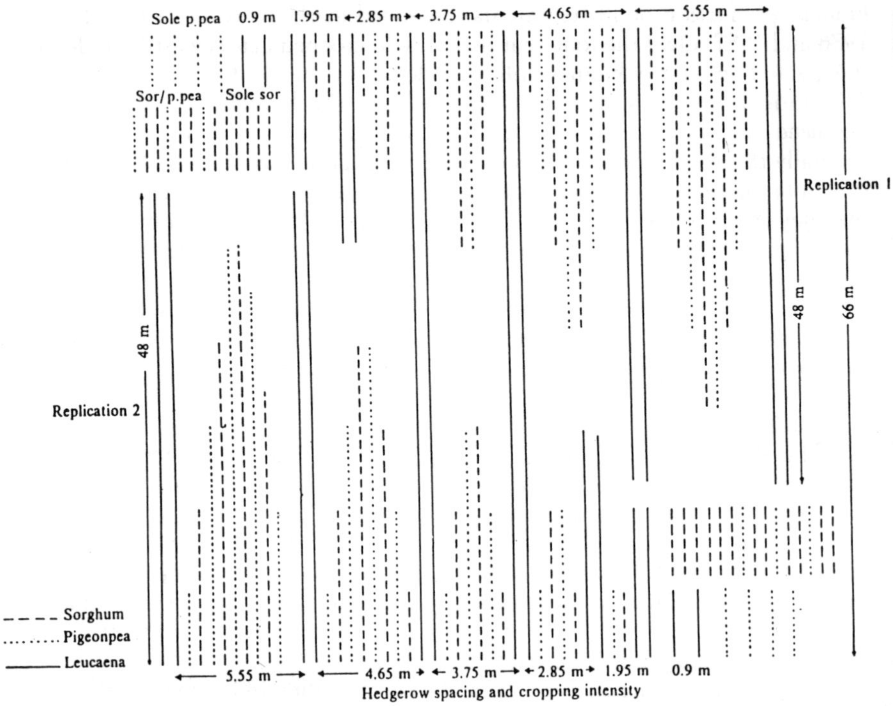

Figure 7.3. Layout of two-way systematic design for agroforestry research on the influence of hedgerow spacing and cropping intensity.

Source: ICRISAT 1986.

the *Leucaena* to grow until late February when fodder demand was high, biomass could reach 25 t/ha due to a combination of greater total intercepted radiation and better value of *e* (table 7.4). This evidence suggests that more biomass would be produced by managing the *Leucaena* optimally as a sole crop rather than by alley cropping.

INTERACTION FOR MOISTURE

Conventional methods for measuring water uptake by agroforestry systems, such as neutron scattering techniques, evaporation models, or diffusion porometry, are extremely tedious and impractical. The latest advances in sap flow techniques have made it relatively straightforward to make direct measurements of transpiration of trees and shrubs, although it is still new to agroforestry (Baker and van Bavel 1987). A simple approach is to calculate the total water uptake, *E*, from equation 2 rewritten as

$$E = \frac{W \times D}{(qD)} \qquad (4)$$

where W is the total biomass produced in kg/m^2, D is the average daily maximum saturation deficit in kPa, and (qD) is the conservative value for each species obtained from the product of q (the dry matter/water ratio) and D. The

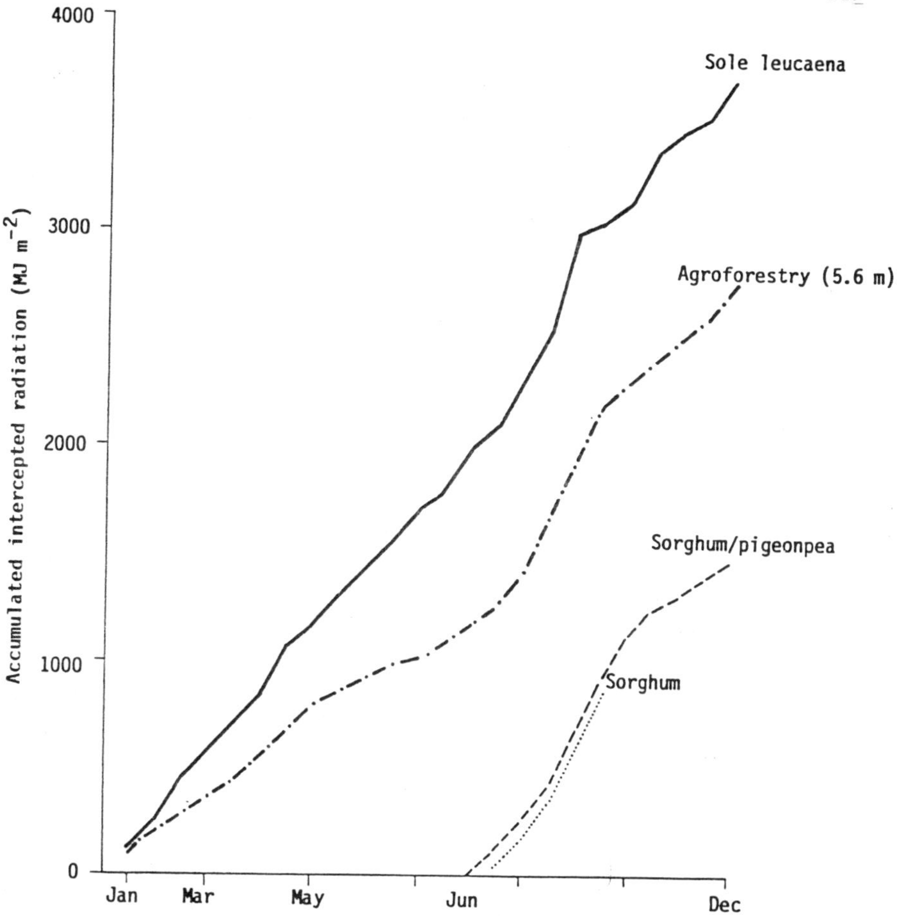

Figure 7.4. Accumulated intercepted solar radiation for agroforestry and cropping systems on Vertic Inceptisols.

Source: ICRISAT 1986.

Table 7.4. Biomass production, intercepted radiation, and radiation conversion efficiency for *Leucaena*, sorghum, and agroforestry combinations on Vertic Inceptisols

	Biomass (t/ha)	Total intercepted radiation (MJ/m^2)	Radiation conversion efficiency (g/MJ)
Sole *Leucaena*			
pruned	11.6	3580	0.32
unpruned*	25.2	4960	0.51
Alley cropping			
3.8 m	11.2	2680	0.42
4.7 m	11.2	2480	0.45
5.6 m	10.6	2740	0.39
Sole sorghum	8.9	890	1.00

* Based on calculations of potential radiation interception and a radiation conversion efficiency of 0.61 g/MJ in the rainy season.
Source: ICRISAT 1986.

value of qD is conservative over a range of tropical environments ranging from 4 g/kg kPa for C_3 species to 8 g/kg kPa for C_4 species like millet and sorghum (Monteith 1988).

For instance, we could estimate E for the sole-cropped *Leucaena* and the alley crop during the dry season in example C. In 1986, sole-cropped *Leucaena* produced 4 t/ha and the alley crop (at an alley width of 5.6 m) produced 1.9 t/ha during the dry season when D was 3 kPa. Calculations (table 7.3) suggest that sole-cropped *Leucaena* would have used 300 mm of water compared to 143 mm by the alley crop. It is probable that greater water uptake by sole-cropped *Leucaena* may be a function of having a deeper root system than the alley crop. This is an aspect of agroforestry research that merits further investigation, especially in conditions where water is the major constraint to productivity.

SUMMARY

This review has highlighted the similarities and differences between intercropping and alley cropping and has pointed out ways in which biophysical relationships can be used to examine the basis for positive or negative interactions between crops for resources. Before attempting to outline future research directions, it is worthwhile to summarize the main differences between the two cropping systems in terms of greater use and more efficient use of resources.

Greater use of resources is probably the most common and best-established basis for higher yields in intercropping over sole cropping. The best examples are from sorghum/pigeon pea, maize/pigeon pea, and pigeon pea/groundnut systems where the LER for grain is consistently higher than for total biomass because of positive interactions in partitioning to grain (table 7.5). The advantage of intercropping over sole cropping, when evaluated in terms of LER, is usually greater than when evaluated in terms of relative productivity (the ratio of the combined biomass of the intercrop per hectare to the biomass of the best sole crop). Where this ratio exceeds 1, it implies that the best sole crop is unable to utilize the full water and nutrient resources, and in some cases, the most appropriate comparison would be a double crop system rather than a sole crop—for example, in the maize/pigeon pea system (Sivakumar and Virmani 1980). At present, there is no evidence that any agroforestry combination has resulted in increased partitioning to grain of the crop, and published evidence suggests that the interaction is mainly negative.

The evidence for more efficient resource utilization is confined to the millet/groundnut system where enhanced partitioning to grain is the primary cause for high LER. Here there is less convincing evidence that intercropping can increase relative productivity per unit of land area because the combined e of the intercrop canopy can never exceed the high value of the C_4 cereal canopy. The data of the *Leucaena*/millet trial (table 7.2) confirmed that alley cropping is unlikely to improve e or increase relative productivity. In general, the most suitable crops for agroforestry are cereals because they are more competitive than legumes, which respond positively to intercropping by allocating more dry matter to grains (table 7.5).

Table 7.5. Summary of resource utilization and relative biomass production

Mechanisms	Species	LER		Relative productivity Intercrop/Best sole	Treatment	Source
		Biomass	Grain			
Greater resource capture	Maize/Ppea	1.80	1.81	1.88	R	Sivakumar and Virmani 1980
	Sorghum/Ppea	1.51 1.23	1.76 1.43	1.44 1.00	P P	Natarajan and Willey 1985
	Millet/Ppea	1.90 1.56	2.35 2.00	1.15 1.36	N N	Natarajan unpublished
	Sorghum/Ppea	1.47 1.57	1.77 1.72	1.04 1.26	N N	Natarajan unpublished
	Gnut/Ppea	1.48	1.58	1.41	R	Willey et al. 1986
	Mean	1.56	1.80	1.28		
More efficient resource utilization	Millet/Gnut	1.28	1.26	1.07	R	Reddy and Willey 1981, Marshall and Willey 1983
	Millet/Gnut	1.29 1.61	1.45 1.36	0.89 1.13	N N	Vorrasoot 1982
	Millet/Gnut	1.43 1.04 0.75	1.67 0.88 1.05	0.30 0.78 0.54	R R R	ICRISAT 1987
	Millet/Gnut	1.02 1.21	1.10 1.84	0.87 1.00	W W	Harris and Natarajan 1987
	Mean	1.20	1.33	0.82		

R = rainfed; P = population; N = fertilizers; W = water regimes.

FUTURE RESEARCH DIRECTIONS

Much of the current attention in agroforestry research is centered on alley cropping with *Leucaena*. This may represent the most extreme agroforestry system because *Leucaena* is highly competitive with crops and because alley cropping is not necessarily the most appropriate system available. Alley cropping was developed for the humid tropics where it was effective in preventing soil erosion and in improving soil fertility through the use of residue mulch and minimum tillage (Kang and Wilson 1987). In many of the cited examples of alley cropping, especially in the arid and semiarid tropics, the use of residue mulch is seldom advantageous because the value of fresh fodder is high and stover and inorganic fertilizers are available at reasonable prices. It is important to realize that the conclusions reached in the above summary are based on alley cropping with *Leucaena*, and future attention should examine the following:

Indigenous Perennial Species

Use of these less competitive species that do not have the aggressive rooting ability of *Leucaena* will enhance partitioning of dry matter to grain when leguminous crops are undersown as in the experience with intercropping. Promising perennial species are *Albizia lebbeck*, perennial pigeon pea (*Cajanus cajan*), and *Acacia ferruginea*.

Simple Experimental Designs

Where the advantages of tree products and potential benefits are well defined, simple experimental designs should be used to explore the range of spatial and population arrangements of trees. Systematic designs are rarely suitable for studies of interaction for physical and chemical resources. As pointed out repeatedly in the discussion, an appropriate sole tree control is essential to test the hypothesis that agroforestry systems are more advantageous than sole-cropping systems.

Yield Advantages of Potential Systems

Evaluations of benefits should also consider soil fertility, water and soil conservation, pest management, and microclimate amelioration.

Regular Measurements

Crop growth and changes in the environment should be measured regularly over a period of time rather than consist of a single final yield measurement. This will promote understanding the causes of differences between treatments and will help extend such generalizations to other locations and situations through modelling.

Below-ground Interaction

This is an important factor, especially as it affects the root distribution of trees and crops.

LITERATURE CITED

Baker, E.F. 1974. Research on mixed cropping with cereals in Nigerian farming systems—a system for improvement. In: *International workshop on farming systems*. Hyderabad, India: ICRISAT. pp. 287–301.

Baker, J.M., and C.H.M. van Bavel. 1987. Measurement of mass flow of water in the stems of herbaceous plants. *Plant Cell Enrivonment* 10: 777–782.

Black, C.R. 1986. *Measurement of transpiration in agroforestry systems*. Consultant's report for ICRISAT. Hyderabad, India. 115 p.

Black, C.R., D.Y. Tang, C.K. Ong, A. Solon, and L.P. Simmonds. 1985. Effects of soil moisture stress on the watrer relations and water use of groundnut stands. *New Phytologist* 100: 313–328.

Corlett, J.E., C.K. Ong, and C.R. Black. 1989. Microclimate modification in intercropping alley cropping systems. In: *The applications of meteorology to agroforestry systems planning and management*. Nairobi: ICRAF. pp 419–430.

Dalal, R.C. 1974. Effects of intercropping maize with pigeon pea on grain yield and nutrient uptake. *Experimental Agriculture* 10: 219–224.

Fuchs, M., G. Asrar, E.T. Kanemasu, and L.E. Hipps. 1984. Leaf area estimates from measurements of photosynthetically active radiation in wheat canopies. *Agricultural and Forest Meteorology* 32: 13–12.

Harris, D., and M. Natarajan. 1987. Physiological basis for yield advantage in a sorhgum/groundnut intercrop exposed to drought. II. Plant temperature, water stress and components of yield. *Field Crop Research* 17: 273–288.

Huxley, P.A. 1983. Some characteristics of trees to be considered in agroforestry. In: *Plant research and agroforestry*. (P. A. Huxley, ed.). Nairobi: ICRAF. pp 3–12.

ICRISAT. 1986 Annual Report. International Crops Research Institute for the Semi-Arid Tropic. Patancheru, Andhra Pradesh, India. 367 p.

ICRISAT. 1987. Annual Report. International Crops Research Institute for the Semi-Arid Tropics. Patancheru, Andhra Pradesh, India. 390 p.

Kang, B.T., H. Grimme, and T.L. Lawson. 1985. Alley cropping sequentially cropped maize and cowpea with leucaena on a sandy soil in southern Nigeria. *Plant and Soil* 85: 267–277.

Kang, B.T., and G.F. Wilson. 1987. The development of alley cropping as a promising agroforestry technology. In: *Agroforestry and a decade of development* (H.A. Stappler and P.K.R. Nair, eds.) Nairobi: ICRAF. pp 227–243.

Kumar Rao, J.V.D.K., P.J. Dart, and P.V.S.S. Sastry. 1983. Residual effect of pigeon pea (*Cajanus cajan*) on yield and nitrogen response of maize. *Experimental Agriculture* 19: 131–141.

Lai, T.M., and Lawton, K. 1962. Root competition for fertilizer phosphorus as affected by intercropping. *Soil Science Society of America Proceedings* 26: 58–62.

Marshall, B., and R.W. Willey. 1983. Radiation interception and growth in an intercrop of pearl millet/groundnut. *Field Crop Research* 7: 141–160.

Marshall, D. C. 1958. Measurement of sap flow in conifers by heat transport. *Plant Physiology* 33: 385–396.

Mcgowan, M., and J.B. Williams. 1980. The water balance of an agricultural catchment. I. Estimation of evaporation from soil water records. II. Crop evaporation: seasonal and soil factors. *Journal of Soil Science* 31: 217–244.

Monteith, J. L. 1988. Steps in crop climatology. In: *Papers presented at international conference on dryland farming, August 1988*, Amarillo, Texas, (in press).

Nair, P.K.R., U.K. Patel, R.P. Singh, and M.K. Kaushik. 1979. Evaluation of legume intercropping in conservation of fertilizer nitrogen in maize culture. *Journal of Agricultural Science* 93: 189–194.

Nambiar, P.T.C., M.R. Rao, M.S. Reddy, C. N. Floyd, P. J. Dart, and R. W. Willey. 1983. Effect of intercropping on nodulation and N_2 fixation by groundnut. *Experimental Agriculture* 19: 79–86.

Natarajan, M., and R.W. Willey. 1980a. Sorghum-pigeon pea intercropping and the effects of plant population density. I. Growth and yield. *Journal of Agricultural Science* 95: 51–58.

Natarajan, M., and R.W. Willey. 1980b. Sorghum-pigeon pea intercropping and the effects of plant population density. II. Resource use. *Journal of Agricultural Science* 95: 59–65.

Natarajan, M., and R.W. Willey. 1985. Effect of row arrangement on light interception and yield in sorghum-pigeon pea intercropping. *Journal of Agricultural Science* 104: 263–270.

Norman, J.M., and P.G. Jarvis. 1974. Photosynthesis in Sitka spruce (*Picea sitchensis* (Bong) Carr.). III. Measurement of canopy structure and interception of radiation. *Journal of Applied Ecology* 11: 375–398.

Ofori, F., and W.R. Stern. 1987. Cereal-legume intercropping systems. *Advances in Agronomy* 41: 41–90.

Reddy, M.S., and R.W. Willey. 1981. Growth and resource use studies in an intercrop of pearl millet. *Field Crop Research* 4: 13–24.

Searle, P.G.E., Y. Comudom, D.C. Sheddon, and R.A. Nance. 1981. Effect of maize + legume intercropping systems and fertilizer nitrogen on crop yields and residual nitrogen. *Field Crop Research* 4: 133–145.

Singh, R.P., C.K. Ong, and N. Saharan. 1989. Microclimate and growth of sorghum and cowpea in alley cropping in semi-arid India. *Agroforestry Systems* (in press).

Singh, R.P., R.J. van den Beldt, D. Hocking, and G.R. Korwar. 1988. Alley cropping in the semi-arid regions of India. In: *Proceedings of workshop on alley cropping, 10–14 March 1988*. Ibadan Nigeria: IITA (in press).

Sivakumar, M.V.K., and S.M. Virmani. 1980. Growth and resource use of maize, pigeon pea and maize/pigeon pea intercrop in an operational research watershed. *Experimental Agriculture* 16: 377–386.

Squire, G.R., C.K. Ong, and J.L. Monteith. 1987. Crop growth in semi-arid environments. In: *Proceedings of the international pearl millet workshop, 7–11 April 1986*, Hyderabad, India: ICRISAT. pp 219–231.

Szeicz, G., J.L. Monteith, and J. dos Santos. 1964. Tube solarimeters to measure radiation among plants. *Journal of Applied Ecology* 1: 169–174.

Vorasoot, N. 1982. *A biological study of the benefits of intercropping in England and India.* Ph D thesis, University of Reading, UK. 184 p.

Willey, R.W. 1979. Intercropping—its importance and research needs. I. Competition and yield advantages. *Field Crops Abstract* 32: 1–10.

Willey, R.W., M. Natarajan, M.S. Reddy, and M.R. Rao. 1986. Cropping systems with groundnut: resource use and productivity. In: *Agrometeorology of groundnut, international symposium proceedings, 21-26 August 1986.* Niamey, Niger: ICRISAT Sahelian Center. pp 193–205.

Willey, R.W., and M.S. Reddy. 1981. A field technique for separating above and below-ground interactions in intercropping: an experiment with pearl millet/groundnut. *Experimental Agriculture* 17: 257–264.

Yamoah, C.F., A.A. Agboola, and G.F. Wilson. 1986. Nutrient contribution and maize performance in alley cropping systems. *Agroforestry Systems* 4: 247–254.

CHAPTER 8

Nitrogen-fixing Plant Interactions in Agroforestry Systems

M. E. *Avery*

The inclusion of nitrogen-fixing plants, primarily legumes, as a component of mixed cropping, traditional bush fallow, and agroforestry systems has led to the belief that nitrogen (N) fixed in the root nodules may be used by the companion crop. Research relating to legume/nonlegume plant interactions has been conducted for decades (Ofori and Stern 1987, Ladd et al. 1986). Yet the results remain inconclusive, and direct evidence of the advantages of intercropping is limited.

The potential for increasing the total N in an agroforestry system by including N-fixing plants (for example, Huxley 1986) is not necessarily an advantage of intercropping. An advantage occurs only if the process of N fixation or the use of fixed N by an associated crop is more efficient than when the crops are grown separately but in some suitable sequence (Willey et al. 1977). By extension, simply using a N-fixing plant in agroforestry is also not necessarily an advantage.

In this chapter, the terms *N-fixing* and *legume* are used somewhat interchangeably for the sake of simplicity, brevity, and deference to the preponderance of data. The purpose of this chapter is to discuss the potential role of N-fixing plants, regardless of the microorganism involved, and not the specific N-fixing symbiosis. Wheeler and colleagues (this volume) provide an excellent review of leguminous (*Rhizobium*) and actinorhizal (*Frankia*) symbioses.

Although we are becoming more proficient at estimating and quantifying the rate of N fixation (see Wheeler et al., this volume), we are only beginning to identify and quantify the benefit of the fixed N to the associated crop. The most commonly offered explanations for the apparent benefit are (1) underground transfer, whether by direct excretion of nitrogenous compounds (Ta and Faris 1987) and/or by root/nodule decay (Ta and Faris 1988, Agboola and Fayemi 1972); (2) stimulation of nonsymbiotic N fixation (Gadgil 1971a, 1971b); (3) more efficient use of nutrients, light, and water (Danso et al. 1987); and (4) the N-sparing effect (Henzell and Vallis 1977). To evaluate the efficiency of supply of fixed N to the associated crop, the mechanisms underlying the assumed transfer must be determined, with an eye to identifying and understanding the environmental and cultural influences on N-fixing/non-N-fixing plant interactions.

The lack of quantitative research reflects inherent difficulties in systems analysis and in identifying the advantages of a given intercropping system. Willey (1985) proposed two distinct objectives in evaluating intercropping advantages: (1) a biological objective to determine the increased biological efficiency of intercropping and (2) a practical objective to determine the advantages that are likely to be obtained by a farmer. The focus of this chapter is on the former, the biological efficiency of intercropping with N-fixing and nonfixing plants within agroforestry systems. Do non-N-fixing plants benefit from being interplanted or mulched with N-fixing plants? Do N-fixing woody perennial legumes differ significantly from annual legumes in the quantity or timing of available N? Do cultural practices or environmental factors significantly increase the amount of N made available by the N-fixing plant?

AVAILABILITY OF FIXED NITROGEN TO THE COMPANION CROP

Topic 1: Do non-N-fixing plants benefit from being interplanted with symbiotic N-fixing plants?

Client Problem

Agroforestry is often recommended for marginal lands and low-input systems. Landowners frequently interplant legumes instead of applying other fertilizers, assuming the N-fixing plants will maintain or enhance soil fertility or will transfer fixed N to the companion crop.

Maintaining site productivity in low-input systems through legume contributions while increasing cropping intensity may not be a viable management practice. LaRue and Patterson (1981) reported that they could find no published evidence that any legume crop satisfied all its N requirements by fixation. The highest percentages (80%) were typical of low-fertility soils or soils artificially made N-poor. In fact, the authors cited results demonstrating the ability of a legume to scavenge soil N and concluded that soybeans, for example, might actually deplete soil N. Thus, increasing the cropping intensity (growing more plants per hectare) through mixed cropping may increase yield in the short term, but in the longer term, it may cause an overall increase in nutrient export (for example, Lulandala and Hall 1987) that is not offset by the legume. The end result may be to decrease site productivity over time, contradicting the view that agroforestry is a sustainable land-use system.

The basic problem is that prevailing management practices are often based on incomplete and conflicting evidence. The client's problem is essentially a lack of information or a gap in knowledge concerning the predicted response of plants in N-fixing/nonfixing agroforestry systems. The client's problem can be redefined as a testable hypothesis.

Hypothesis

Clearly stated hypotheses require the researcher to focus on the underlying scientific theory being tested. The process of formulating a hypothesis is illustrated below, but the reader is referred to the first book in the series *A Handbook on the Management of Agroforestry Research*, chapters 3 and 4, for a more complete explanation. Depending on the focus of the research, hypotheses may take several forms, as in the following example.

> Total N content significantly increases in nonleguminous plants raised in association with legumes.

The hypothesis stated above is a generic hypothesis, expressing the underlying scientific principle. It must be tailored to individual research by including specific species, time period, and if possible, predicted mechanism(s).

> Total N content significantly increases in sorghum raised in association with the woody perennial legume *Leucaena leucocephala* during the first growing season, as a result of below-ground processes such as root/nodule excretion and/or root/nodule turnover.

Identifying the mechanism(s) is one of the most useful and difficult aspects of biological research. Although it may seem obvious, it is important that the materials and methods selected to test the hypothesis do, in fact, test the hypothesis. After deciding on the appropriate and available method(s), the researcher may be required to redefine the hypothesis. For the hypothesis stated above, the researcher may not have a method for specifically identifying the mechanism, in which case the hypothesis could be restated.

> Total N content significantly increases in sorghum raised in association with the woody perennial legume *Leucaena leucocephala* during the first growing season, as a result of below-ground processes.

The phrase "as a result of below-ground processes" is less precise, but it does indicate that above-ground processes (for example, green manuring, litter fall, plant death) are not part of the experiment.

Given that N transfer occurs, both the quantity and timing of N transfer from N-fixing to nonfixing plants are important in agroforestry systems. Annual plants grow exponentially and require N rapidly at critical stages. Research is needed to determine whether significant N transfer occurs naturally, in a timely manner, or whether it is effective only when the trees are cut for green manure.

Materials and Methods

A benefit is usually measured as an increased growth response, such as an increase in height, dry weight, total N, and/or biomass. The underlying assumption is that the N-fixing plant increases the amount of N available to the nonfixing plant. Actual measurement (for example, Kjeldahl determination, ^{15}N assay) of N in the nonfixing plant is required to test this hypothesis. Increases in plant height and dry matter, for example, although they suggest that available N increases, do not prove that total N increases.

Initially, yield comparisons were used to ascertain the benefit of legume/nonlegume intercrops in agronomy, forestry, and agroforestry systems. Although increases in yield may in fact be the response that is most important to the client, yield comparisons alone are not a direct measure of the interaction between N-fixing and nonfixing plants. Yield comparisons as such will not be discussed in this chapter (see Ong, this volume).

Two methods, total N determination and stable isotope analysis, are widely used for measuring N in both plants and soils. Determining the amount of total

N, especially by using Kjeldahl procedures, does not require sophisticated equipment and thus is universally applicable. Stable isotope analysis, although it is becoming more widely used, requires access to a mass spectrometer or an emission spectrometer and thus is less available. The following segments briefly characterize these methods and refer the reader to specific literature for more detailed descriptions. This methods section is equally applicable to all four topics discussed in this chapter and will not be repeated.

Total nitrogen determinations

Kjeldahl procedures are commonly used to analyze total N in plant material (Bremner and Mulvaney 1982, Nelson and Sommers 1973) and soil samples (Bremner and Mulvaney 1982) because these procedures are relatively simple and inexpensive. Essentially, the Kjeldahl method converts organic N to NH_4^+ –N by digestion with concentrated sulfuric acid (H_2SO_4). The digest is then distilled by using an alkali to liberate NH_3. The distillate is analyzed for NH_4^+ –N either through titrimetric (Bremner and Mulvaney 1982) or colorimetric procedures (Weatherburn 1967). If inorganic N (such as NO_3^-, NO_2^-) is also predicted to be in the sample, then pretreatment of the sample or another variation in the procedure is required (Bremner and Mulvaney 1982). The total N data generated through these procedures provides a quantitative measure of plant N but does not distinguish between the different sources of available N.

A comparison of total N accumulation in the treated plants and the control plants would either support or refute the hypothesis, but it relies on the unproven assumption that the legume derives all of its N through N fixation (refer to LaRue and Patterson 1981). The importance of selecting appropriate control species and managing the control optimally has been demonstrated by numerous authors (see Ong, this volume).

Stable isotope analysis

The stable isotope ^{15}N is frequently used as a tracer in plant and soil studies. Nitrogen tracer techniques have been used to directly measure biological N_2 fixation and to distinguish the proportions of N in the crop components of mixed cropping systems. Nitrogen-15 tracer techniques have been used extensively to determine whether and how much N was transferred. Tracer techniques permit direct measurement of soil N transformations, which otherwise can be studied only by using indirect measures. Nitrogen-15 detected in the nonfixing plant can be attributed to transfer from the N-fixing plant, if an appropriate control is used (see Wheeler et al., this volume).

Nitrogen tracers can be defined as materials containing an unusually high or low content of ^{15}N. Usually, ^{15}N-enriched materials are used, but ^{15}N-depleted N can also be used as a tracer (Bremner 1977). Mass differences are used to identify the tracers. The natural environment has a ^{15}N abundance of 0.37%. The ^{15}N in the sample is expressed as the atom % ^{15}N excess over the natural abudance of 0.37%. The determination of ^{15}N abundance requires either a mass spectrometer or an emission spectrometer. The use of stable isotopes as tracers is described in detail by Wheeler and colleagues (this volume), Vose (1980), and Bremner (1977).

Critical Review

The renewed interest in biologically efficient systems has produced numerous research articles pertaining to N-fixing/nonfixing plant interactions in multiple cropping systems. Research has focused on the components individually or the system collectively. Although a considerable amount of descriptive research exists on this topic, quantitative research is more elusive. Additionally, the majority of the research uses indirect measurement methods that imply a benefit but do not directly measure the role of N-fixing plants. This is an important distinction in trying to identify and quantify the potential advantage of N-fixing plants in agroforestry systems.

Lulandala and Hall (1987) measured the fodder and wood production of *Leucaena leucocephala* intercropped with maize or beans over a three-year period. The control was clean-weeded *Leucaena*. Mean annual fodder yields (kg/ha) showed no significant difference between management regimes (*Leucaena* at various spacings and intercropped). However, plots intercropped with beans maintained greater fodder yields that culminated in a significantly higher value at 26 months. The authors attributed the increase in *Leucaena* yield to supplemental N fixing by the beans, even though both species were presumed to be N fixing. It would be easy to infer from this report that N was transferred from the beans to the *Leucaena*. The authors' nutrient data did not include measurements at 26 months, and therefore it was impossible to use N content to support their conclusion. However, they did report annual nutrient (N, P, K, Ca, Mg) removals (see table 8.1). As with fodder yields, there was no significant difference in annual nutrient levels due to management regimes.

Table 8.1. Nutrient removals through fodder harvest from *Leucaena* hedgerows

Management regime	Mean annual nutrient removal (kg/ha)				
	N	P	K	Ca	Mg
Clean weeded	88	4	42	18	8
Beans intercrop	95	5	58	20	8
Maize intercrop	80	5	37	23	9

Adapted from: Lulandala and Hall 1987.

Interpretation of these results is further confounded by the application of fertilizer to all plots with crops, but not to the control plots of clean-weeded *Leucaena*. One cannot distinguish whether the increased N in fodder intercropped with beans (versus the control) reflects applied fertilizer, biological N, or some other interaction.

In the case described by Lulandala and Hall (1987), the N-fixing beans would be less competitive for soil N than maize and thereby increasing the amount of soil N available to the *Leucaena*. In this study, for example, ^{15}N tracer studies could distinguish between the sources of available N, but total N determinations could not. The use of a tracer could distinguish between fertilizer N and fixed N but could not identify beans or *Leucaena* as the source of fixed N. This research was not specifically designed to address the issue of N transfer, but the conclusions imply a fixed N benefit to the companion crop from intercropping with legumes, a benefit that the reported nutrient data does not support.

In another study, in which N was determined not to be limiting, Ahimana and Maghembe (1987) described the response (height, volume, biomass) of *Eucalyptus tereticornis* to five treatments: weeded, unweeded, or intercropped with beans, maize, or sorghum. At the end of three years, the authors reported that *Eucalyptus* weeded and intercropped with beans had significantly greater yields than *Eucalyptus* intercropped with maize or sorghum. Nitrogen accumulation by above-ground parts of *Eucalyptus*, however, indicated no significant difference between weeded monoculture and intercropped beans, although the N level in the weeded *Eucalyptus* was slightly higher than in *Eucalyptus* intercropped with beans (table 8.2). These treatments had significantly ($P < 0.05$) higher accumulations of N than *Eucalyptus* unweeded or intercropped with sorghum or maize. The results are similar to those reported by Lulandala and Hall (1987), although in this case, N was apparently not the limiting factor. If Ahimana and Maghembe (1987) had reported only their yield data, it would have been tempting to infer a N benefit from the beans.

Table 8.2. Total accumulation of nutrients by above-ground parts of *Eucalyptus tereticornis* under various management regimes as compared to plant-available soil reserves at Mafiga, Morogoro, Tanzania

	Nutrients (kg/ha)*					
Management regime	N	P	K	Ca	Mg	Na
Unweeded monoculture	36.9a	5.0a	40.5a	132a	10.1a	5.4a
Weeded monoculture	188.7b	24.7b	203.1bc	648.0b	49.9b	27.0b
Intercropped beans	175.9b	23.4bc	181.8c	630.7b	56.9b	26.3b
Intercropped maize	147.6c	21.4cd	229.8b	485.5c	39.4c	19.4c
Intercropped sorghum	140.3c	19.2d	201.9c	473.7c	36.8c	19.3c
Estimated available soil reserves**	2200	18	601	3384	652	92

* Values in a single column followed by the same letter are not statistically different at $P < 0.05$.
**Based on a 30 cm-ha furrow slice.
Adapted from: Ahimana and Maghembe 1987.

Ahimana and Maghembe (1987) also evaluated the total above-ground nutrient content for the different treatments in relation to the estimated total plant "available" soil reserves. For most nutrients, and especially N, Ca, and Mg, the amount of nutrients held in the biomass was negligible compared to the estimated available soil reserves. A proportionately higher amount of P and, to a lesser extent, K was held in the biomass. The authors concluded that problems related to P and K nutrition might arise in future rotations. This study illustrates the advantage of estimating a nutrient budget for the agroforestry system.

The interaction between the N-fixing and nonfixing plants may occur primarily in two ways: above-ground through litter fall, seed shed, and plant death or below-ground through transfer of nitrogenous compounds and root and/or nodule sloughing. The above-ground processes are discussed in topic 2.

Investigations into below-ground processes have produced a myriad of results. Research involving increased N content in woody perennials raised with legumes was reported by DeBell et al. (1989), Funk et al. (1979), and Gadgil (1971a, 1971b). In contrast, many others (for example, Rao et al. 1987, Sprent 1983) reported little, if any, N transfer by actively growing legumes.

Avery and Rhodes (1990) described the growth characteristics and total N content of *Leucaena leucocephala* and the companion crop *Sorghum bicolor* grown in a greenhouse with N-limiting nutrient medium in sole and mixed cropping. Plant height, dry weight, and total N content increased significantly ($P < 0.01$) in sorghum intercropped with nodulated *Leucaena* over the control, sole sorghum. Given an intercropping mixture of three sorghum and two *Leucaena*, sorghum grown in mixed culture gained an estimated 5.6 mg N per plant over 185 days. Although it is statistically significant, this 40% increase in N content of intercropped sorghum represented less than 1% of the *Leucaena* total N budget during this period and was inadequate to sustain normal physiological development of sorghum.

The small, but significant, increase in total N content in sorghum intercropped with nodulated *Leucaena* suggested that fixed N was transferred from the legume to the sorghum roots via the soil. Two possible sources of legume N are root/nodule turnover and root/nodule exudates. Characterization of the nitrogenous solutes might be used to discriminate between root/nodule turnover or excreted nitrogenous compounds. In either case, these metabolites could be additional sources of N for the sorghum. The total N input into agroforestry systems by either of these processes is predicted to be relatively insignificant. Extrapolating from greenhouse experiments to the field must be done with caution. Further greenhouse and field studies are required to address the precise mechanisms of N transfer below-ground and environmental factors or cultural practices that may enhance the process(es).

Not unexpectedly, below-ground processes appear to be susceptible to a variety of influences. The quantity of N supplied will depend on the effectiveness of the N fixers (see Wheeler et al., this volume), the distribution of nodules within the soil horizons, phosphorus supply (Hansen and Pate 1987), and availability of other nutrients. Environmental factors that influence N fixation appear to affect not only the source but also the transfer. Ta and Faris (1988) assessed the effects of environmental conditions on the fixation and transfer of N from alfalfa to associated timothy by using ^{15}N dilution techniques. High light intensity, long days, and cool temperatures (20°/16°C day/night) were the best conditions for high N fixation by alfalfa and high N transfer from alfalfa to the associated timothy. This result is particularly interesting because it suggests that environmental conditions conducive to N fixation are also the most conducive to N transfer. This contrasts with the theory discussed under topic 4, where aging or stressed legumes are perceived as the most likely source of N for associated species.

In an earlier study, Ta and Faris (1987) suggested that species variation was one cause of variation in N benefit of legumes to associated grasses. In addition to differences in legumes' ability to fix N, they reported that N absorption by grasses also varied with species. The implication is that specific matching of legume and nonlegume, in terms of N fixation and N adsorption, might enhance the N benefit to an agroforestry system.

In terms of biological efficiency, the research literature currently available does not support convincingly the use of N-fixing plants to increase the availability or transfer of fixed N to the companion crop through below-ground processes. The quantity produced by natural processes other than plant death (see topic 2) appears to be relatively insignificant in the physiological development of the companion crops. The time lapse prior to a noticeable response in the companion crop suggests that the recorded benefits might be residual effects and therefore, by definition, not an advantage of intercropping.

Recognizing the complexity of the interactions between plants and the environment, enough tantalizingly positive results exist to warrant more, carefully designed research. This research should be clearly focused to quantify the N budgets in agroforestry systems and to identify mechanisms underlying the interactions. Mechanistic research would allow us to predict plant response instead of relying on site-specific trials.

Topic 2: Do green manures using N-fixing plants, primarily legumes, contribute more N to the associated crop than green manures using nonlegumes?

Background

Green manuring, the practice of incorporating plant material into the soil while it is green or shortly after maturity, has received the most attention in agroforestry research (Kang et al. 1981, Yamoah et al. 1986, Ssekabembe 1985). This research typically involves evaluating crop response to leguminous green manures as substitutes for inorganic fertilizers. The results consistently indicate that additional N applications are needed to optimize yield (Kang et al. 1985, Yamoah et al. 1986).

The species most frequently selected for green manures are legumes. The underlying assumption appears to be that legumes contribute more N to the system than nonlegumes and that this benefit outweighs the inherent metabolic cost incurred by N-fixing plants. Ssekabembe (1985), for example, identified N-fixing capacity as one of the desirable characteristics when choosing tree species for hedgerow intercropping (green manuring).

It is also common to equate a plant's N-fixation rate or amount of N fixed with the amount of N available to the companion crop through green manuring. Clearly, this is dangerous, as N availability to the associated crop is not directly correlated with the amount of N fixed. The nutrient content of the green manure is important, but also important are the rates of decomposition and nutrient release, and the demand by the associated crop.

The purpose of this section is not to review the advantages and disadvantages of green manuring per se, but rather to address the issue of whether N-fixing plants offer any measurable biological advantage over nonfixing plants in green manuring. Aspects of green manuring and the use of legumes for green manures are also discussed in the chapters by Ong, Wheeler and colleagues, and Young in this volume.

Client Problem

The shortened fallow period and the high price or lack of inorganic fertilizers in many regions has encouraged the use of legumes as green manures to maintain or enhance soil fertility. In agroforestry systems, leguminous trees are commonly selected for this purpose. Growing plants for green manure requires farmers to allocate space to this "fertilizer" crop, to increase labor, and to forego other products (such as fodder and fuel) that the tree might have provided.

Hypothesis

Green manures composed of N-fixing plants contribute significantly greater amounts of N to the associated crop than do non-N-fixing plants.

Critical Review

Most of the agroforestry research on green manuring involves alley cropping with leguminous trees. Different levels of prunings from various tree or shrub species are applied to the crop growing between the rows of trees as a source of nutrients, especially N. The exclusive use of legumes in this type of agroforestry research suggests that legumes contribute more N to the associated crop than nonlegumes, although this was not specifically tested in the papers by Kang et al. (1981, 1985) or Yamoah et al. (1986).

The reader usually assumes that species in the Leguminosae family are nodulated and N fixing. This assumption is not always warranted, especially for tropical legumes. It is well documented that nodulation may be poor or absent on leguminous trees because of adverse soil effects, the absence of appropriate strains of *Rhizobium*, or the species' resistance to nodulation (Wheeler, personal communication). Agroforestry research reports without measurements of nodulation or the rate of N fixation make it impossible for the reader to accurately assess whether the legume was, in fact, capable of N fixation.

Wheeler (personal communication) provided an example of this potential pitfall by reviewing the alley-cropping experiments reported by Yamoah and associates (1986). They compared the effect of returning repeated prunings of three species, *Gliricidia sepium*, *Cassia siamea*, and *Flemingia congesta*, to the intercropped maize. Maize yields were greatest when intercropped with *Gliricidia* and *Cassia*. The total N content (versus percent N) in the prunings of both species was similar, although *Cassia* produced the greater biomass. The reader might assume that the N fixation by the trees was partly responsible for sustaining crop yield, although the authors alluded only indirectly to this possibility. Such an assumption may well be ill founded, for Allen and Allen (1981) noted that *Cassia siamea* belongs to the poorly nodulated subgenus *Fistula* and has been reported as non-nodulated in Pakistan, South Africa, Java, the Philippines, and Hawaii (Lim and Burton 1982). It is possible, therefore, that Yamoah and colleagues inadvertently tested whether green manures of N-fixing plants contribute significantly more N to the system than nonfixing plants, in which case, their results indicate they do not. Yamoah and colleagues (1986) suggested that improving soil physical properties might prove more important in many alley-cropping systems than supplying nutrients.

Agroforestry practices predicated on the idea of increasing the amount of N available to the companion crop or to the soil organic matter through the application of *Leucaena* prunings alone appear to be unjustified, given the results reported by Kang and colleagues (1981). They reported that *Leucaena* prunings yielded large quantities (180 kg to 240 kg/ha/yr) of N, but that the efficiency with which the associated maize crop could use the N from the prunings was low compared to its use of N [inorganic] fertilizer. This was confirmed in subsequent experiments where, despite the high N yield from the prunings, the application of low levels of [inorganic] N was needed to obtain high maize yield (Kang et al. 1985). In addition, the 5 t to 8 t/ha of dry top prunings added annually did not significantly increase the soil organic matter level. Kang and colleagues (1981) attributed this, in part, to the rapid decomposition of *Leucaena* leaves because of their low C/N ratio.

There is no evidence to date that using leguminous (versus nonleguminous) woody species as green manures increases the amount of N available to the companion crop in agroforestry systems. To assess the effects of green manures solely on the basis of potential N contribution is, admittedly, too simplistic. This is, however, the main reason offered for selecting a N-fixing versus nonfixing plant for green manure. Decades of agricultural research have shown that green manuring benefits a soil's organic matter content, structure, and permeability (Toech et al. 1980) and thus provides a nutrient source for current or subsequent cropping. Part of the effect can be attributed to the N fixed when a legume is used for green manure, although the effect also occurs with nonlegume green manures and with nutrients other than N (Toech et al. 1980).

In many regions, prunings are too valuable for other products to be used as green manure. Quantitative research addressing the long-term nutrient budgets of agroforestry systems is required to ascertain whether agroforestry practices are consistent with sustainability in low-input systems.

WOODY PERENNIALS AS A SOURCE OF BIOLOGICALLY FIXED NITROGEN

Topic 3: Do leguminous woody perennials differ significantly from annual legumes in the quantity or timing of available N?

Background

One of the main advantages attributed to agroforestry is the possibility of improving the fertility of agricultural lands because the tree legume component might add additional amounts of N to the system over time (Huxley 1986, Nair 1984). The topic under discussion is whether N-fixing trees used in agroforestry systems confer an advantage to soil fertility or to the companion crop that is significantly different from that of an annual N-fixing plant. The potential contribution of woody legumes has been described by several authors (Nair et al. 1984, Vergara 1982, Brewbaker and Hu 1981), but their particular role is ill defined, and experimental data is limited.

Client Problem

The mixture of crops and trees that is characteristic of agroforestry systems requires more intensive management than do sole-cropping systems in either forestry or agriculture. In agroforestry, one can choose which component will be the N-fixing species. Trees are, by definition, woody and perennial; therefore, selecting trees as one of the components increases the opportunity costs for the land manager. Including trees for biological purposes has longer-term consequences than including annual or herbaceous perennial plants. Thus the need for information is heightened.

Hypothesis

Nitrogen-fixing woody perennials contribute significantly greater quantities of N, over a longer period of time, than do consecutive plantings of annual N-fixing crops in agroforestry systems.

Critical Review

There have been relatively few studies of perennial N-fixing plants individually or in combination with other crops, especially in comparison with annual or short-lived perennial agricultural legumes. The design of agroforestry systems usually requires extrapolation of experience gained from agricultural mixed-cropping systems (see Ong, this volume), forest ecosystems (Hansen et al. 1987b), or forest plantations (DeBell et al. 1989, Turvey and Smethurst 1983).

Comparing leguminous woody perennials to agricultural legumes is admittedly somewhat artificial. Tree physiology usually causes N-fixation rates for individual plants to be less than those of agricultural legumes; plant densities are usually less in tree than in herbaceous plant systems; and agroforestry systems and forestry plantations are usually relegated to less productive sites than agricultural systems. Nonetheless, to evaluate the potential advantages of intercropping with a leguminous tree in agroforestry systems, quantitation is required. Unfortunately, no direct comparison between annual legumes and tree legumes has been reported.

A number of studies have investigated the role of leguminous trees and shrubs (primarily *Acacia* species) in the native *Eucalyptus* forest ecosystems of Australia. These forests typically occupy marginal sites with distinct wet and dry seasons, analogous to many of the sites proposed for agroforestry systems. The general conclusion from these studies is that forest legumes are effective in accumulating and turning over soil N but grossly ineffective in contributing fixed N to the system.

Depending on the species and plant densities, the estimated N accretion (versus fixation) rates from various woody legumes are 0.05 kg to 1.0 kg/ha/yr of N in undisturbed *Eucalyptus tetrodonta* open forest (Langkamp et al. 1981), 0.10 kg to 1.6 kg/ha/yr of N in jarrah (*E. marginata*) forests (Hansen et al. 1987b), and 2.2 kg/ha/yr of fixed N in dense stands of *Acacia pulchella* (Monk et al. 1981). Higher rates of 12 ± 4 kg/ha/yr of N were recorded by Langkamp and colleagues (1979) for fertilized plantations of *A. holosericea* (1,111 trees per hectare), but these are still low in comparison to rates for agricultural legumes. Stevenson (1986), citing data recorded by Moore in 1966, provided the following figures for gains in soil N through biological N fixation. The N gain in soils amended with crop residues was 15 kg to 78 kg/ha/yr of N, and in field plots under sodlike crops, it was 14 kg to 56 kg/ha/yr of N. Most of the agricultural literature reports N-fixation rates for agronomic legumes (for example, average 150 kg to 200 kg/ha; Power and Doran 1984) that are not synonymous with N accretion or N gains to a system.

Because planting a legume as one of the component crops is not in itself an advantage of intercropping (Willey et al. 1977), the data cited above suggests that additional research is needed to substantiate the potential fixed N contribution of woody legumes to agroforestry systems. One method for measuring accretion is to estimate N budgets on a chronosequence of trees.

The above studies reported a decreasing dependence on N fixation with increasing plant age. Proportional dependencies of the species on fixed N_2 were relatively high (13% to 61%) in the first-year seedlings, then declined markedly to 1.1% to 3.4% in the second year, to 0.3% to 1.6% in the third year, and, with one exception, to well below 1% in the fourth and sixth years (Hansen et al. 1987b). This decline in N fixation of field-grown plants suggests that leguminous woody perennials will not necessarily contribute more N to a system over time than herbaceous legumes. Sprent (1983) characterized N fixation in woody perennials as following the law of diminishing returns as the plants age. The total N fixed per unit of ground area increases over time, but the rate of fixation per unit of biomass drops steadily. The perennating organs act as stores of N as well as of carbon. There is no net gain of N after a season's growth (Sprent 1983).

Strongly seasonal nodulation and nodule activity was exhibited by *Acacia* species in the Australian field studies. Hansen and Pate (1987) concluded that this seasonality was a result of water stress rather than high temperature through comparison with fully symbiotic plants raised in the greenhouse. Low nutrient availability, particularly of P, was also indicated to limit symbiotic N fixation under natural conditions. Both of these conditions, moisture stress and limited nutrient availability, have implications for agroforestry systems. It is difficult to predict the net effect. Nodule sloughing may increase the amount of available N in the system, but in turn, it will decrease the N-fixing capability of the plant.

Available N and extractable P are the nutrients most likely to limit plant growth on many of the sites proposed for agroforestry. DeBell and colleagues (1989) reported that concentrations of N and P in *Eucalyptus* foliage increased as the percentage of *Albizia* increased in *Eucalyptus-Albizia* plantations in Hawaii. They attributed the higher concentrations of N and P in the foliage to enhanced root growth or increased rates of nutrient cycling associated with the presence of *Albizia*, to additions of enriched *Albizia* litter to the soil in mixed plantings, or to both conditions. The positive responses in *Eucalyptus* growth and foliage nutrient concentrations were recorded from a wet coastal site. In contrast, in a companion test conducted on a much drier site, *Albizia* grew very poorly and provided no advantage to the *Eucalyptus* plantations. The results cited above emphasize the importance of moisture deficits. Careful matching of species to site may have significant effects on the availability of N and other nutrients within a given system.

No direct comparison between annual legumes and woody perennial legumes has been reported in terms of the amount or timing of N made available. The Leguminosae family is composed mainly of woody perennials, so the opportunities to use leguminous trees in agroforestry systems are many. Research on the N budgets of N-fixing trees is limited. To date, there is no obvious advantage, in terms of biologically fixed N availability, in selecting N-fixing trees as the legume component. There might be an ecological advantage, however, in that trees may survive stress that annual plants may not.

The research does not indicate that N-fixing trees will be major contributors to the N economy of agroforestry systems, given the current objectives and practices of most agroforestry systems. A number of forest plantation and ecosystem studies have recorded increases in N fertility when N-fixing plants are included. These increases usually require major contributions by above-ground parts (for example, plant death—topic 2) and long time periods. Whether this data is applicable to designing agroforestry systems has yet to be determined.

Topic 4: Do cultural practices (such as defoliation and grazing) or environmental factors (such as shading and moisture deficits) significantly increase the amount of N made available by the N-fixing woody perennial to the nonfixing plant?

Background

As with annual legumes, the aging or stressed perennial legume might act as a significant N source for associated species. Any factor in the environment, natural or man-made, that affects the growth of the host plant usually affects the development and functioning of its root nodules. The implication is that changes in nodule function (and thus in N fixation) will in turn affect the amount of N available to the nonfixing plant.

Cultural practices or environmental factors that effectively stress the woody legume may increase the availability of N to the associated crop. This may prove useful in agroforestry systems, where leguminous trees are frequently cut for a variety of products and to reduce shading of the intercrop. It has yet to be determined whether substantial changes in N availability within agroforestry systems can be produced through different management strategies.

Client Problem

Agroforesters usually recommend repeated cutting of woody species for a variety of reasons. The predicted effect of this practice on tree development and growth is known (see Cannell, this volume). The effect on N availability within the agroforestry system, however, is virtually unknown. Developing management strategies with the potential to enhance the amount or timing of N availability in the system is impossible without this information.

Hypothesis

Repeated cutting of leguminous woody perennials significantly increases the amount of total N absorbed by the companion crop within the current growing season (without green manuring).

Critical Review

A review of the current agroforestry literature revealed no research specifically addressing the effect of different cultural practices, except for green manuring (refer to topic 2) and harvesting (Ahimana and Maghembe 1987), on the amount of available N in agroforestry systems. It is well documented that environmental stresses tend to decrease the amount of N available to the host plant (Sprent and Minchin 1983), but whether these stressed plants can increase the amount of N available to the associated crops remains controversial. In theory, two probable sources of increased plant-generated N are: (1) increased nodule and/or root turnover and (2) increased excretion of nitrogenous compounds. Additionally, the availability of soil N to the companion crop may increase as the woody legume becomes less competitive for soil N.

The implications for agroforestry are described by Sprent (1983). Although no direct evidence exists, pruning *Leucaena* is likely to cause some nodule decay, releasing N into the soil. Pruning, which reduces the available photosynthate, essentially makes nodules a luxury. The nodules' total mass is small (1% to 5% of the total dry matter of young plants), but their N content is high (5% of nodule dry weight). The total N input via nodule breakdown alone, however, appears to be relatively insignificant. When it is supplemented by root death and leaf fall, Sprent (1983) concluded, appreciable amounts of N may become available. Whether this actually occurs with leguminous woody perennials is unknown.

The physiological basis for this theory is described in an excellent review of the effects of environmental factors on nodulation and N fixation by Sprent and Minchin (1983). The common agroforestry practice of pruning leguminous trees, for example, represents various defoliation or shading treatments. Changing the quantity of light a plant intercepts affects photosynthesis and thus the carbohydrate supply to the nodules. The general effect of shading is on the growth and metabolism of the host plants, but the severity of the light loss will determine the mode of expression. Nodule mass, rather than nitrogenase activity, will be affected during moderate shading, while severe shading can curtail both nodule development and N fixation. Defoliation represents severe shading, where a new supply of carbohydrates becomes available after regrowth. Thus, the initial result of defoliation is a decrease in both nodule mass and fixation, followed by a slow recovery (Sprent and Minchin 1983).

Working with forested ecosystems in New Zealand, Gadgil (1971a, 1971b) conducted a series of glasshouse experiments on the influence of damaged and undamaged lupins (*Lupinus arboreus*) on N uptake in *Pinus radiata* seedlings. She reported that damage caused by sudden shading, defoliation, and drought inhibited root growth, nodulation, and probably the rate of N fixation and caused some degree of lupin root decomposition. Interestingly, Gadgil (1971b) found no evidence that nodule sloughing was associated with the treatments, although changes occurred within the nodules. The results suggested that damage to lupin plants could increase N availability to *P. radiata* in the field by causing decomposition of lupin root material.

In contrast to the prevailing theory, Gadgil's (1971a) experiments with undamaged lupins indicated that the tree crop could benefit from the lupins' presence even if the lupins were not under stress. Undamaged lupins also increased N uptake in pine seedlings through litter and lupin seedling exudate. Surprisingly, lupin tops placed on the soil surface did not significantly alter the amount of N absorbed by the pines. As with the damaged lupins, Gadgil concluded that the increased N uptake in pines was due either to direct N transfer or to stimulation of nonsymbiotic N fixation.

Cultural practices can mimic or ameliorate environmental stresses. Successive harvesting has been reported to increase the amount of N absorbed by grasses in grass/legume mixtures (Ta and Faris 1987) as well as the interspecies distance and the legume/grass ratio (Brophy et al. 1987). The idea of being able to manipulate N availability through different cultural practices is appealing, but the research to either support or refute it is lacking. More research is needed to determine the effects of different cultural practices on woody legume species. This is especially true regarding the role of woody legumes in meeting the objectives of agroforestry systems.

CONCLUSIONS

It is apparent and not surprising that a review of the current literature on the role of N-fixing plants in agroforestry systems follows the trend of agronomic mixed-cropping systems; the literature is inconclusive and conflicting, descriptive rather than quantitative. In terms of biological efficiency, the research literature reports conflicting results about the ability of N-fixing plants to increase the availability or transfer of fixed N to the companion crop through below-ground processes. The assumed benefit of using legumes (versus nonlegumes) for green manures appears to be unsubstantiated, as do the purported advantages of using woody perennial rather than annual legume species. Too little is known about the effect of cultural practices on legumious trees to predict the trees' response in terms of increasing the amount of N available in agroforestry systems. The current literature provides no direct evidence of benefit, biological or economical, conferred by the mere use of species capable of nodulation.

Admittedly, agroforestry research is relatively new. We must learn from the experience of other systems and concentrate on understanding mechanisms and quantifying nutrient budgets. The complexity of biophysical interactions and the inherent variability of sites clearly indicate that empirical research will never adequately address the research needs required for designing agroforestry systems. Further research to determine the quantity and timing of additional N availability from N-fixing trees is needed to clarify their biological contribution to agroforestry systems and their actual contribution to management objectives.

ACKNOWLEDGEMENT

I thank Karen Haugen for her editorial contributions.

LITERATURE CITED

Agboola, A. A., and A. A. A. Fayemi. 1972. Fixation and excretion of nitrogen by tropical legumes. *Agronomy Journal* 64: 409–412.

Ahimana, C., and J. A. Maghembe. 1987. Growth and biomass production by young *Eucalyptus tereticornis* under agroforestry at Morogoro, Tanzania. *Forest Ecology Management* 22: 219–228.

Allen, O. N., and E. K. Allen. 1981. *The Leguminosae: a source book of characteristics*. London: McMillan. 812 p.

Avery, M., and D. Rhodes. 1990. Growth characteristics and total nitrogen content of a *Leucaena*/sorghum agroforestry system. *Plant and Soil* 127: 259–267.

Bremner, J. M. 1977. Use of nitrogen-tracer techniques for research on nitrogen fixation. In: *Biological nitrogen fixation in farming systems of the tropics* (A. Ayanaba and P. J. Dart, eds.). Chichester, UK: John Wiley and Sons. pp 335–352.

Bremner, J. M., and C. S. Mulvaney. 1982. Nitrogen-total. In: *Methods of soil analysis. II. Chemical and microbiological properties*. Agronomy Monograph 9, 2nd ed. Madison, Wisconsin: American Society of Agronomy, pp 595–624.

Brewbaker, J. L., and T. W. Hu. 1981. *Nitrogen-fixing trees of importance in the tropics*. Paper for U. S. National Academy of Science workshop, September 1981. Mimeo. 15 p.

Brophy, L. S., G. H. Heichel, and M. P. Russelle. 1987. Nitrogen transfer from forage legumes to grass in a systematic planting design. *Crop Science* 27: 753–758.

Danso, S. K. A., F. Zapata, and G. Hardarson. 1987. Nitrogen fixation in fababeans as affected by plant population density in sole or intercropped systems with barley. *Soil Biology and Biochemistry* 19: 411–415.

DeBell, D. S., C. D. Whitesell, and T. H. Schubert. 1989. Using N_2–fixing *Albizia* to increase growth of *Eucalyptus* plantations in Hawaii. *Forest Science* 35(1): 64–75.

Funk, D. T., R. Schlesinger, and F. Ponder Jr. 1979. Autumn-olive as a nurse plant for black walnut. *Botanical Gazette* 140 (Suppl.): S110–S114.

Gadgil, R. L. 1971a. The nutritional role of *Lupinus arboreus* in coastal sand dune forestry. I. The potential influence of undamaged lupin plants on nitrogen uptake by *Pinus radiata*. *Plant and Soil* 34: 357–367.

Gadgil, R. L. 1971b. The nutritional role of *Lupinus arboreus* in coastal sand dune forestry. II. The potential influence of damaged lupin plants on nitrogen uptake by *Pinus radiata*. *Plant and Soil* 34: 575–593.

Hansen, A. P., and J. S. Pate. 1987. Comparative growth and symbiotic performance of seedlings of *Acacia* spp. in defined pot culture or as natural understorey components of a eucalypt forest ecosystem in S. W. Australia. *Journal of Experimental Botany* 38(186): 13–25.

Hansen, A. P., J. S. Pate, and C. A. Atkins. 1987a. Relationships between acetylene reduction activity, hydrogen evolution and nitrogen fixation in nodules of *Acacia* spp.: experimental background to assaying fixation by acetylene reduction under field conditions. *Journal of Experimental Botany* 38(186): 1–12.

Hansen, A. P., J. S. Pate, A. Hansen, and D. T. Bell. 1987b. Nitrogen economy of post-fire stands of shrub legumes in jarrah (*Eucalyptus marginata* Donn ex Sm.) forest of S. W. Australia. *Journal of Experimental Botany* 38(186): 26–41.

Henzell, E. F., and I. Vallis. 1977. Transfer of nitrogen between legumes and other crops. In: *Biological nitrogen fixation in farming systems of the tropics* (A. Ayanaba and P. J. Dart, eds.). Chichester, UK: John Wiley and Sons. pp 73–88.

Huxley, P. A. 1986. The prediction of biological productivity and sustainability of tree-crop mixtures. *Tropical Agriculture* 63: 68–70.

Kang, B. T., H. Grimme, and T. L. Lawson. 1985. Alley cropping sequentially cropped maize and cowpea with *Leucaena* on a sandy soil in southern Nigeria. *Plant and Soil* 85: 267–277.

Kang, B. T., G. F. Wilson, and L. Sipkens. 1981. Alley cropping maize (*Zea mays* L.) and leucaena (*Leucaena leucocephala* Lam) in southern Nigeria. *Plant and Soil* 63: 165–179.

Ladd, J. N., J. H. A. Butler, and M. Amato. 1986. Nitrogen fixation by legumes and their role as sources of nitrogen for soil and crop. In: *Biological Agriculture and Horticulture*, 3(2/3): 269–286.

Langkamp, P. J., G. K. Farnell, and M. J. Dalling. 1981. Acetylene reduction rates by selected leguminous and non-leguminous plants of Groote Eylandt, Northern Territory. *Australian Journal of Botany* 29: 1–9.

Langkamp, P. J., L. B. Swinden, and M. J. Dalling. 1979. Nitrogen fixation (acetylene reduction) by *Acacia pellita* on areas restored after mining at Groote Eylandt, Northern Territory. *Australian Journal of Botany* 27: 353–361.

LaRue, T. A., and T. G. Patterson. 1981. How much nitrogen do legumes fix? In: *Advances in agronomy*, Vol. 34, New York: Academic Press. pp 15–39.

Lim, G., and J. C. Burton. 1982. Nodulation status of the Leguminosae. In: *Nitrogen fixation*, Vol. 2, *Rhizobium* (W. J. Broughton, ed.). Oxford: Clarendon Press. pp 1–34.

Lulandala, L. L. L., and J. B. Hall. 1987. Fodder and wood production from *Leucaena leucocephala* intercropped with maize and beans of Mafiga, Morogoro, Tanzania. *Forest Ecology and Management* 21: 109–117.

Monk, D., J. S. Pate, and W. A. Loneragan. 1981. Biology of *Acacia pulchella* R. Br. with special reference to symbiotic nitrogen fixation. *Australian Journal of Botany* 29: 579–592.

Nair, P. K. R. 1984. Role of trees in soil productivity and conservation. In: *Soil productivity aspects of agroforestry*, Nairobi: ICRAF. pp 29–49.

Nair, P. K. R., E. C. M. Fernandes, and P. N. Wambugo. 1984. Multipurpose leguminous trees and shrubs for agroforestry. *Agroforestry Systems* 2: 145–163.

Nelson, D. W., and L. E. Sommers. 1973. Determination of total nitrogen in plant material. *Agronomy Journal* 65: 109–112.

Ofori, F., and W. R. Stern. 1987. Cereal-legume intercropping systems. In: *Advances in agronomy*, Vol. 41. New York: Academic Press. pp 41-90.

Power, J. F., and J. W. Doran. 1984. Nitrogen use in organic farming. In: *Nitrogen in crop production* (R. D. Hauck, ed.). Madison, Wisconsin: American Society of Agronomy. pp 585–598.

Rao, M. R., T. J. Rego, and R. W. Willey. 1987. Response of cereals to nitrogen in sole cropping and intercropping with different legumes. *Plant and Soil* 101: 167–177.

Sprent, J. I. 1983. Agricultural and horitcultural systems: implications for forestry. In: *Biological nitrogen fixation in forest ecosystems: foundations and applications* (J. C. Gordon and C. T. Wheeler, eds.). The Haque: Martinus Nijhoff/Dr. W. Junk. pp 213–232.

Sprent, J. I., and F. R. Minchin. 1983. Environmental effects on the physiology of nodulation and nitrogen fixation. In: *Temperate legumes* (D. G. Jones and D. R. Davies, eds.). Boston: Pitman Advanced Publishing Program. pp 269–317.

Ssekabembe, C. K. 1985. Perspectives on hedgerow intercropping. *Agroforestry Systems* 3: 339–356.

Stevenson, F. J. 1986. The nitrogen cycle in soil: global and ecological aspects. In: *Cycles of soil*, New York: John Wiley and Sons. pp 106–154.

Ta, T. C., and M. A. Faris. 1987. Species variation in the fixation and transfer of nitrogen from legumes to associated grasses. *Plant and Soil* 98: 265–274.

Ta, T. C., and M. A. Faris. 1988. Effects of environmental conditions on the fixation and transfer of nitrogen from alfalfa to associated timothy. *Plant and Soil* 107: 25–30.

Toech, F. R., J. A. Hobbs, and R. L. Donahue. 1980. Cropping systems. In: *Soil and water conservation for productivity and environmental protection*. New Jersey: Prentice-Hall. pp 233–273.

Turvey, N. D., and P. J. Smethurst. 1983. Nitrogen fixing plants in forest plantation management. In: *Biological nitrogen fixation in forest ecosystems: foundations and applications* (J. C. Gordon and C. T. Wheeler, eds.). The Hague: Martinus Nijhoff/Dr. W. Junk. pp 233-259.

Vergara, N. T. 1982. *New directions in agroforestry: the potential of tropical tree legumes*. Honolulu: Environment and Policy Institute, East-West Center. 52 p.

Vose, P. B. 1980. Stable isotopes as tracers: mainly the use of ^{15}N. In: *Introduction to nuclear techniques in agronomy and plant biology*. New York: Pergamon Press. pp 151–176.

Weatherburn, M. W. 1967. Phenol-hypochlorite reaction for determination of ammonia. *Analytical Chemistry* 39: 917–974.

Willey, R. W. 1985. Evaluation and presentation of intercropping advantages. *Experimental Agriculture* 21(2): 119–133.

Willey, R. W., B. A. Krantz, M. R. Rao, M. S. Reddy, M. Natarajan, and L. P. A. Oyen. 1977. *Cropping systems research at ICRISAT*. Hyderabad, India: ICRISAT. 7 p.

Yamoah, C. F., A. A. Agboola, and K. Mulongoy. 1986. Decomposition, nitrogen release and weed control by prunings of selected alley cropping shrubs. *Agroforestry Systems* 4: 239–246.

CHAPTER 9

Soil Microorganisms in Agroforestry Systems

C. T. Wheeler
I. M. Miller
R. Narayanan
D. Purushothaman

Much research in agroforestry systems is concerned with increasing the biological input of nutrients to trees and to the crops grown concurrently or consecutively with them and with determining how and in what quantities these nutrients become available. The exchange of nutrients among the plants of the agroforestry system result largely from the activity of appropriate soil microorganisms. Associative or symbiotic microorganisms are responsible for nitrogen input and for the availability of other minerals, especially phosphorus, in the ecosystem. Other bacteria make available the nutrients of dead and decaying plants for uptake by the root systems of crop species. Because of their importance in cropping systems and the emphasis rightly placed on them in research programs, this chapter primarily focuses on the symbiotic microorganisms *Rhizobium* and *Frankia* and on ecto- and endomycorrhizas. Also briefly considered are the potential roles of such free-living microorganisms as associative nitrogen fixers, plant growth regulating and phosphate-solubilizing bacteria, and those organisms concerned with nutrient transformations of decaying plant material.

SYMBIOTIC NITROGEN FIXATION IN AGROFORESTRY SYSTEMS

Selecting the Symbioses Best Suited to Agroforestry

The important nitrogen-fixing symbioses are (1) those between many legume tree species and *Rhizobium* or *Bradyrhizobium* and (2) those between *Frankia* and woody species within the eight nonleguminous plant families (the so-called actinorhizal plants—see Dixon and Wheeler 1986) that are nodulated by this nitrogen-fixing actinomycete. For temperate and warm temperate conditions, the most important of the *Frankia* associations are with *Alnus* (Betulaceae) or *Elaeagnus* and *Hippophaë* (Elaeagnaceae), and in the tropics and subtropics, with members of the Casuarinaceae. This last family has been subject to recent taxonomic revision and is divided into four genera—*Casuarina*, *Allocasuarina*, *Gymnostoma*, and *Ceuthostoma*. The most promising candidates for agroforestry are in the first two genera. Some species, for example *Casuarina* and legumes such as some of the *Acacias* and *Prosopis* species, are drought resistant and clearly should be considered for use in areas subject to water shortage. Employing trees that periodically shed their leaves may increase nutrient cycling, and this may also improve penetration of light to intercrops. However, informed selection of species for many situations requires much more research information than is currently available.

Specificity of Nodulation

How specific the interactions between host plant and microsymbiont are is essential information when deciding whether to inoculate an exotic species before introduction into a new area or when it is believed that the native microbial population is not effective. The specificity of nodulation of the host plant by the microsymbiont varies considerably; in general, *Frankia* is of broader host specificity than *Rhizobium*.

A system based on the specificity of infection of host legume genera has been used to classify *Rhizobium* for well over 50 years. Specificity of infection has many practical attractions for those concerned with the applications of *Rhizobium* technology, even though it is imperfect because numerous strains of rhizobia are able to infect across specificity groups and because there is new evidence of affinities from chemotaxonomic and numerical taxonomic data. It remains an important criterion for speciation of the genus in *Bergey's Manual of Systematic Bacteriology*, with modifications incorporating new taxonomic data (Jordan 1984).

The genus *Rhizobium* now includes fast-growing rhizobia that produce acid on yeast mannitol agar (YMA) and are most frequently of temperate origin. There are three species in the genus: *R. leguminosarum* (biovars *viceaea*, *trifolii*, and *phaseoliif*), *R. meliloti*, and *R. loti*. This last species includes rhizobia that are capable of nodulating *Leucaena* and *Mimosa*. The genus *Bradyrhizobium* consists of slow-growing bacteria that do not produce acid on YMA and that most commonly infect tropical legumes. A strain of *Bradyrhizobium* is also responsible for nodulation of the woody nonlegume *Paraspnia* (Trinick and Galbraith 1980). The nodule bacteria in this genus are a heterogeneous group within which texonomic relationships have not yet been resolved. Only one species, *B. japonicum*, is recognized (Jordan 1984).

Frankia strains vary considerably in their ability to infect actinorhizal plant species from different genera. On the basis of cross-inoculation studies with isolates from a range of species, Baker (1987) suggested that isolates fall into at least four cross-inoculation groups: (1) strains that nodulate *Alnus* and *Myrica*; (2) strains that nodulate *Casuarina* and *Myrica*; (3) strains that nodulate the Elaeagnaceae (*Elaeagnus*, *Hippophaë*, and *Shepherdia*) and *Myrica*; and (4) strains that nodulate only the Elaeagnaceae. These groups of *Frankia* strains can be defined partly in terms of their method of infection.

Groups 1 and 2 represent strains that nodulate by the traditional root hair infection mechanism. In this infection pathway, *Frankia* penetrates deformed root hairs, and the hyphae grow intracellularly down through the root hair and into the root cortex. Strains in group 4 infect host Elaeagnaceous plants by intercellular penetration (Miller and Baker 1986). Root hairs are not involved in infection. *Frankia* hyphae enter the root tissue by penetration through the middle lamella between two epidermal cells and then colonize the intercellular spaces of the root cortex. Group 3 represents a small number of so-called flexible strains. These strains infect *Myrica* by root hair infection and the Elaeagnaceae by intercellular penetration (Miller and Baker 1986). It should be noted that many legumes, particularly tree species, have epidermal routes of infection and penetrate the root cortex intercellularly (Sprent and de Faria 1988).

Special Requirements for Microsymbiont Isolation

Both rhizobia and *Frankia* are difficult to isolate directly from soil, and consequently, most isolates are obtained from nodules. Isolation and culture of rhizobia normally poses few problems (Vincent 1970). Most strains will grow readily on a variety of defined media or on YMA. The main problem is adequate surface sterilization of the plant material to reduce contaminants before crushing and plating out of the bacterial suspension. Inability to absorb Congo Red from the nutrient agar is a useful indicator for rhizobia, but the absolute test is nodulation when the isolate is inoculated back onto the host plant.

The slow growth of *Frankia* presents problems for its isolation. However, isolation and culture of isolates from many species, particularly alders, has become fairly routine over the decade since the first isolates were obtained (Callaham et al. 1978). Most isolates now are obtained by subculture of outgrowths on isolation plates from surface sterilized nodule fragments or from endophyte fractions that are separated from surface sterilized nodule homogenates, for example, by microfiltration through 20 μm mesh nylon screen (Diem et al. 1982, Benson 1982). Media containing propionate and Tween 80 as carbon sources are most suited for culture of a wide range of *Frankia* strains, although pyruvate or glucose may sometimes be used (Burggraaf and Shipton 1983, Malcolm et al. 1985).

Most problems during isolation are caused by other microbial contaminants overrunning culture plates before *Frankia* colonies develop. Fungal contaminants can be controlled by the inclusion of cycloheximide in the media. The growth of many bacteria is suppressed by propionate, which makes this compound particularly useful as a component of culture media. Because most *Frankia* strains so far isolated prefer this organic acid as a carbon source, its use in isolation media is recommended.

It should be noted that numbers of *Frankia* strains have been isolated, particularly from *Casuarina* nodules, that are unable to reinfect their host species, although they may be infective on other species. The reasons for this are unknown.

Tolerance and Adaptability of Microsymbionts

Microbial populations are genetically diverse, and consequently, it is not surprising that symbiotic nitrogen fixers differ widely in their ability to nodulate and fix nitrogen in different host plant species. Strains vary in their tolerance of a wide range of environmental and soil factors such as temperature, pH, water stress, mineral nutrition, and salinity. However, extreme soil conditions will reduce the nodulation and nitrogen fixation due to effects on both the microsymbiont and the host plant. Often, the host plant is more sensitive to the constraint under investigation than the microsymbiont. It is essential to ensure that the strain(s) of microsymbiont and varieties of host plant species are properly matched, that they are suited for survival in the environment to which they are to be introduced, and that the introduced bacteria are fully competitive with the native microbial strains in the soil. Clearly, parallel breeding programs are needed for selection of microsymbionts and host plant species tolerant of environmental stress.

Legume tree species tolerant of semiarid, waterlogged, and saline conditions have been identified (Felker 1984, Tomar and Gupta 1985), but selective breeding programs have been pursued seriously only with *Leucaena* and *Prosopis* (Burley et al. 1986). Selection of rhizobia for superior nodulation of woody legumes is relatively recent, as appreciation of the importance of these plants has increased. For example, Sanginga and colleagues (1986) identified two superior *Rhizobium* strains from a number of isolates from *Leucaena*, *Sesbania*, *Tephrosa*, and *Acacia* that gave good nodulation and nitrogen fixation in *Leucaena leucocephala* in pot trials and in the field. From 211 strains isolated from Brazilian legume trees, da Silva and Franco (1984) selected 19 that grew at pH 4.6 and were promising for nodulation of trees on acid soils. Halliday (1984) has provided a useful account of appropriate selection procedures for isolation, selection, and testing of legume tree rhizobia. Detailed coverage is available in many reviews (for example, Alexander 1985).

Frankia strains also vary widely in their effectivity, that is, in their ability to fix symbiotic nitrogen (Normande and Lalonde 1982). Differences result from the influence of the microsymbiont on specific nitrogen-fixing activity (nitrogen fixed per unit weight of nodules) and on nodulation (weight of nodules per plant) (Hooker and Wheeler 1987). The occurrence of *Frankia* infective for particular host plant species in soils is variable—*Frankia* capable of nodulating *Alnus* species are widespread in soils of many temperate countries (Huss-Danell and Frej 1986), whereas *Casuarina* Frankiae may be restricted in adverse soils (Lawrie 1982). Much of the work on survival and adaptation of *Frankia* in adverse soil conditions has been carried out in temperate regions. However, *Frankia* strains tolerant of adverse conditions such as pH, moisture stress, or salinity can be selected for incorporation into inoculum preparations (Shipton and Burggraaf 1983, Faure-Raynaud et al. 1986). There is a great need for more extensive research in the tropics, particularly on Frankiae nodulating the Casuarinaceae (Dawson and Gibson 1987).

A special feature of *Frankia* that affects its infectivity and effectivity is the ability to sporulate in vivo. Although most *Frankia* strains form sporangia in culture, in the field *Frankia* produces sporangia [sp(+)] in some nodules but not in others [sp(–)]. Spore production may be observed frequently in nodules of *Alnus* or *Myrica*, but is less common in *Casuarina* or Elaeagnaceous nodules. Genetic differences between *Frankia* strains are believed to be responsible for the sporulation response in vivo (Torrey 1987). These observations have practical implications because sp(+) crushed nodule inoculum can be several times more infective than sp(–) inoculum. However, the effectivity of nitrogen fixation may be less in sp(+) than in sp(–) nodules. *Frankia* strains introduced into the soil can survive for many months, and there is some evidence that sp(–) Frankiae can grow saprophytically in soil (Smolander et al. 1987).

There has also been relatively little study of actinorhizal plants, so there is enormous scope for improvement and selection for growth in particular environments (Hopmans et al. 1983, Tomar and Gupta 1985, Hennessey et al. 1985). Vegetative and micropropagation methods for the rapid multiplication of *Casuarina* (Lundquist and Torrey 1984) and *Alnus* (Perinet and Lalonde 1983) are available.

Strain Selection for Improving the Competitivity of the Microsymbionts

As in any agricultural system, it is clearly essential that the microbes introduced into agroforestry sites be competitive with the native soil microorganisms for nodulation of the plant of interest. The introduced strains should also remain in the soil for several years so that as high a percentage as possible of the new roots formed on the tree crop are nodulated by the introduced microorganism. Techniques for identifying inoculated rhizobial strains are well established and usually include introduction of, or selection for, antibiotic resistance and/or serotyping, for example by ELISA on re-isolation from nodules (Alexander 1984, Mortenssen et al. 1987). Application of such techniques has shown that while inoculation with superior strains may successfully establish competitive populations in some soils, particularly when numbers of native rhizobia are low, it is common for a majority of the nodules formed to contain native rhizobia. This problem may be lessened to some extent by introducing inoculum with bacterial numbers greatly in excess of those in the soil. Weaver and Frederick (1974) suggest that soybean inoculum should contain 1,000 times the soil population of *Bradyrhizobium japonicum* for the introduced strain to produce 50% of the nodules. However, native populations of rhizobia are often large and highly competitive with the introduced strains. Dowling and Broughton (1986), in their review of competitivity and survival of introduced rhizobia, note that "any attempt to predict the outcome of field inoculations requires knowledge of not only the size but also the nature of the indigenous rhizobial population." These authors quote evidence for genetic exchange between *Rhizobium* strains in the laboratory and suggest that it may occur in the field. If it does, or if it could be encouraged, then desirable genetic traits may spread from the indigenous to the introduced rhizobia, and vice versa. It is conceivable that the overall effectiveness of the rhizobial population could be increased by gene flow in the population.

What currently hinders most study of the ability of *Frankia* to compete in agroforestry or other ecosystems is the lack of markers by which strains may be recognized and recovered from the environment into which they are introduced. Antibiotic resistance or serological techniques, which readily permit re-isolation of specific rhizobial strains, are not sufficiently specific for use as field tools. Instead, reliance on more time-consuming and technically difficult techniques, such as strain protein patterns or DNA homology, has been suggested (Lechevalier 1984). Further development, particularly of serological techniques, is required if critical studies of *Frankia*'s ability to compete, infect, and survive under field conditions are to be carried out.

Allelopathic Effects on Interactions between Host Plant and Microsymbiont

Organic compounds from plants and soil can either promote or inhibit the growth of microsymbionts by influencing infection, nodule function, and plant growth. For example, Mallik and Tesfai (1987) demonstrated stimulation of *Bradyrhizobium* growth by allelochemicals from some grasses and herbs. The naphthoquinone juglone, produced by black walnut, will inhibit the growth of interplanted *Elaeagnus* and *Alnus*. Nodulation is inhibited by this compound, as is the growth of *Frankia* in culture (Vogel and Dawson 1985). Vogel and Dawson obtained evidence that *Frankia* strains varied widely in their tolerance of this compound so that strain selection provides one possible avenue for overcoming allelopathic effects on the nodulation of actinorhizal plants. Actinorhizal plants may also influence pathogenesis. For example, Li and colleagues (1969) showed that phenolics contained in the roots of *Alnus rubra* inhibited the growth of the root parasite *Poria weirii*.

There is much controversy about adverse allelopathic effects of some agroforestry trees on associated crops. For example, some authors claim *Casuarina* litter is harmful to crop growth, whereas others find no effect. Clearly, further research is required to resolve these conflicts. It should be noted that nodulation by *Frankia* may help prevent allelopathic growth inhibition of *Alnus crispa* by phenolics from interplanted *Populus* (Jobidon and Thibault 1982) and that mycorrhizal infections can influence allelopathy (see below). Toxic effects may be reduced by soil inoculation with microbial strains that degrade allelopathic residues—Dommergues and colleagues (1979) found that inoculation of a sandy soil containing phytotoxic root material with *Enterbacter cloacae* restored its fertility.

Cross-Infection between a Legume Crop and the Tree Component of an Agroforestry System

Of particular interest in agroforestry is the improved nodulation that may occur due to cross-infection between rhizobia nodulating a preceding or concurrent legume crop and the introduced tree legume. Most work on the diversity of *Rhizobium* strains that effectively nodulate both herbaceous and woody species has been carried out with *Leucaena*. Thus Trinick (1968) reported the isolation of fast-growing isolates from *Leucaena* that sometimes nodulate *Vigna* species effectively. In a subsequent paper (Trinick 1980), effective cross-inoculations were demonstrated with fast-growing isolates from woody species of *Leucaena*, *Mimosa*, *Acacia*, and *Sesbania*, and from the herbaceous *Lablab*. Again, many of the isolates were effective on *Vigna* and some other legumes. Slow-growing *Bradyrhizobium* isolated from a range of herbaceous and shrubby legumes were unable to form nodules on any of these legumes except *Lablab*. The range of cross-inoculations for *Leucaena* has been extended by Jarvis (1983), who found effective nodulation by fast-growing strains isolated from 11 plant species, including *Lotus*, *Onobrychis*, *Astragalus*, and *Coronilla*. Some slow-growing isolates from some tree species may also be cross-infective. For example, Basak and Goyal (1980) obtained isolates from *Albizia lebbeck*, *Dichrostachys nutans*, *Prosopis cineraria*, and three species of *Acacia* that all nodulated cowpea effectively. Research is now required to establish in the field the advantages to plant productivity that may result from growing appropriate legume tree species with herbaceous legumes inoculated with *Rhizobium* strains that are effective on both herbaceous and woody legumes. The inoculated legumes may be grown either prior to planting the tree species or as an intercrop with them.

Benefits of Microbial Inoculation

Formulation of a stable inoculum is essential for the commercial distribution of improved strains to be used where appropriate strains of microsymbionts are absent or are not effective. Rhizobia in logarithmic growth in shake cultures or fermentors are mixed into peat or one of the other carrier materials, such as coals and lignite, coconut coir dust, maize cob compost, vermiculite, charcoal, or silt. Novel carriers, such as alginates and acrylamide gels, have also been used successfully (Mugnier et al. 1982). However, it is usually possible to devise a successful inoculum carrier using local materials. Further details concerning the production, storage, and use of inoculants are available (Vincent 1970, Bergersen 1980, Alexander 1984).

Using *Frankia* strains to nodulate nursery stock that show superior symbiotic effectivity can improve early seedling growth (Wheeler et al. 1986). Small-scale inoculation of nursery stock can be accomplished readily by watering onto the seedlings an aqueous suspension of washed culture, infection being improved if the surface layers of the seed bed are kept moist following inoculation. Inoculation can also be achieved with *Frankia* mixed in alginate gels (Sougoufara et al. 1989). Large-scale inoculation techniques for greenhouse culture of actinorhizal seedlings have also been developed in which *Frankia* is applied with a greenhouse spray (Perinet et al. 1985).

Problems of Observation of Nodulation

Many reports of non-nodulation based on field observations carried out in one country or in a single area are due to environmental conditions that may inhibit nodulation. Poor nodulation of many of the common agricultural legumes in tropical regions is also the result of high soil temperature and low soil moisture during the growing season.

In addition, various characteristics of the root systems of woody legumes make observation and quantification of nodules more difficult than with the herbaceous legumes, especially for mature trees. Although nodule clusters may survive for several years, soil compaction and the development of surface roots can make exploration for nodules difficult without significantly damaging the tree. On many sites, nodules may be found at a depth of 0 cm to 30 cm in the soil, but some species growing in arid conditions form nodules at considerably greater depth and could be mistakenly categorized as non-nodulating from superficial investigation. Definite proof of how well a particular species nodulates may have to be obtained in glasshouse experiments. For example, Felker and Clark (1982) found that many reports of non-nodulation of mesquite (*Prosopis* spp.) in the field were based on searches of the top meter of the soil. The ability of these plants to nodulate was shown clearly in controlled experiments—plentiful nodulation was found at 3.2 m in a soil column in which the top 0.5 m was allowed to dry, but in which moisture was maintained at depth. Field nodulation of this genus can be good when adequate moisture is available. Other environmental features such as soil acidity, aeration, mineral availability, and plant age can all affect the occurrence of nodules—Hogberg (1986) took five days to find nodules on *Xeroderris stuhlmannii* in the Tanzanian savanna!

Conversely, care must be taken not to make false records of nodulation.

Nodules must not be confused with insect galls, *Agrobacterium* tumors, or mycorrhizal roots. Careful morphological, anatomical, and physiological investigation can resolve the nature of questionable material. The root system of adjacent legumes can become so intertwined with the roots of the tree under investigation that nodulation of the legume may be mistakenly attributed to the tree: there are suggestions that reports of root nodulation of the nonlegume *Rubus ellipiticus* by *Frankia* may have resulted from confusion with nodules on adjacent bushes of *Myrica rubra*. The acetylene reduction assay can provide a quick and easy method of confirming the nitrogen-fixing ability of root tubercles. However, very low levels of acetylene reduction should not be accepted as evidence of symbiotic nitrogen fixation, for such activity often is due to nonsymbiotic, nitrogen-fixing microorganisms.

Quantification of Nodulation in the Field

Many estimates of nitrogen fixation attempt to relate rates of fixation to levels of nodulation in the field. However, accurate quantification of nodulation per tree or area can present major problems. It is often not possible to excavate whole trees, and consequently, the experimental area has to be subsampled, usually by taking cores with an auger. Field nodulation is not necessarily random, so it is important to sample both laterally and at different depths to determine the pattern of nodule distribution. The work of Hogberg and Kvarnstroem (1982) provides a guide to the procedure that might be followed. To determine the biomass of root nodules and fine roots in a four-year-old stand of *Leucaena leucocephela*, they took 30 random 7 cm diameter auger samples at three soil depth levels (0 to 10 cm, 10 to 30 cm, and 30 to 50 cm). An additional sample was excavated from a depth of 50 to 70 cm at 15 of the 30 sampling sites. A rather uniform, lateral distribution of nodules was found throughout the stand, with three-quarters of the nodules in the top 30 cm of the soil profile. Other investigations suggest a more random distribution. From soil cores taken in a Mexican coffee plantation where *Inga jinicuil* was used as a shade tree, Roskoski (1981) noted that fine roots and nodules were concentrated close to the coffee plants. By contrast, Lindblad and Russo (1986) found the greatest biomass of nodules within a radius of 50 cm of the stems of *Erythrina poeppigiana* in a coffee plantation. Nodules were confined to the upper (12 cm) soil layer, and their incidence decreased markedly with distance from the stem. This variation in nodule distribution at different study sites shows that it is not possible to predict patterns of nodule distribution from experience gained elsewhere. Techniques of vegetation analysis must be applied carefully at new sites if meaningful data are to be obtained.

With smaller shrub legumes, it may be possible to excavate whole plants within quadrats distributed randomly over the study area, as has been done for relatively uniform stands of *Cytisus scoparius* (Wheeler et al. 1987). With larger tree legumes, a rational approach would be to establish the root distribution patterns both horizontally and vertically around specimen trees and then to sample with a soil borer at specific depths along transects radiating from the stem of the tree. This approach should permit statistical analysis of nodule distribution in the ecosystem. There is, however, a great need for a full comparison of sampling procedures that may be applied to both natural and agroforestry systems to determine the best approach for quantifying woody plant nodulation.

Techniques for Measuring Nitrogen Fixation

Several techniques have been employed to measure rates of fixation in the field. Some depend on accurate measurement of changes in the quantity of nitrogen in different parts of the ecosystem. These techniques are laborious but require little sophisticated equipment. Others, using ^{15}N-labeled materials, are much more expensive and require the use of a mass spectrometer to determine isotope enrichment. Another technique determines the reduction of acetylene to ethylene by the nodulated plants. While this is a relatively simple technique, requiring only the use of a basic gas chromatograph (cheap, portable instruments can be made or purchased), the results obtained give only general rates of fixation and cannot be used except in a most general sense to quantify nitrogen fixation (Minchin et al. 1983). A recent account of an attempt to calibrate the acetylene assay for field use by comparing total nitrogen accumulation with acetylene reduction is provided by Hansen et al. (1987). The other principal techniques considered briefly here are nitrogen balance and isotope techniques.

Nitrogen balance techniques

In order to compensate for the various sources of nitrogen input and loss in the ecosystem, accurate measurement requires a detailed knowedge of nitrogen cycling within the ecosystem. The accumulation of nitrogen that results from symbiotic fixation may then be calculated from a balance sheet of nitrogen changes. The major changes over a given period of time are described by the following equation (Herridge 1982):

$$N_{sf} = \Delta SN + \Delta VN + N_h + N_l + N_g - [N_{af} + N_{fe} + N_p + N_{am}]$$

where

N_{sf} = symbiotic nitrogen fixation
SN = soil nitrogen
VN = nitrogen in non-nodulated vegetation
N_h = nitrogen harvested
N_l = nitrogen lost through leaching
N_g = gaseous loss of nitrogen (denitrification)
N_{af} = nonsymbiotic nitrogen fixation
N_{fe} = fertilizer or manure nitrogen added
N_p = rain or dust nitrogen
N_{am} = ammonia fixation by plants

Some of these components, such as N_{af}, N_p, or N_{am}, will probably be small and may usually be ignored, although their potential contribution to the balance sheet should be considered during planning.

Accurate measurement of the different components, especially of components with major nitrogen content such as the soil and vegetation associated with nitrogen-fixing trees, is essential if meaningful results for symbiotic nitrogen fixation are to be obtained. Measurement errors of the nitrogen content of associated vegetation may be particularly large in intercropped agroforestry systems. If the intercrop is a legume, then it will be very difficult to assign accurate values for symbiotic nitrogen fixation to the associated nitrogen-fixing tree species by nitrogen budgeting methods. Unfortunately, methods for measuring denitrification are imprecise and are likely to introduce substantial errors in estimates of nitrogen fixation.

Another N balance method involves comparing rates of nitrogen accretion in the nitrogen fixer with that of either non-nondulated nonlegumes, or a non-nodulating isoline of the same plant. The problem with the former comparison is that it is necessary to select a reference species with growth and rooting habits similar to those of the legume that will explore the nitrogen content of a similar volume of soil. The latter comparison relies on the availability of non-nodulating mutants, which also undoubtedly will show different growth characteristics from the nodulated plant on nitrogen-poor soil. Non-nodulating isolines of species other than soybean and peanut are rare. Woody legumes and non-leguminous nitrogen fixers have rarely been screened for this trait.

Isotope techniques

The most precise method for determining nitrogen fixation is with glasshouse plants that can be sealed in containers for assay. An atmosphere of ^{15}N is provided so that subsequent measurement of the isotope in the plant tissue may be made by mass spectrometry. Experimental details are provided by Bergersen (1980) and Silvester (1983). This technique is not practical for field use, particularly with woody species of large size. Also, the effects of handling are likely to negate the accuracy of measurements carried out on excised, nodulated roots.

The most satisfactory alternative method is to determine nitrogen fixation by N isotope dilution. This technique requires a control group of non-nodulated plants grown in soil similar to that of the nodulated plants under investigation. Mineral nitrogen, as ammonium or nitrate, is then added to the soil in amounts (< 3 kg/ha) that do not materially affect nitrogen fixation. The nitrogen fixed by the nodulated plant is given by the following expression:

$$\frac{\%\ ^{15}\text{N excess [control} - \text{N-fixing plant]}}{\%\ ^{15}\text{N excess control plant}} \times \text{N content of N-fixing plant}$$

The major problem is again to ensure that the root systems of the control and nodulated plants are sufficiently similar so that the same soil volume is explored by both. Exploitation of mineral nitrogen at lower levels in the soil by a deep-rooting control species will greatly affect the isotope dilution rates obtained. Some authors have used several plants as a check on their suitability as reference controls but if different plants give different estimates of enrichment with ^{15}N from soil uptake, then the problem of selecting the most suitable reference control remains. Screening methods for evaluating the suitability of reference plants have been proposed (Ledgard et al. 1985).

A less satisfactory isotope technique involves measurement of the variation in natural abundance of ^{15}N in plant tissue caused by discrimination in biological systems in favor of ^{14}N. Soil microbiological reactions, in particular denitrification, tend to increase the ^{15}N abundance of soil nitrogen compared with atmospheric nitrogen. Consequently, plant nitrogen assimilated from soil has more ^{15}N than nitrogen fixed directly from the atmosphere (Silvester 1983).

However, problems arise because it is necessary to make very precise measurements on the mass spectrometer of small variations in ^{15}N levels—a small variation in δ $^{15}N\%$ corresponds to a large difference in nitrogen fixation. Also, the factors that control isotope discrimination in the soil and uptake by the plant are not fully understood (Broadbent et al. 1980). Other plant associations, such as mycorrhizas, may affect isotope fractionation. It is not surprising, therefore, that estimates of nitrogen fixation in the field using the technique have often proved unsatisfactory (Binkley et al. 1985). However, in some situations, such as with the deep-rooted nodulation of *Prosopis*, the measurement of nitrogen natural abundance ratios may be the only practical technique for the estimation of nitrogen fixation (Shearer et al. 1983). Up to 60% of the nitrogen of *Prosopis* is fixed at depth, and depending on stand density, 40 kg to 135 kg of nitrogen per hectare may be deposited in the surface soils of mesquite woodlands (Virginia 1986).

The above observations will indicate to the reader that the comments of Beringer, made at the Fifth International Conference of N Fixation in 1984, still largely hold true concerning our ability to accurately measure nitrogen fixation in the field. He noted that large amounts of useless data have been, and continue to be, collected. He encouraged researchers to continue to make field measurements while being aware of the limitations of the techniques they employ. These should be refined continually as constraints upon their interpretations become apparent.

Nitrogen Cycling and Crop Productivity in Agroforestry Systems

There are many reports implying that symbiotic nitrogen fixation sustains soil fertility in agroforestry systems. However, there are relatively few experiments to prove this beyond a doubt and to quantify the contributions that nitrogen fixation can make to improve crop production. In many studies, accurate observations of the nodulation of the tree species are lacking, even though it is well documented that nodulation of legume trees can be poor or absent (see above). Estimates by de Faria and colleagues (1989) from the published literature suggest that 97% of the Papilionoideae are nodulated, compared with 90% of Mimosoideae and 23% of Caesalpinoideae. Extensive surveys have now been made of nodulation of tree legumes in some countries such as Brazil, and the NIFTAL project at the University of Hawaii has produced a catalog of known nodulated legume trees (de Faria et al. 1989). Fortunately, more of the common legume trees used for agroforestry have at least a potential for nodulation, although even among these, both nodulated and non-nodualted plants may be found in particular situations. Without basic information on nodulation, it is not possible, as some authors have been tempted to do, to assign to nitrogen fixation the benefits for crop yield that may arise from the cycling of nitrogen from the tree species.

Given that measurements made with the acetylene reduction assay are likely to be very imprecise, accurate data for the input of fixed nitrogen and its subsequent cycling in agroforestry systems are conspicuously absent. A flow diagram for nitrogen cycling in a coffee plantation shaded by *Erythrina* or *Inga*, adapted from Aranguren and associates (1982), is shown in figure 9.1. The authors presented details of the nitrogen content and its transformation for a number of the pool, but wisely left unquantified input due to fixation

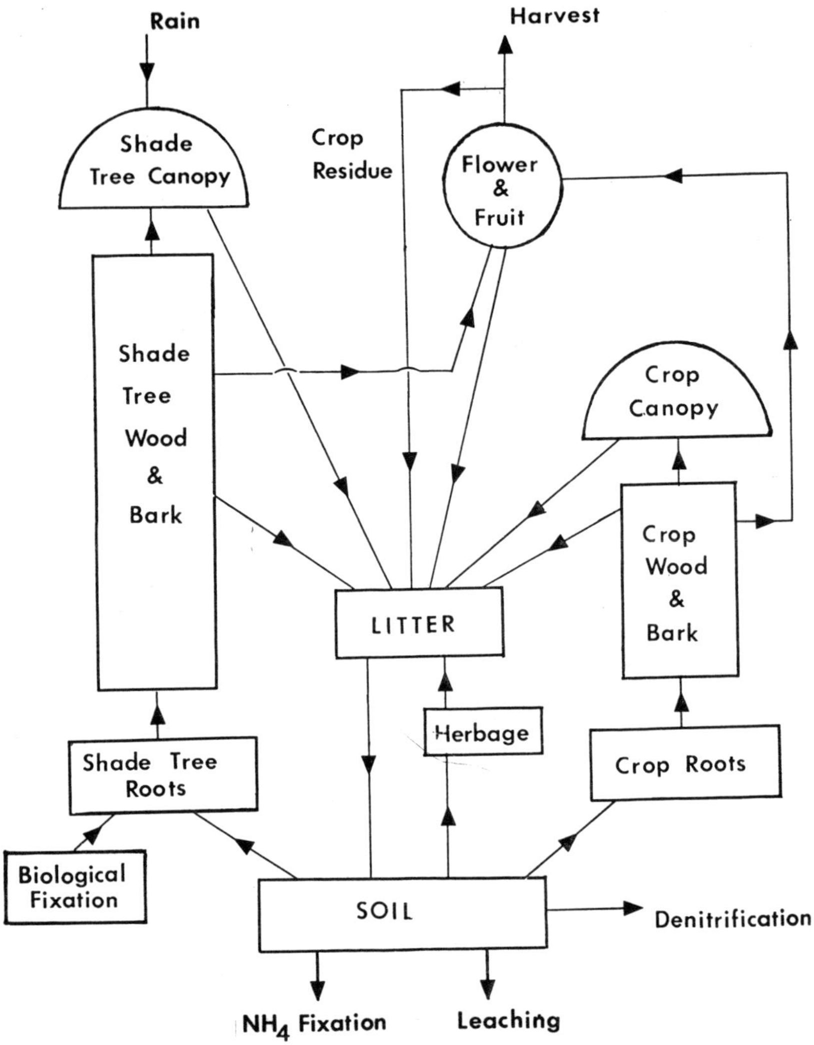

Figure 9.1. Nitrogen cycling in a plantation comprising a crop such as coffee with nitrogen-fixing shade trees.
Adapted from: Aranguren et al. 1982.

and losses from denitrification. Very careful estimates of nitrogen input and output over several years should enable the levels of nitrogen that are contributed by biological fixation to be estimated. These can then be compared with data obtained by other techniques.

As far as actinorhizal plants are concerned, it is generally accepted that nitrogen fixation rates of well-nodulated trees are at least equivalent to that of legumes. In forestry, both experimental and commercial use has been made of actinorhizal species as "nurses" and as sources of nitrogen for associated tree species. Positive effects on growth have been shown—for example, for *Populus-Alnus* mixes where success depends on the use of *Populus* clones that do not inhibit nitrogen fixation by outgrowing and shading the alder too rapidly

(Heilman and Stettler 1983). Freidrich and Dawson (1984) showed not only that mixed plantings with *Elaeagnus*, *Alnus*, and the legume *Robinia* increased soil nitrogen and growth of black walnut (*Juglans nigra* L.) in North America but also that other factors such as weed control and soil shading contributed much to the promotion of growth.

Other factors such as the fertility of the site and the density of the stand are important in determining the effects that nitrogen-fixing trees have on plantation productivity. On fertile soils, the incorporation of a nitrogen-fixing species may have no effect, or it may even inhibit productivity of the non-nitrogen-fixing crop tree (Binkley 1984, Malcolm et al. 1985).

While data is gradually accumulating, particularly for alders, on nutrient cycling in pure or mixed tree stands (Coté and Camiré 1987), information concerning nutrient effects on intercropped species is scarce. Actinorhizal trees can enhance the productivity of understory pasture as shown by a 40% to 60% increase in yield of herbage under *Alnus glutinosa* and *Alnus rubra*, planted on a neglected pasture, in which much of the additional nitrogen from alder root and litter decomposition was incorporated into the grass cover (Wheeler et al. 1986). Such benefits to food crop species may be expected when they are interplanted with *Casuarina*, for example. But basic information is only emerg-ing slowly from which the benefits and drawbacks of cycling nitrogen and other nutrients in such systems may be analyzed critically.

ROLE OF MYCORRHIZAS IN AGROFORESTRY SYSTEMS

It is now well documented that mycorrhizas, which are a range of symbiotic associations between soil fungi and plant roots, improve the uptake of a range of macronutrients, especially immobile phosphorus. This increased uptake has been attributed to factors such as reduced spatial diffusion of the nutrient to the plant, increased rate of absorption and concentration at the absorbing surface, and chemical alteration of the nutrient making it more available for uptake (Harley and Smith 1983). The mycorrhizal association also improves uptake of micronutrients such as zinc and copper (Gilmore 1971). Mycorrhizas can enhance nitrogen fixation by *Rhizobium* and *Frankia* (Daft et al. 1985) and have been associated with the reduction of disease susceptibility. They can also transfer materials between plants of the same and different species (Francis and Read 1984).

In view of these benefits, it is vital to explore the relevance of this association to the development of successful agroforestry systems, particularly because the mycorrhizal fungi occur in all plant taxa and because little host specificity is apparent (Redhead 1980, Smith 1980).

Two Important Mycorrhizal Associations for Agroforestry

Five types of mycorrhizas are now recognized (Harley and Smith 1983), but only two are relevant to agroforestry—ectomycorrhizas and endomycorrhizas. The first type, which occurs almost exclusively in tree species, is characterized by the absence of intracellular penetration of root cortical cells by the fungus. The fungul mycelium develops on the surface of short roots that lack secondary thickening and forms a mantle around the root. Changes in root color and morphology that result can be used to distinguish the different kinds of ectomycorrhizas. The mycorrhizal roots may be either extensively branched or undivided, depending on the host species. The structure of the fungal sheath is a function of the infecting fungus and may have a smooth surface with very few emanating hyphae, or may form an external mycelial network. Within the fungal mantle, the hyphae grow in the intercellular spaces of the host root, forming a network called the Hartig net (Harley and Smith 1983). The fungi involved are mostly higher basidiomycetes (such as *Boletus*, *Suillus*, *Amanita*, *Lactarius*, *Tricholoma*, *Pisolithus*, *Scleroderma*, and *Rhizopogon*) some ascomycetes (*Tuber*), and zygomycetes.

The second type, the vesicular-arbuscular mycorrhizas (VAM), is predominant. It is characterized by less fungal development on the root surface and haustorial penetration of the cortical cells. Infection does not induce readily discernible change in root morphology, and VAM are detectable only by staining. The distinguishing features of VAM are the arbuscles and vesicles. Arbuscles are haustoria-like structures that develop due to repeated dichotomous branching of the invading hyphae and are ephemeral structures with a lifespan of one to three weeks. Vesicles are terminal sac-like swellings of the hyphae and are primarily storage organs of the fungus. VAM does not form a sheath around the root, but the extra-matrical hyphae can extend from the root for several centimeters into the soil. The VAM fungi mostly belong to the family Endogonaceae (*Glomus*, *Sclerocystus*, *Gigaspora*, *Acaulaspora*). They are obligate symbionts and cannot be cultured axenically.

Special Problems for Isolation and Culture

In vitro culture studies have shed light on the basic biology and physiology of the fungal symbiont and can provide specific isolates for artificial inoculation of seedlings (Marx 1980, Trappe 1977). Techniques are available for culture of ectomycorrhizas, but none have been developed for endomycorrhizas.

Ectomycorrhizas. While ectomycorrhizas can be isolated from sclerotia, rhizomorphs, and surface-sterilized ectomycorrhizal roots (Schenck 1982), sporocarp isolation is generally favored because it does not involve pretreatment of the fungus. Most species can be isolated readily, but some are difficult (for example, *Camphedious*). The major impediment in the isolation of some species is the lack of precise information regarding their nutrient requirements.

Young sporocarps are preferred for direct isolation, and isolation is performed immediately after collection (Moser 1958). For hypogeous fungi such as *Rhizopogon*, the center of the gleba is ideal for isolation. For *Scleroderma*, the interior of the peridial tissue gives better results. If isolation is delayed, the sporocarp should be refrigerated after collection. The tissue explants are transferred onto nutrient agar in tubes or plates; most species proliferate profusely from the tissue explants. Stock cultures are commonly stored on nutrient agar slants in test tubes under refrigeration. Long storage may result in loss of viability, although several species have been stored for over three years (Marx and Daniel 1976).

VAM Isolation and Culture. Monoculture of VAM on artificial media has been unsuccessful, necessitating collection of spores or infected roots from field soil, augmentation on host plants in the greenhouse, and recollection of spores from the culture soil (Mosse 1973). Progress has come with the discovery that fungal fructifications (sporocarps) and large, distinctive resting spores found near or attached to mycorrhizal roots in the soil will produce typical VAM infections when used to inoculate seedlings grown in sterilized soil and in axenic culture. A range of such resting spores, differing in morphological characteristics and life histories, has now been recovered from soils using a variety of techniques (Daniels and Skipper 1982). The spores can be multiplied in the presence of a host plant, and within four to six months they will give thousands of new spores of the same kind. This makes it possible to categorize the spores as originating from distinct fungi, although the normal taxonomic criteria of species identification by growth and reproduction in monoaxenic culture cannot be applied. The process of spore development and the range of variability within a progeny can be studied in open pot cultures containing the host plant.

Use of Cultures for Plant Inoculation

Ectomycorrhizas. Use of pure mycelial cultures has been advocated as the most biologically efficient method for plant inoculation (Trappe 1977, Marx 1980). However, the limited availability of pure mycelial cultures is the major deterrent to large-scale nursery applications. Procedures for producing pure mycelial inoculum have been developed in various parts of the world. A commercial source of vegetative inoculum of *Pisolithus tinctorius* has been developed by IMRD in Athens, Georgia, and by Abbotts Laboratory in Chicago (Marx et al. 1982).

Tree seedlings with ectomycorrhiza have been used as inoculum for new seedlings. Similarly, excised mycorrhizas have also been used as inoculum. The simplest and easiest method of inoculation is with soil. The main problems of this approach are the transport of large quantities of soil and the lack of control over the specific fungi present in the mixture. Sporophores of fungi such as *Pisolithus tinctorius* and *Rhizopogon luteolus* have been dried or chopped into small pieces and incorporated successfully into soil inoculum, but fresh sporophores enhance the infective capacity of the soil. Use of sporophores essentially involves inoculation with spores, since the vegetative matrix of the sporophore soon disintegrates after incorporation into the soil. The practices for mycorrhizal inoculation of nursery seedlings in tropical and subtropical countries are discussed by Mikola (1973), and in temperate regions by Marx and colleagues (1982).

VAM. Either a mixed inoculum containing chlamydospores, soilborne vesicles, mycelium, and infected root fragments or a pure mycelial inoculum of spores separated from soil may be used for plant inoculation. Though each method has its merits and demerits, mixed inoculum is credited with faster and surer infection. Storing dried inoculum, even for two weeks, results in reduced infection rates. Spore inoculum, however, is easier to quantify and permits regulation of inoculum dosage.

Though numerous techniques for recovering VAM propagules are available, the most basic involves wet sieving and decanting (Gerdemann and Nicolson 1963) followed by collection of spores by differential sedimentation in gelatin columns (Mosse and Jones 1968). The plate methods (Smith and Skipper 1979) and the adhesion/flotation method (Sutton and Barron 1972) do not require prior sieving and decanting. For VAM spore extraction, density gradient centrifugation is also used (Mertz et al. 1979).

Influence of the Environment on Strain Selection

Ectomycorrhizas

The genotype of the fungus is of paramount importance for exploitation of the mycelial association. A majority of the ectomycorrhizal fungi have a broad host spectrum, but some are host specific. Thus *Pisolithus tinctorius* forms ectomycorrhizas with nearly 73 tree species, but *Suillus grevilli* has a limited host range. By the same token, some tree species such as *Alnus* have a limited range of ectomycorrhizal partners, while others have a broad range. The importance of isolate selection for infection and effectivity has been demonstrated by several workers (Maroneck and Hendrix 1980).

That environmental and soil factors can affect the tree crop and symbiont specificity is evident from studies of *Pisolithus tinctorius*. This ectomycorrhizal fungus infected ponderosa and Scots pine better than other fungi on poor soils, while the reverse was true on better soils (Riffle and Tinus 1982). The temperature tolerance of fungal strains also varies considerably (Trappe 1977). *Pisolithus tinctorius* can form mycorrhizal associations at high temperatures, while *Suillus variegatus* is adapted to low temperatures (Marx and Bryan 1971, Slankis 1974).

Endomycorrhizas

VAM are reported to have worldwide distribution (Mosse 1973). Their occurrence in lowland humid tropics or on tropical hosts has been little surveyed, but observations are available from all three tropical regions of the world: Asia, Africa, and the neotropics (Thappar and Khan 1973, Redhead 1980, Thomazini 1974). In India, Thapper and Khan (1973) noted that 22 forest tree species were all infected by endogonaceous fungi. However, in investigations in Puerto Rico (Edminston 1970) and Brazil (St. John 1980), almost a third of the species inspected lacked the association, contradicting the concept of the universal occurrence of VAM.

The VAM fungi show less host specificity than the ectomycorrhizal fungi, and any potential host species can be infected by any fungal species (Mosse 1973, Hayman 1978). However, the fungi differ in their effectiveness, which depends more on the soil and plant system they colonize than on the host plant itself (Borea and Azcon-Aguilar 1983). Low infection intensity in several species of annual crops in the humid tropics has been attributed to inhospitable soil environment after land clearing or to lack of mycorrhizal inoculum (Janos 1980).

Although light and air temperature are reported to influence the development of VAM (Furlan and Fortin 1977, Hayman 1978, Moawad 1980), generally, information on the effect of soil factors on the ecophysiology of VAM associated with tropical plants is dismally modest. The observance of a decline in mycorrhizal infection at a soil temperature of 35°C (Meyer 1973) indicates that mycorrhizal growth may be depressed in areas subject to high irradiation and emphasizes the need for selecting thermophilic fungal species like certain *Gigaspora* species. The soil type also affects the occurrence of VAM. Endogonaceous spores are more abundant in loamy than sandy soils in dry areas of Pakistan (Saif et al. 1977), but in the humid conditions of Jamaica, VAM developed well in sandy soils (Black 1980). Although VAM improve water transport in plants, they do not compensate for water loss in semiarid regions (Menge et al. 1978, Safir et al. 1972).

The Effects of Biological and Allelopathic Interactions on Mycorrhizal Associations

Mycorrhizas release various metal chelates that directly or indirectly alter nutrient availability, break down primary materials, and inhibit root pathogens (Cromack et al. 1979, Schroth and Hancock 1982). The most potent of these, especially in pathology, are the siderophores that occur less in soils of clear-cut areas than in forest soils (Perry et al. 1984, Leong 1986). Mycorrhizal fungi can also detoxify allelochemicals produced by grasses and herbs, and thus allow trees to compete successfully (Perry et al. 1987).

Ectomycorrhizas

Improvement in the resistance of trees to root infections when mycorrhizas are well developed is extensively reported. For example, association with *Pisolithus tinctorius* increased the survival of *Pinus tideae* seedlings infected with the root pathogen *Rhizoctonia solani* (Wingfield 1968). Similarly, seedlings of *Picea mariana* inoculated with *Suillus granulatus* grew well in the presence of the root pathogen *Mycelium radicis attrovirens* (Richard et al. 1971). Addition of basidiospores of *Lactaria laccata* to nursery soil containing *Fusarium oxysporum* decreased the mortality of Douglas fir seedlings (Stack and Sinclair 1975). Over one hundred species of ectomycorrhizal fungi are known to produce antibiotics, which are variable in their biological spectra, being antifungal, antibacterial, and antiviral (Marx 1972). The host roots can also produce inhibitors that contribute to pathogen resistance as a response to ectomycorrhizal infection.

There are many reports on the susceptibility of ectomycorrhizas to phytotoxins from litter (Chu-Chou 1978, Schoenberger and Perry 1982), and inhibition of ectomycorrhizas by products of other microorganisms such as *Streptomyces* species has also been reported. The magnitude of the inhibition varies with the nature and concentration of the allelochemical and with the fungal species (Rose et al. 1983).

Endomycorrhizas

Host-specific interactions have demostrated that the benefits of interactions among VAM, plant pathogen, and host plant may vary with differing species or strain combinations (Mosse 1973). Disease severity may be increased (Ross 1972), decreased (Schonbeck and Dehne 1977), or unaffected by endomycorrhizal infection (Menge et al. 1978). Most studies have been with the VAM *Glomus mossae*. How the VAM affects the pathogen is not known, although increased concentration of arginine and reducing sugars have been observed in the mycorrhizal plants. Use of surplus carbohydrates, secretion of antibiotics, provision of a physical barrier, and protection of rhizosphere organisms have all been suggested as possible mechanisms against infection by pathogens.

Interaction of VAM with viral pathogens has been reported by Schonbeck and Schinger (1972) and by Daft and Okusanya (1973). In both reports, viral infection was greater in mycorrhizal than nonmycorrhizal plants. This greater susceptibility was attributed to increased phosphate levels.

Much less work has been done to study allelopathic effects on VAM than on ectomycorrhizas. That such effects do occur is illustrated by the poor development of VAM associations on hardwood seedlings growing under ponderosa pine where fungal spores were absent from the soil (Kovacic et al. 1984).

Nematode Infections

Parasitic nematodes on feeder roots sometimes directly inhibit ectomycorrhizal development (Sutherland and Fortin 1968). Several, such as lance (*Hoplolaimus coronatus*) and pine cystoid (*Meloidodera floridensis*), feed on ectomycorrhizas formed by *Pisolithus tinctorius* and *Thelophora terrestris* on short leaf and loblolly pine (Ruehle and Marx 1971). However, mycorrhizal infection may in some instances help reduce the incidence of nematode attack. For example, Roncadori and Hussey (197) showed that mycorrhizal colonization by *Gigaspora margarita* inhibited the attack of cotton by *Meloidogyne incognita*.

OTHER MICROORGANISMS

The tree crop in an agroforestry system has major effects on microbial population of the soil beneath and adjacent to the canopy. The accumulation of organic carbon, derived from litter or root exudates, acts as an energy source for microbial metabolism and growth. In turn, this affects both the chemical properties of the soil and such physical properties as compaction, water infiltration, and aggregate structure (Virginia 1986).

Nitrogen-fixing bacteria that associate with plant roots, such as *Azospirillum*, have been the subject of intensive study for many years, but a signficant role in the nitrogen nutrition of crop species remains to be demonstrated. Growth effects undoubtedly do occur in some instances, and these have been variously ascribed to the production of growth substances or to effects on soil nutrient availability (Okon and Kapulnik 1986). Associations of such bacteria with tree roots in tropical regimes have been studied little—in temperate forests the general consensus of experimental evidence again is that the influence on nitrogen nutrition is insignificant. Other free-living, nitrogen-fixing organisms are undoubtedly significant to the ecology of an area over a long period of time but are unlikely to be important to crops in short rotations.

Some bacteria are known to influence nodulation. For example, *Pseudomonas cepacia* improves nodulation of *Alnus rubra* by *Frankia* (Knowlton and Dawson 1983), and *Pseudomonas putida* improves nodulation of *Phaseolus vulgaris* by *Rhizobium*, possibly through the production of phosphate-solubilizing compounds (Grimes and Mount 1984). Demonstration of similar events and their significance for plant productivity in tropical agroforestry systems await investigation.

Other microorganisms, such as phosphate-solubilizing fungi and bacteria, may affect plant growth through effects on mineral availability (Thomas et al. 1985). Nitrogen fixation often results in net acidification of soil (Dixon and Wheeler 1983); and reduced pH, even at restricted soil microsites, may change the availability of phosphate or iron (Sprent 1986). Siderophore production by some microorganisms may be important in aiding iron uptake by crop plants (Neilands and Leong 1986).

Finally, mention must be made of the central role bacteria play in various transformations of the nitrogen cycle in the soil. Factors such as the carbon to nitrogen ratios of the organic material, temperature, water supply, pH, soil disturbance and aeration, and the availability of other nutrients can all affect rates of nitrogen mineralization. The pattern of change in soil microorganisms in a stand of deciduous nitrogen-fixing trees, such as *Alnus glutinosa*, follows a predictable pattern. Leaves and litter decompose rapidly during the six months after leaf fall, with the numbers of ammonifying bacteria being highest soon after leaf fall and decreasing towards the end of the decomposition period. The numbers of denitrifying bacteria also increase after litter fall (Kjoller et al. 1985). In a study of a number of agricultural legumes, Frankenburger and Abdelmagid (1985) noted that the composition of ligno-protein complexes in the residues was an important factor influencing nitrogen mineralization rates. Clearly, this will have even greater effect on the mineralization of residues from tree crops.

Nitrifiers in soils have a great survival capacity. Nitrification rates probably are largely determined by soil organic content, temperature, and moisture content (Berg and Rosswall 1985). Again, little is known of the functioning of these bacteria in ammonium nitrification in agroforestry systems, although formation of the more readily leached nitrites and nitrates is clearly important, especially in areas subject to periods of high rainfall. Further information on techniques for the study of nitrifying bacteria may be obtained from Prosser (1988).

CONCLUSIONS

While there is an enormous literature concerning the role of microorganisms in the maintenance of soil fertility in conventional agriculture and forestry, information obtained specifically for agroforestry is only gradually becoming available. Discussions in this chapter have been directed mainly to the two elements, nitrogen and phosphorus, that most commonly govern plant productivity. Integrated studies on the availability of these elements in the environment of agroforestry systems are rare. Further studies to fill this gap in our knowledge are essential if the potential of agroforestry is to be fully realized.

LITERATURE CITED

Alexander, M. 1984. *Biological nitrogen fixation; ecology, technology and physiology*. New York: Plenum Publishing Co. 258 p.

Alexander, M. 1985. Ecological constraints on nitrogen fixation in agricultural ecosystems. *Advances in Microbial Ecology* 8: 163–178.

Aranguren, J., G. Escalante, and R. Herrera. 1982. Nitrogen cycle of tropical perennial crops under shade trees. *Plant and Soil* 67: 259–269.

Baker, D.D. 1987. Relationships among pure-cultured strains of *Frankia* based on host specificities. *Physiologica Plantarum* 70: 245–248.

Basak, M.K., and S. Goyal. 1980. Studies on tree legumes. IV. Characterization of the symbiont and direct and reciprocal cross-inoculation studies with tree legumes and cultivated legumes. *Plant and Soil* 56: 39–51.

Benson, D.R. 1982. Isolation of *Frankia* strains from alder actinorhizal root nodules. *Applied Environmental Microbiology* 44: 461–465.

Berg, P., and T. Rosswall. 1985. Ammonium oxidiser numbers, potential and actual oxidation rates in two Swedish arable soils. *Biology and Fertility of Soils* 1: 131–140.

Bergersen, F.J. 1980. *Methods for evaluating biological nitrogen fixation*. Chichester, UK: John Wiley and Sons.

Binkley, D. 1984. Importance of size-density relationships in mixed stands of Douglas fir and red alder. *Forest Ecology and Management* 9: 81–85.

Binkley, D., P. Sollins, and W.A. McGill. 1985. Natural abundance of nitrogen-15 as a tool for tracing alder-fixed nitrogen. *Soil Science Society of America Journal* 49: 444–447.

Black, R. 1980. The role of mycorrhizal symbiosis in the nutrition of tropical plants. In: *Tropical mycorrhiza research* (P. Mikola, ed.). Oxford: Clarendon Press. pp 191–202.

Borea, J.M., and C. Azcon-Aguilar. 1983. Mycorrhizas and their significance in nodulating nitrogen fixing plants. *Advances in Agronomy* 36: 1–54.

Broadbent, F.E., R.S. Rauschkolb, K.A. Lewis, and C.Y. Chang. 1980. Spatial variability of nitrogen-15 and total nitrogen in some virgin and cultivated soils. *Soil Science Society of America Journal* 44: 524–527.

Burggraaf, A.J.P., and W.A. Shipton. 1983. Studies on the growth of *Frankia* isolates in relation to infectivity and nitrogen fixation (acetylene reduction). *Canadian Journal of Botany* 61: 2774–2782.

Burley, J., C.E. Hughes, and B.T. Styles. 1986. Genetic systems of tree species for arid and semi-arid lands. *Forest Ecology and Management* 16: 317–344.

Callaham, D., P. Deltredici, and J.G. Torrey. 1978. Isolation and cultivation *in vitro* of the actinomycete causing root nodulation in *Comptonia*. *Science* 199: 899–902.

Chu-Chou, M. 1978. Effects of root residues on growth of *Pinus radiata* seedlings and a mycorrhizal fungus. *Annals of Applied Biology* 90: 407–416.

Coté, B., and C. Camiré. 1987. Tree growth and nutrient cycling in dense plantings of hybrid poplar and black alder. *Canadian Journal of Forest Research* 17: 516–523.

Cromack, K., P. Sollins, W.C. Graustein, K. Seidel, A.W. Todd, G. Spycher, Y.L. Ching, and R.L. Todd. 1979. Calcium oxalate accumulation and soil weathering in mats of the hypogeous fungus *Hysterangeum crassum*. *Soil Biology and Biochemestry* 11: 463–468.

Daft, M.J., and B.O. Okusanya. 1973. Effect of *Endogone* mycorrhiza on plant growth. V. Influence of infection on the multiplication of viruses in tomato, petunia and strawberry. *New Phytologist* 72: 975–983.

Daft, M.J., D.M. Clelland, and I.C. Gardner. 1985. Symbiosis with endomycorrhizas and nitrogen-fixing organisms. *Proceedings of the Royal Society of Edinburgh* 85: 263–282.

Daniels, B.A., and H.D. Skipper. 1982. Methods for the recovery and quantitative estimation of propagules from soil. In: *Methods and principles of mycorrhizal research* (N.C. Schenck, ed.). St. Paul, Minnesota: American Phytopathological Society. pp 29–36.

da Silva, G.G., and A.A. Franco. 1984. Selecaode estirpes de *Rhizobium* spp. de leguminosas florestais em meio de cultura tolerantes a acidez e a toxidez do Al. *Pesquisa Agropecuaria Brasileira* (Brasilia) 19: 169–173.

Dawson, J.O., and A. H. Gibson. 1987. Sensitivity of selected *Frankia* isolates from *Casuarina*, *Allocasuarina* and North American host plants to sodium chloride. *Physiologica Plantarum* 70: 272–278.

de Faria, S.M., G.P. Lewis, J.I. Sprent, and J.M. Sutherland. 1989. Occurrence of nodulation in the leguminosae. *New Phytologist* 111: 607–619.

Diem, H.G., D. Gauthier, and Y.R. Dommergues. 1982. Isolation of *Frankia* from nodules of *Casuarina equisetifolia*. *Canadian Journal of Microbiology* 18: 526–530.

Dixon, R.O.D., and C.T. Wheeler. 1983. Biochemical, physiological and environmental aspects of symbiotic nitrogen fixation. In: *Biological nitrogen fixation in forest ecosystems: foundations and applications* (J.C. Gordon and C.T. Wheeler, eds.). The Hague: Martinus Nijhoff/Dr. W. Junk. pp 108–172.

Dixon, R.O.D., and C.T. Wheeler. 1986. *Nitrogen fixation in plants*. Glasgow: Blackie.

Dommergues, V.R., H.G. Diem, and F. Ganry. 1979. The effect of soil microorganisms on plant productivity. In: *Soils research in agroforestry* (H.O. Mongi and P.A. Huxley, eds.). Nairobi: ICRAF. pp 205–242.

Dowling, D.N., and W.J. Broughton. 1986. Competition for nodulation of legumes. *Annual Review of Microbiology* 40: 131–157.

Edminston, J. 1970. *Survey of mycorrhizas and root nodules in the E1 Verde forest. PF15-F20 in a tropical rain forest* (H.T. Odum and R.F. Pidgeon, eds.). Oak Ridge, Tennessee: US Atomic Energy Commission.

Faure-Reynaud, M., M.A. Bonnefoy-Poirier, and A. Moiroud. 1986. Influence de pH acides sur la viabilite d'isolates de *Frankia*. *Plant and Soil* 96: 347–358.

Felker, P. 1984. Legume trees in semi-arid and arid areas. *Pesquisa Agropecuaria Brasileira* (Brasilia) 19s/n: 47–59.

Felker, P., and P.R. Clark. 1982. Position of mesquite (*Prosopis* spp.) nodulation and nitrogen fixation (acetylene reduction) in 3m long phraetophytically simulated soil columns. *Plant and Soil* 64: 297–305.

Francis, R., and D.J. Read. 1984. Direct transfer of carbon between plants connected by vesicular/arbuscular mycorrhizal mycelium. *Nature* 307: 1.
Frankenburger, W.T., and H.M. Abdelmagid. 1985. Kinetic parameters of nitrogen mineralisation rates of leguminous crops incorporated into the soil. *Plant and Soil* 87: 257–271.
Freidrich, J.M., and J.O. Dawson. 1984. Soil nitrogen concentration and *Juglans nigra* growth in mixed plots with nitrogen-fixing *Alnus*, *Elaeagnus*, *Lespedesa* and *Robinia* species. *Canadian Journal of Forest Research* 14: 864–868.
Furlan, V., and J.A. Fortin. 1977. Effects of light intensity on the formation of vesicular-arbuscular endomycorrhizas on *Allium cepa* by *Gigaspora calospora*. *New Phytologist* 79: 335–340.
Gerdemann, J.W., and T.H. Nicolson. 1963. Spores of mycorrhizal *Endogone* species extracted from soil by wet sieving and decanting. *Transactions of the British Mycological Society* 46: 235–246.
Gilmore, A.E. 1971. The influence of endotrophic mycorrhizae on the growth of peach seedlings. *Journal of American Society of Horticultural Science* 96: 35–38.
Grimes, R., and M.S. Mount. 1984. Influence of *Pseudomonas putida* on nodulation of *Phaseolus vulgaris*. *Soil Biology and Biochemistry* 16: 27–30.
Halliday, J. 1984. Integrated approach to nitrogen fixing tree germplasm development. *Pesquisa Agropecuaria Brasileira* (Brasilia) 19s/n: 91–117.
Hansen, A.P., J.S. Pate, and C.A. Atkins. 1987. Relationships between acetylene reduction activity, hydrogen evolution and nitrogen fixation in nodules of *Acacia* spp.; experimental background to assaying fixation by acetylene reduction under field conditions. *Journal of Experimental Botany* 38: 1–12.
Harley, J.L., and S.E. Smith. 1983. *Mycorrhizal symbiosis*. London: Academic Press.
Hayman, D.S. 1978. Endomycorrhizae. In: *Interactions between non-pathogenic soil micro-organisms and plants* (Y.R. Dommergues and S.W. Krupa, eds.). Amsterdam: Elsevier. pp 401–442.
Heilman, P.C., and R.F. Stettler. 1983. Phytomass production in young mixed plantations of *Alnus rubra* Bong. and cottonwood in western Washington. *Canadian Journal of Microbiology* 29: 1007–1013.
Hennessey, T.C., L.K. Bair, and R.W. McNew. 1985. Variation in response among three *Alnus* spp. clones to progressive water stress. *Plant and Soil* 87: 135–141.
Herridge, D.F. 1982. A whole system approach to quantifying biological nitrogen fixation by legumes and associated gains and losses of nitrogen in agricultural systems. In: *Biological nitrogen fixation for tropical agriculture* (P.H. Graham and S.C. Harris, eds.). Cali, Colombia: CIAT. pp 593–608.
Hogberg, P. 1986. Nitrogen fixation and nutrient relations in savannah woodland trees. *Journal of Applied Ecology* 23: 675–688.
Hogberg, P., and M. Kvarnstroem. 1982. Nitrogen fixation by the woody legume *Leucaena leucocephala* in Tanzania. *Plant and Soil* 66: 21–28.
Hooker, J.E., and C.T. Wheeler. 1987. The effectivity of *Frankia* for nodulation and nitrogen fixation in *Alnus rubra* and *Alnus glutinosa*. *Physiologica Plantarum* 70: 333–341.
Hopmans, P., L.A. Douglas, P.M. Chalk, and S.C. Dalbridge. 1983. Effects of soil moisture, mineral nitrogen and salinity on nitrogen fixation (acetylene reduction) by *Allocasuarina verticillata* (Lam.) L. Johnson seedlings. *Australian Forest Research* 13: 189–196.
Huss-Danell, K., and A.K. Frej. 1986. Distribution of *Frankia* in soils, forest and afforestation sites in northern Sweden. *Plant and Soil* 90: 407–417.
Janos, D.P. 1980. Mycorrhizae influence tropical succession. *Biotropica* 12(Supp.): 56–64.
Jarvis, B.D.W. 1983. Genetic diversity of *Rhizobium* strains which nodulate *Leucaena leucocephala*. *Current Microbiology* 8: 153–158.
Jobidon, T., and J.R. Thibault. 1982. Allelopathic growth inhibition of nodulated and unnodulated *Alnus crispa* var. *mollis* Fern. seedlings by *Populus balsamifera*. *American Journal of Botany* 69: 1213–1223.
Jordan, D.C. 1984. III. Rhizobiaceae. In: *Bergey's manual of systematic bacteriology*, 9th ed. Baltimore, Maryland: Williams and Wilkens. 964 p.
Kjoller, A., S. Struwe, and K. Vestberg. 1985. Bacterial dynamics during decomposition of alder litter. *Soil Biology and Biochemistry* 17: 463–468.
Knowlton, S., and J.O. Dawson. 1983. Effects of *Pseudomonas cepiacea* and cultural factors on the nodulation of *Alnus rubra* roots by *Frankia*. *Canadian Journal of Botany* 61: 2877–2882.
Kovacic, D.A., T.V. St. John, and M.I. Dyer. 1984. Lack of vesicular/arbuscular mycorrhizal inoculum in a ponderosa pine forest. *Ecology* 65: 1755–1759.
Lawrie, A.C. 1982. Field nodulation in nine species of *Casuarina* in Victoria. *Australian Journal of Botany* 30: 447–460.
Lechevalier, M.P. 1984. The taxonomy of the genus *Frankia*. *Plant and Soil* 78: 1–6.

Ledgard, S.F., J.R. Simpson, J.R. Freney, F.J. Bergersen, and R. Morton. 1985. Assessment of the relative uptake of added and indigenous soil nitrogen by nodulated legumes and reference plants in the nitrogen-15 dilution measurement of nitrogen fixation: glasshouse application of method. *Soil Biology and Biochemistry* 17: 323–328.

Leong, J. 1986. Siderophores: their biochemistry and possible role in the biocontrol of plant pathogens. *Annual Review of Phytopathology* 24: 187–209.

Li, C.Y., K.C. Lu, E.E. Nelson, W.B. Bollen, and J.M. Trappe. 1969. Effect of phenolics and other compounds on growth of *Poria weirii* in vitro. *Microbios* 3: 305–311.

Lindblad, P., and R. Russo. 1986. Acetylene reduction by *Erythrina poepigiana* in a Costa Rican coffee plantation. *Agroforestry Systematics* 4: 33–37.

Lundquist, R., and J.G. Torrey. 1984. The propagation of *Casuarina* species from stem cuttings. *Botanical Gazette* 145: 378–384.

Malcolm, D.C., J.E. Hooker, and C.T. Wheeler. 1985. *Frankia* symbiosis as a source of nitrogen in forestry: a case study of symbiotic nitrogen fixation in a mixed *Alnus-Picea* plantation in Scotland. *Proceedings of the Royal Society of Edinburgh* 85B: 263–282.

Mallik, M.A.B., and K. Tesfai. 1987. Stimulation of *Bradyrhizobium japonicum* by allelochemicals from green plants. *Plant and Soil* 103: 227–231.

Maroneck, D.M., and J.W. Hendrix. 1980. Synthesis of *Pisolithus tinctorius* ectomycorrhizae on seedlings of four woody species. *Journal of American Society of Horticulture Science* 105: 823–825.

Marx, D.H. 1972. Ectomycorrhizae as biological deterrants to pathogenic root infections. *Annual Review of Phytopathology* 10: 429–454.

Marx, D.H. 1980. Ectomycorrhizal fungus inoculation; a tool for improving forestation practices. In: *Tropical mycorrhiza research* (P. Mikola, ed.). London: Oxford University Press. pp 13–71.

Marx, D.H., and W.C. Bryan. 1971. Influence of ectomycorrhizae on survival and growth of aseptic seedlings of loblolly pine at high temperature. *Forest Science* 17: 37–41.

Marx, D.H., and W.J. Daniel. 1976. Maintaining cultures of ectomycorrhizal and plant pathogenic fungi in sterile water cold storage. *Canadian Journal of Microbiology* 22: 338–341.

Marx, D.H., J.L. Ruehle, D.S. Kennedy, C.E. Cordell, J.W. Riffle, R.J. Molina, W.H. Pawuk, S. Navratil, R.W. Tinus, and O.C. Goodman. 1982. Commercial vegetative inoculum of *Pisolithus tinctorius* and inoculation techniques for development of ectomycorrhizae on container-grown tree seedlings. *Forest Science* 28: 339–366.

Menge, J.A., E.L.V. Johnston, and R.G. Platt. 1978. Mycorrhizae dependency of several citrus cultivars under three nutrient schemes. *New Phytologist* 81: 553–559.

Mertz, S.M., J.J. Heithaus, and R.L. Bush. 1979. Mass production of axenic spores of the endomycorrhizal fungus. *Transactions of the British Mycological Society* 72: 167–169.

Meyer, F.H. 1973. Distribution of ectomycorrhiza in native and man-made forests. In: *Ectomycorrhizae: their ecology and physiology* (G.C. Marks and T.T. Koslowski, eds.). London: Academic Press. pp 79–106.

Mikola, P. 1973. Application of mycorrhizal symbiosis in forestry practice. In: *Ectomycorrhizae: their ecology and physiology* (G.C. Marks and T.T. Koslowski, eds.). London: Academic Press. pp 383–411.

Miller, I.M., and D.D. Baker. 1986. Nodulation of actinorhizal plants by *Frankia* strains capable of both root hair infection and intercellular penetration. *Protoplasma* 131: 82–91.

Minchin, F.R., J.F. Witty, J.E. Sheehy, and M. Muller. 1983. A major error in the acetylene reduction assay. I. Decreases in nodular nitrogenase activity under assay conditions. *Journal of Experimental Botany* 34: 641–649.

Moawad, M. 1980. Ecophysiology of vesicular/arbuscular mycorrhiza. In: *Tropical mycorrhiza research* (P. Mikola, ed.). London: Oxford University Press, pp 203–205.

Mortenssen, A.M., J.G. Gustafssen, and H.D. Ljunggren. 1987. Competition in soil between effective and ineffective strains of *Rhizobium leguminosarum* biovar *trifolii* in the nodulation of the red clover-*Triboilum pratense* studied with ELISA. *Journal of General Microbiology* 133: 3467–3472.

Moser, M. 1958. Die Kunshiche Mykorrhizaimpfung and Forstpflanzen. I. Erfahrungen bei der Reinkulture von Mykorrhizapilzen. *Forstwissenschaft Centralbibliothek* 77: 32–40.

Mosse, B. 1973. Advances in the study of vesicular/arbuscular mycorrhiza. *Annual Review of Phytopathology* 11: 171–196.

Mosse, B., and G.W. Jones. 1968. Separation of *Endogone* species from organic soil debris by differential sedimentation on gelatin columns. *Transactions of the British Mycological Society* 51: 604–608.

Mugnier, G.J., H.G. Diem, and Y.R. Dommergues. 1982. Polymer entrapped *Rhizobium* as an inoculant for legumes. *Plant and Soil* 65: 219–231.

Neilands, J.B., and S.A. Leong. 1986. Siderophores in relation to plant disease and growth. *Annual Review of Plant Phytopathy* 37: 187–208.

Normande, P., and M. Lalonde. 1982. Evaluation of *Frankia* strains isolated from provenances of two *Alnus* species. *Canadian Journal of Microbiology* 28: 1133–1142.

Okon, Y., and Y. Kapulnik. 1986. Development and function of *Azospirillum* inoculated roots. *Plant and Soil* 90: 3–16.

Perinet, P., J.G. Brouillette, J.A. Fortin, and M. Lalonde. 1985. Large scale inoculation of actinorhizal plants with *Frankia*. *Plant and Soil* 87: 175–183.

Perinet, P., and M. Lalonde. 1983. In vitro propagation and nodulation of the actinorhizal plant *Alnus glutinosa* (L.) Gaertn. *Plant Science Letter* 29: 9–17.

Perry, D.A., S.L. Rose, D. Pilz, and M.M. Schoenberg. 1984. Reduction of natural ferric iron chelators in disturbed forest soils. *Soil Science Society of America Journal* 48: 379–382.

Perry, D.A., R. Molina, and M.P. Amaranthus. 1987. Mycorrhizas, mycorrhizospheres and reforestation: current knowledge and research needs. *Canadian Journal of Forest Research* 17: 929–940.

Prosser, J. 1988. *Nitrification*. *Society of General Microbiology Special Publications*. Vol. 20. London: Oxford University Press. 228 p.

Redhead, J.F. 1980. Mycorrhiza in natural tropical forests. In: *Tropical mycorrhiza research* (P. Mikola, ed.). London: Oxford University Press. pp 127–142.

Richard, C., J.A. Fortin, and A. Fortin. 1971. Protective effect of and ectomycorrhizal fungus against the root pathogen *Mycelium radicis atrovirens*. *Canadian Journal of Forest Research* 1: 246–251.

Riffle, J.W., and R.W. Tinus. 1982. Ectomycorrhizae characteristics, growth and survival of artificially inoculated ponderosa and Scots pine in a greenhouse and plantation. *Forest Science* 28: 646–660.

Roncardori, R.W., and R.S. Hussey. 1977. Interaction of the endomycorrhizal fungus *Gigaspora margarita* and root-knot nematode on cotton. *Phytopathology* 67: 1507–1511.

Rose, S.L., D.A. Perry, D. Pilz, and M.M. Schoenberg. 1983. Allelopathic effects of litter on the growth and colonization of mycorrhizal fungi. *Journal of Chemical Ecology* 9: 1153–1162.

Roskoski, J.P. 1981. Nodulation and nitrogen fixation by *Inga jinicuil*, and woody legume in coffee plantations. I. Measurement of nodule biomass and field acetylene reduction rates. *Plant and Soil* 59: 201–206.

Ross, J.P. 1972. Influence of *Endogone* mycorrhiza on *Phytophthora* rot of soybean. *Phytopathology* 62: 896–897.

Ruehle, J.L., and D.H. Marx. 1971. Parasitism of ectomycorrhizae of pine by lance nematode. *Forest Science* 17: 31–34.

Safir, G.R., J.S. Boyer, and J.N. Gerdemann. 1972. Nutrient status and mycorrhizal enhancement of water transport in soybean. *Plant Physiology* 49: 700–703.

Saif, S.R., I. Ali, and A.A. Zaidi. 1977. Vesicular-arbuscular mycorrhiza in plants and endogonaceous spores in soil of northern areas of Pakistan. *Pakistan Journal of Botany* 9: 129–148.

Sanginga, N., K. Mulongoy, and A. Ayanaba. 1986. Inoculation of *Leucaena leucocephala* (LAM) de Wit. with *Rhizobium* and its contribution to a subsequent maize crop. *Biological Agriculture and Horticulture* 3: 347–352.

Schenck, N.C. 1982. *Methods and principles of mycorrhiza research*. St. Paul, Minnesota: American Pathology Society. 244 p.

Schoenberger, M.M., and D.A. Perry. 1982. The effect of site disturbances on growth and ectomycorrhizae on Douglas fir and western hemlock seedlings. A greenhouse bioassay. *Canadian Journal of Forest Research* 12: 343–353.

Schonbeck, F. and H.W. Dehne. 1977. Damage to mycorrhizal and non-mycorrhizal cotton seedlings by *Thielaviopsis basicola*. *Plant Desease Reporter* 61: 266–267.

Schonbeck, F., and V. Schinger. 1972. Investigations on the influence of endotrophic mycorrhiza on TMV lesion formation in *Nicotiana tabacum* L. var. *xanthine*. *Phytopathology* 73: 78–80.

Schroth, M.N., and J.G. Hancock. 1982. Disease-suppressive soil and root colonizing bacteria. *Science* 216: 1376–1381.

Shearer, G., D.H. Kohl, R.A. Virginia, B.A. Bryan, J.L. Skeeters, E.T. Nilsen, M.R. Sharifi, and P.W. Rundel. 1983. Estimates of nitrogen fixation from variation in the natural abundance of nitrogen-15 in a Sonoran desert ecosystem. *Oecologia (Berlin)* 56: 365–373.

Shipton, W.A., and A.J.P. Burggraaf. 1983. Aspects of the cultural behavior of *Frankia* and possible ecological implications. *Canadian Journal of Botany* 61: 2783–2792.

Silvester, W.B. 1983. Analysis of nitrogen fixation. In: *Biological nitrogen fixation in forest ecosystems: foundations and applications* (J.C. Gordon and C.T. Wheeler, eds.). Dordrecht, The Hague: Martinus Nijhoff/Dr. W. Junk. pp 173–212.

Slankis, V. 1974. Soil factors influencing formation of mycorrhizae. *Annual Review of Phytopathology* 12: 437–457.

Smith, G.W., and H.D. Skipper. 1979. Comparison of a method to extract spores of vesicular-arbuscular mycorrhizal fungi. *Soil Science Society of America Journal* 43: 722–725.

Smith, S.E. 1980. Mycorrhizas of autotrophic higher plants. *Biological Reviews of Cambridge Philosophical Society* 55: 475–510.

Smolander, A., C. Van Dijk, and V. Sundman. 1987. Survival of *Frankia* strains introduced into soils. *Plant and Soil* 106: 65–72.

Sougoufara, B., H.G. Diem, and Y.R. Dommergues. 1989. Response of field-growth *Casuarina equisetifolia* to inoculation with *Frankia* strain ORS 021001 entrapped in alginate beads. *Plant and Soil* 118: 133–137.

Sprent, J.I. 1986. Nitrogen fixation in a sustainable agriculture. *Biological Agriculture and Horticulture* 3: 87–94.

Sprent, J.I., and S. de Faria M. 1988. Mechanisms of infection of plants by nitrogen fixing organisms. *Plant and Soil* 110: 157–165.

Stack, R.W., and W.A. Sinclair. 1975. Protection of Douglas fir seedlings against *Fusarium* root rot by a mycorrhizal fungus in the absence of mycorrhiza formation. *Phytopathology* 65: 468–472.

St. John, T.V. 1980. A survey of mycorrhizal infection in an Amazonian rain forest. *Acta Amazonica* 10: 527–533.

Sutherland, J.R., and J.A. Fortin. 1968. Effect of the nematode *Aphelenchus avenae* on some ectotrophic mycorrhizal fungi and on red pine mycorrhizal relationship. *Phytopathology* 58: 519–523.

Sutton, J.C., and G.L. Barron. 1972. Population dynamics of *Endogone* spores in soil. *Canadian Journal of Botany* 50: 1909–1914.

Thapper, H.S., and S.N. Khan. 1973. Studies on endomycorrhiza in some forest species. *Proceedings of the Indian National Science Academy* 39: 687–694.

Thomas, G.V., M.V. Shantaran, and N. Saraswathy. 1985. Occurrence and activity of phosphate-solubilizing fungi from coconut plantaiton soils. *Plant and Soil* 87: 357–364.

Thomazini, L.I. 1974. Mycorrhiza in the plants of the "Cerrado." *Plant and Soil* 41: 707–711.

Tomar, O.S., and R.K. Gupta. 1985. Performance of some forest tree species in saline soils under shallow and saline water table conditions. *Plant and Soil* 87: 329–335.

Torrey, J.G. 1987. Endophyte sporulation in root nodules of actinorhizal plants. *Physiology of Plants* 70: 279–288.

Trappe, J.M. 1977. Selection of fungi for ectomycorrhizal inoculation in nurseries. *Annual Review of Phytopathology* 15: 203–222.

Triniek, M.J. 1968. Nodulation of tropical legumes. I. Specificity in the *Rhizobium* symbiosis of *Leucaena leucocephala*. *Experimental Agriculture* 4: 243–253.

Trinick, M.J. 1980. Relationships among the fast-growing rhizobia of *Lablab purpureus*, *Leucaena leucocephala*, *Mimosa* spp., *Acacia Farnesiana Sesbania grandiflora* and their affinities with other rhizobial groups. *Journal of Applied Bacteriology* 49: 39–53.

Trinick, M.J., and J. Galbraith. 1980. The *Rhizobium* requirement of the non-legume *Parasponia* in relation to the cross-inoculation group concept of legumes. *New Phytologist* 86: 17–26.

Vincent, J.M. 1970. A manual for the practical study of root nodule bacteria. *International Biological Programme Handbook 15*. Oxford: Blackwell Scientific. 164 p.

Virginia, R.A. 1986. Soil development under legume tree canopies. *Forest Ecology Management* 16: 69–79.

Vogal, C.S., and J.O. Dawson. 1985. Effect of juglone on growth in vitro of *Frankia* isolates and nodulation of *Alnus glutinosa* in soil. *Plant and Soil* 87: 79–89.

Weaver, R.W., and L.R. Frederick. 1974. Effect of inoculum rate on competitive nodulation of *Glycine max* L. Merrill. II. Field Studies. *Agronomy Journal* 66: 233–236.

Wheeler, C.T., O.T. Helgersen, D.A. Perry, and J.C. Gordon. 1987. Nitrogen fixation and biomass accumulation in plant communities dominated by *Cytisus scoparius* L. in Oregon and Scotland. *Journal of Applied Ecology* 24: 231–237.

Wheeler, C.T., J.E. Hooker, A. Crowe, and A.M.M. Berrie. 1986. The improvement and utilization in forestry of nitrogen fixation by actinorhizal plants with special reference to *Alnus* in Scotland. *Plant and Soil* 90: 393–406.

Wingfield, E.B. 1968. *Mycotrophy in loblolly pine. I. The role of* Pisolithus tinctorius *and* Rhizoctonia solani *in survival of seedlings. II. Mycorrhiza formation after fungicide treatment.* Ph D thesis, Virginia Polytechnic Institute and State University, Blacksburg, Virginia.

CHAPTER 10

Allelopathy

Stephen B. Horsley

Interference occurs when one plant species fails to germinate, grows more slowly, shows symptoms of damage, or does not survive in the presence of another plant species. Interference can result from competition, allelopathy, or other indirect influences. *Competition* is the phenomenon by which one plant removes a limited resource (such as light, water, or nutrients) from the environment, thereby reducing the survival or growth of a neighboring plant. *Allelopathy* is the phenomenon in which a plant or microorganism releases a natural product into the environment that subsequently reduces or enhances the survival or growth of neighboring plants. Interference also occurs indirectly, for example, when one plant provides shelter for herbivores that selectively browse or trample another plant.

Agroforestry intentionally combines woody perennials with agricultural crops or pasture plants in a variety of spatial or temporal arrangements, thus the choice of species combinations may dramatically influence the productivity and ultimate success of some agroforestry systems. The challenge in plant interference work is identifying which of these various factors causes the associated plant response. Allelochemicals originating in foliage leachings, root products, or mulches of crops or woody plants may result in reduced productivity or death of companion plants. For example, black walnut (*Juglans nigra*) is widely known for its allelopathic interference with field crops and conifers (Rietveld 1982). Contact with walnut foliage leachings or roots results in the death of many species. Plant residue mulches commonly used in agroforestry systems to protect soil from erosion, conserve moisture, and supply nutrients such as nitrogen may be the source of allelochemicals that interfere with crop productivity.

The concept of allelopathy is at least 2,000 years old, though the term was not coined until 1937 (Willis 1985). During the past 25 years, there has been a significant effort to understand the role of allelopathy in ecological processes. This chapter discusses methods for evaluating whether allelopathy is an important component of interference in agroforestry systems. Grodzinsky (1965) and Rice (1984) have published comprehensive monographs on allelopathy, and since 1983, there have been six major symposia or monographs on this topic (Chou and Waller 1983, Putnam and Tang 1986, Thompson 1985, Waller 1987). The proceedings of the North American Symposium on Allelopathy was published in 1983 as a special issue of the *Journal of Chemical Ecology* (Vol. 9, No. 8) and *Plant and Soil* published a special allelopathy edition (Vol. 98, No. 3) in 1987.

SEPARATING ALLELOPATHY FROM COMPETITION AND OTHER INFLUENCES

Demonstrating allelopathy requires evidence that interference occurs and that the interference is not due to competition or other indirect influences. The complexity of interference phenomena and the potentially large number of interacting factors makes separating allelopathy from competition and other influences especially difficult. Thus an unequivocal demonstration has rarely been achieved in practice.

Besides releasing phytotoxic natural products, one plant can influence another in other ways: reducing light intensity or changing its quality; taking up limited water; changing humidity; absorbing limited nutrients; changing the ratio of soil NH_4/NO_3 (Horsley 1988); changing the soil level through accumulation of organic matter; changing soil reaction; favoring or reducing pathogenic activity; or sheltering or excluding predators that selectively browse or trample vegetation or that defecate or urinate in the area (Harper 1977). Further, none of these factors is independent. For example, a shortage of water will reduce the production and transport of nitrate nitrogen. Thus, the study of interference phenomena is difficult, and to demonstrate a cause and effect relationship requires the study not only of allelopathy, but also of competition and other influences.

Despite its complexity, research on interference can be approached experimentally. In 1890, German physician Robert Koch faced difficulties in proving that microorganisms were responsible for causing disease. In formulating his now famous postulates, Koch described the symptoms of disease in a particular plant. Then, in turn, the potentially causal organism was isolated from the diseased plant, grown in culture and characterized, inoculated into a healthy plant to produce the original disease symptoms, and finally re-isolated from the diseased plant and shown to have the same identifying characteristics found in the first culture. These same procedures can be used to evaluate interference.

Client Problem

Because agroforestry is a relatively new field, little work has been conducted on species compatibility (Wood 1988). Some species currently used in agroforestry systems reportedly have allelopathic properties (Watanabe et al. 1988). For example, allelopathic interference from *Eucalyptus* foliage leachings and volatiles (del Moral and Muller 1970) and plant residues (Suresh and Rai 1987) has been described. Also, residue mulches of *Leucaena leucocephala* reportedly have allelopathic properties.

General Methods

Koch's postulates can be adapted to evaluate interference phenomena (Harper 1977, Fuerst and Putnam 1983). The first step is to demonstrate interference between plants and quantify it to the extent possible. The circumstances in which interference occurs should be described, including the stage of plant development affected (seed germination, growth on cotyledonary reserves, growth after the cotyledonary reserve phase) and the symptomology. Symptoms of interference should be as specific as possible because they will be used later to eliminate alternative causes of interference or to bioassay allelochemicals. Symptoms such as foliage coloration, tissue nutrient content, loss of geotropic behavior, change in water potential, wilting, change in the CO_2 exchange rate, and epinasty are useful descriptors. Gross measures of growth that are influenced by many factors, such as a reduction in seed germination, plant height, dry weight accumulation, or survival, often are not specific enough to be used in separating components of interference or to act as discriminators in bioassays of fractions of natural products.

Techniques such as replacement series experiments offer an opportunity to quantify interference and symptomology (Dekker et al. 1983, Radosevich 1987). In experiments of this type, two plants are grown together, either at a fixed plant density with varying proportions of each species (100:0, 75:25, 50:50, 25:75, 0:100), or with varying total plant density and species proportion (double replacement). The effects of density and proportion of each species on total yield or other plant attributes can be measured and quantified. Given interference and symptomology, additional experiments can be conducted that concentrate on reducing the number of alternative causes of the symptoms (Horsley 1977a, 1986, 1987).

Proving that competition is operative requires measuring the supply of each important resource and demonstrating increased utilization of the resource by the presumed aggressor species or decreased utilization by the presumed susceptible species. Demonstrating that symptoms of interference are caused by a change in supply of the resource within the range of change caused by the aggressor, but without presence of the aggressor, is an important element in determining competition. Lack of symptom development with a change in a resource is evidence that the resource is not a key factor controlling interference. By contrast, symptom development suggests involvement of the resource in interference but does not rule out interaction with other resources or allelopathy. For example, the addition of organic compounds to the soil may result in microbial immobilization of nitrogen.

Experiments that restore the level of the presumed limiting resource, for example, through fertilization, irrigation, or supplemental lighting, in the presence of the presumed aggressor are useful in evaluating competition. Also valuable are manipulative experiments that restrict access of the aggressor to one resource but not to others. Horsley (1987, 1988) used fiberglass rod and string barriers to restrain the foliage of aggressor hayscented fern (*Dennstaedtia punctilobula*) from susceptible black cherry (*Prunus serotina*) seedlings without interfering with the use of below-ground resources by the fern. Harper (1977) and Fuerst and Putnam (1983) have suggested other useful methods for studying competition.

Five steps are required to demonstrate that allelopathy is the cause of interference:

1) One or more phytotoxic natural products must be released from the aggressor plant.
2) Phytotoxic natural products must be identified.

3) The presumed allelochemicals must be transported and accumulated in the environment in sufficient quantities to account for the documented interference.

4) Allelochemicals must be taken up by the susceptible plant.

5) Allelochemicals must reproduce the symptoms (physiological action) of interference in the susceptible plant.

While satisfying these conditions may seem a simple matter, there are enormous methodological problems, and few reports to date are free of deficiencies. The remainder of this chapter discusses some specific agroforestry practices and critically reviews methods used to study such problems.

ALLELOPATHY AND INTERCROPPING

Allelopathic interference can result from natural products in intercrop foliage leachings, root products, and volatiles. There are four ways in which these chemicals are released into the environment:

leaching
root exudation
volatilization
microbial decomposition of plant remains

Each mode of allelochemical release is affected by a number of factors that may change the quantity or quality of the natural products obtained.

Client Problem

When woody plants are grown with either crop or pasture plants, there is opportunity for allelopathic interaction between species. Rain water passing through the foliage of a woody intercrop may leach allelopathic natural products from the foliage and carry them to the crop plants below. Allelopathic natural products exuded from the roots of the crop, pasture, or woody plants may interfere with the companion species. Finally, volatile natural products released from plant tissue may interfere with the growth of nearby plants.

Hypothesis

Allelopathic natural products in foliage leachings, root washings, or plant volatiles are a key factor regulating productivity of intercrops containing woody plants and crop or pasture plants.

This generic hypothesis actually combines three hypotheses addressing the first three of the four ways allelochemicals are released into the environment (given above). These are discussed and reviewed below; the fourth, microbial decomposition of plant remains, is discussed later in this chapter.

Leaching

Leaching is the removal of substances from plants by aqueous solutions such as rain, dew, mist, and fog. Radioisotope labeling of plant tissue before leaching has shown that large quantities of both inorganic elements and many classes of organic natural products are leached from plant tissue (Tukey 1970). Both the quantity and quality of leachable natural products differ greatly with species, physiological age of tissue, stage of plant development, plant health, lighting, temperature, nutritional conditions, and the intensity and volume of the leaching solution (Tukey 1970). Vigorous, young, waxy leaves subjected to high-intensity precipitation events of short duration occurring at low temperatures tend to produce the smallest quantity of leachates. Old, easily wet, senescent or dead leaves, and long, low-intensity precipitation events (drizzle, dew, mist, fog) occurring at high temperatures tend to produce the largest quantity of leachates.

Methods and Critical Review

In initial studies to determine the presence of allelopathic activity, the protocol for leachate preparation should be as similar as possible to that prevailing naturally during symptom development.

In studies of allelopathy, extracts or leachates of varying amounts of live or dead (fresh or dried) tissue usually have been made by grinding, soaking, or misting whole or chopped tissue with different amounts of water or organic solvents for several hours to several days. Often the amounts of tissue or the leaching conditions that occur in the natural environment have not been considered when protocols for leachate preparation have been developed. The leachate-preparation protocol used can have a dramatic effect on the quantity and quality of natural products obtained. Obviously, larger amounts of foliage produce larger quantities of natural products. The quantity of leaching water affects solute concentration for a given drop size and leaching duration. Conditions such as chopping or grinding of tissue, lengthy soaking, or unrefrigerated storage of leachates (leading to microbial contamination) affect tissue enzyme activity, which may qualitatively change some of the natural products in the leachate. Organic solvents extract qualitatively different natural products than water, and hot or boiling solvents extract natural products of different quantity and quality than solvents at low or room temperature.

Horsley (1986, 1987) investigated the possibility that foliage leachates of hayscented fern interfered with the growth of black cherry seedlings. He determined the number of fern fronds within a 15 cm radius of natural black cherry seedlings, since biomass of the ferns changed during the growing season. Zero, one-half, one, and two times this number of fronds were used in a concentration series to prepare leachates each week. The amount of leaching water (distilled water adjusted to the pH of local rainwater) and frequency of leachate application were determined from 25-year records of average rainfall for June, July, and August near the study site. Leaching water was applied as a mist from an atomizer or as a 16-hour soaking solution at 4°C. Stored leachates were frozen for up to four days at –20°C. Thus the leachate-preparation protocol attempted to simulate natural conditions and to minimize changes in natural products in the leachates. Del Moral and Muller (1969) simulated natural conditions in fog-drip preparation from *Eucalyptus*, and Ramos and colleagues (1983) used realistic techniques for collecting natural leachates of coffee trees.

Root Exudation

Root exudation is the release of substances into the surrounding medium by healthy, intact plant roots. A variety of natural products has been found in plant root exudates, though in comparison with leaves, the amounts of organic materials are much smaller (Rovira 1969). Many factors can affect the quantity and quality of natural products obtained. These include plant species, plant age or stage of development, temperature, lighting and nutritional conditions, soil microorganisms, root supporting medium, soil moisture, and root damage. For example, Tubbs (1973) found that an inhibitor from sugar maple roots was detectible in his bioassay only when maple root growth was at a maximum. And Boulter and colleagues (1966) found that greater amounts of amino acids were exuded into sand by pea roots than into solution culture. Similarly, exudation in soils can be expected to vary with soil physical and chemical properties. Root exudation usually is increased greatly by wilting conditions and root damage (Clayton and Lamberton 1964). Soil microorganisms modify root-cell permeability and root metabolism, and rhizosphere organisms may absorb or excrete qualitatively different natural products than plant roots.

Methods and Critical Review

Studies of allelopathic activity of aggressor plant roots have not distinguished among natural products originating from root exudates, dead root tissue, and microbial rhizosphere products. Root washings have been collected with a "stairstep apparatus," or soil has been extracted directly. In the stairstep technique, a four-step staircase is constructed (Bell and Koeppe 1972, Lovett and Jokinen 1984). Pots of the presumed aggressor plant are placed on steps one and three, with pots of the presumed susceptible plant on steps two and four. Sand or other artificial materials are the usual potting media. Nutrient solutions are repetitively circulated sequentially from pot one through four so that the effluent from pot one waters the plant in pot two, and so on. Soil rarely is used as a potting medium because its physical or chemical properties may limit percolation of the solution or may fix or liberate elements such as phosphorous, aluminum, or manganese that are sensitive to soil reaction (Horsley 1977a). The control in stairstep experiments is a staircase with susceptible plants on all four steps. Allelopathy is indicated if susceptible plants on steps two and four of the series containing aggressor plants grow less than those on steps two and four of the control series. By contrast, if susceptible plants on step four of the series containing the aggressor grow less than the plants on step two of the same series with respect to the controls on steps four and two, competition is indicated. While the stairstep technique is useful, it is difficult to ensure that composition of the nutrient solution does not change during its passage down the staircase. Such a change could result in an erroneous conclusion (Harper 1977, Horsley 1977a). Also, the use of semisterile, artificial media, without native soil microorganisms, organic matter, or fine material, provides no information on whether any natural product in the root washings would be effective in the soil.

Tang and Young (1982) used Amberlite XAD-4 hydrophobic styrene-divinyl benzene copolymer to trap hydrophobic natural products from intact root systems of plants. Inorganic nutrients and hydrophilic organic molecules such as sugars and amino acids pass through XAD-4. Trapped natural products are readily extracted with methanol or acetone. In initial studies to determine the presence of phytotoxic root washings, presumed aggressor plants can be grown in containers and the pot effluent collected in bulk. Hydrophobic natural products, but not inorganic nutrients, are removed from half of the effluent with XAD-4. The remaining half of the effluent contains inorganic nutrients and hydrophobic natural products. Susceptible plants grown in soil can be treated with these two watering solutions. Growth differences and symptom development are attributable to the presence of hydrophobic natural products.

Coll and colleagues (1982) described a trapping system similar to the XAD-4 technique for use in aqueous systems. In their system, C18 silica gel was used as the adsorbent. Tension lysimeters offer the opportunity to collect natural soil solutions, but they have not been used in studies of allelopathy.

To extract natural products from soil in which aggressor plants have grown is difficult because soils have a great capacity for adsorbing organic substances on both organic and inorganic surfaces. Water, organic solvents, neutral EDTA, sodium pyrophosphate, calcium oxide, and aqueous or alcoholic sodium hydroxide have been used to extract soil allelochemicals. Some of these extractants are not suitable for removing allelochemicals from soil because they alter the structure of natural products. Kaminsky (1980) showed that sodium hydroxide caused qualitative changes in the natural products extracted and recommended against its use in studies of allelopathy. Moreover, sodium hydroxide solubilizes organic material not available to plants (Dalton et al. 1983). The organic chemistry of natural products prohibits use of harsh methods—strong acids or bases or high temperatures—when isolating unknown chemicals.

Volatilization

Volatilization is the release of natural products into the atmosphere. A variety of plants either secrete or excrete metabolic products into special structures such as tricomes and glands, into intercellular spaces and canals, or onto leaf surfaces. In hot, dry weather, natural products with high vapor pressure are released into the atmosphere where they may be absorbed directly by plants or adsorbed onto soil surfaces.

Methods and Critical Review

Much of the work leading to the recognition of plant volatiles as allelopathic agents has been done by Cornelius Muller and associates in semiarid regions of southern California with plants in the genera *Artemisia*, *Eucalyptus*, and *Salvia* (Muller 1969). In his studies, seeds of susceptible plants were germinated in closed containers. Whole foliage or foliage extract was placed inside the container without physical contact with the seeds. Control containers contained no foliage or extracts. Differences in seed germination were attributed to the presence of volatile natural products. Most of the natural products identified in Muller's work using this technique were terpenoids, primarily monoterpenes and sesquiterpenes. More recently, Bradow and Connick (1988a, 1988b) have used similar procedures to evaluate methyl ketones and alcohols from *Amaranthus* residues. Volatile natural products often can be quantified directly from air samples by gas chromatography. A technique for trapping volatile insect pheromones using an ethyl vinyl benzene-divinyl benzene copolymer (Porapak-Q gas chromatography column packing) has been developed (Cross et al. 1976). This may be useful in studies of allelopathy.

ALLELOPATHY AND MULCHING

Allelopathic interference can result from natural products released from mulches of plant residue.

Client Problem

To improve nitrogen nutrition of crop plants, plant-residue mulches, particularly of nitrogen-fixing species, are commonly used in agroforestry systems. These plant residues may in fact result in allelopathic interference and decreased crop production.

Hypothesis

Productivity of crop plants is regulated by allelopathic natural products in plant-residue mulches or their transformation products.

Background

Mulching and conservation tillage, which leave plant residues on the soil surface or incorporate them into the soil, result in the liberation of large quantities of water-soluble and partially water-soluble natural products as the residues decompose. When plant tissues age and die, cell membrane integrity is lost. Substances such as enzymes that normally are compartmented in living tissue are released into the immediate surroundings and react with other natural products, resulting in qualitative changes in some of these products. For example, in *Prunus* species, when tissues senesce or are disrupted, enzymes with B-glucosidase activity react with the cyanogenic glycosides, prunasin and amygdalin, producing mandelonitrile and sugar. Subsequent reactions of mandelonitrile yield hydrogen cyanide, benzaldehyde, and benzoic acid. Thus natural products not found in living tissue may be created by enzymatic activity during tissue senescence.

Precipitation leaches natural products from surface residues over a period of days or weeks (Guenzi et al. 1967). Turner and Rice (1975) found that ferulic acid was released from hackberry (*Celtus laevigata*) leaves for nearly one year on the forest floor. Once natural products enter the soil in leachates or as incorporated plant residues, additional qualitative changes occur as a result of physiochemical action of the soil and the activities of resident soil microorganisms. Hydrolysis of glycosides, producing free sugar and aglycone, occurs readily in acid soils and as a result of microbial activity. Further, microbial use of natural products as carbon sources results in the transformation of one compound to another along a degradative route (Dalton et al. 1983). For example, the phenolic acid ferulic acid is transformed by bacteria and fungi to vanillic acid, then to protocatechuic acid. The aromatic structure subsequently disappears with the formation of aliphatic organic acids. The speed and completeness of these transformations depend on microbial species and factors that influence their activity, including soil properties such as texture, moisture, oxygen, organic matter, metal and nitrogen content (C/N ratio), soil reaction, species resistance to microbial attack, and concentration in the soil (Dalton et al. 1983, Haider and Martin 1975, Martin and Haider 1979).

In addition to degradation, soil microorganisms synthesize new natural products from the carbon sources released during residue decomposition, for example, the antibiotic patulin that is produced by *Penicillium urticae* following amendment of the soil with plant residues (Norstadt and McCalla 1968). Thus soil toxicity may develop at several times during residue decomposition, initially when natural products are leached from surface residues and later during microbial transformation and synthesis (Dalton et al. 1983, Kimber 1973, Martin et al. 1972).

Methods and Critical Review

Bioassays are the principal method for testing whether allelochemical activity is present. Several kinds of bioassays have been used to evaluate allelochemical activity resulting from the decomposition of plant residues:

1) Field soils or residues are collected at several times after natural or intentional residue deposition or residue incorporation (Barnes and Putnam 1983, Cheng 1983, Chou and Kuo 1986, Chou et al. 1987, Cochran et al. 1977). Then bioassays are conducted by germinating test seeds or growing test plants in the soil, or by mulching field plots or greenhouse pots containing test seeds or plants with residues.

2) Leachates of residues are prepared and used as a moisture source for seed germination, plant growth, or microbial growth (Kuo et al. 1983, Liang et al. 1983).

3) Soil extracts are prepared and used as a moisture source for seed germination, plant growth, or microbial growth (Chou 1983, Guenzi and McCalla 1966a, 1966b).

4) Volatiles from residues are bioassayed in closed systems (Bradow and Connick 1988a, 1988b).

Initial experiments to determine the involvement of allelochemicals arising from residue decomposition should concentrate on simulating field conditions as closely as possible. For example, the same quantity, quality, and age of residue documented during symptom description should be used; soil moisture and aeration conditions also should be similar.

It is clear from the previous discussion that soil properties and soil microbial populations play an important role in determining the array of natural products present in the soil. Experiments that use artificial media lacking active microbial populations may give results of little value in determining the cause of inhibition in field situations. Further, appropriate controls are essential because treated plants are compared with the control to determine whether interference has occurred. Introducing organic carbon sources into the soil increases the use of nitrogen by microbes (immobilization) and may result in a deficient nitrogen supply for higher plants. Because plant residues differ considerably in decomposibility, using inert organic material in controls that differs significantly in decomposibility and microbial nitrogen use from residue treatments is inappropriate and may lead to erroneous conclusions (Martin et al. 1972). Horticultural peat commonly has been used as a control in studies of allelopathy. Martin and colleagues (1972) showed that this material is extremely resistant to microbial attack compared with a residue such as sawdust. Similarly, comparisons of residue treatments with controls lacking added organic material can cause bias in interpreting results. Organic materials that appear useful for controls in studies of allelopathy include glucose, chromatography cellulose, and excelsior (Barnes and Putnam 1983, Rose et al. 1983).

Bioassays

Bioassays are useful in demonstrating the involvement of allelochemical activity in aggressor plant leachings, root washings, volatiles, or decomposition products. The initial definition of interference is in terms of symptoms of plant injury, and these symptoms must be preserved in all phases of the work. Inappropriate bioassay methods often have been used to suggest the involvement of allelopathy in field interference phenomena. Many investigators have concluded that allelopathy was operative on the basis of laboratory studies of plant or soil extract toxicity, without field testing. (This identification of allelopathy often is called allelopathic potential).

Typically, extracts or leachates are bioassayed on seed germination or some aspect of seedling growth of non-native, standard species (such as lettuce) under semisterile (such as petri dish or sand culture) conditions without involvement of soil components or soil microorganisms. For example, Stowe (1979) used a variety of laboratory bioassays to evaluate allelopathy as an explanation for plant distribution in an old field. He found little correlation between allelopathic potential identified in laboratory bioassays and the actual field distribution of plants. Similar results were found by del Moral and Cates (1971) and Chou (1983). Studies of allelopathic potential seldom have relevance in understanding interference phenomena in natural systems and should be avoided.

Identification of allelochemicals

Isolating and identifying allelochemical natural products from plant tissue and soil is a critical barrier to proving that allelopathy is operating. A wide array of natural products has been identified in allelopathic phenomena. These include simple phenols, benzoic acids and aldehydes, acetophenones, cinnamic acids, cinnamic acid derivatives and coumarins, phenylpropenes, quinones, flavonoids, tannins, monoterpenes, sesquiterpenes and sesquiterpene lactones, cyanogenic compounds, acetylenes, aliphatic organic acids, alkaloids, amino acids, and simple lactones (Horsley 1977b, Putnam 1985, Rice 1984). The literature on allelopathy suggests that in temperate ecosystems, the phenolic benzoic and cinnamic acids and coumarins are the primary allelochemical agents, while in arid and semiarid environments, terpenoid compounds are most important.

Since 1960, there has been a revolution in the methodology used in natural products organic chemistry. Mass spectroscopy and ^{1}H-nuclear magnetic resonance (NMR) became available about 1960; ^{13}C-nuclear magnetic resonance and single crystal X-ray crystallography came into common use about 1970. High-performance liquid chromatography became increasingly available during the 1980s, and the array of analytical techniques for identifying allelochemicals is expanding annually. With these instrumental methods, both the structure and the stereochemistry of many natural products has been discovered.

Research on allelopathy has had its greatest increase in history during this same period, yet few studies of allelopathy have included the use of these techniques. It has been common when an inhibitory plant or soil extract was found to look only for phenolic inhibitors, and because many phenolic acids are water soluble and almost universally distributed in higher plants, they usually are found. By contrast, it has been uncommon to use up-to-date methods to search for and identify pure active compounds through repeated fractionation of an inhibitory extract, bioassaying each fraction. Work by Fischer and Quijano (1985) is an exception. Presumed allelochemicals infrequently have been isolated from soil and quantified. In some studies where this has been attempted, the harsh methods discussed earlier have been used. Again, phenolics often are the only compounds investigated. Unlike plant pathology and entomology, which interface with organic chemistry, few natural products organic chemists or biologists with natural products training have been involved with studies of allelopathy. Thus, the methodology for identifying allelochemicals should be critically evaluated before being accepted as authoritative. Moreover, biologists who lack training in natural products organic chemistry or who lack the instrumentation required to identify natural products should seek collaboration with a qualified scientist.

Once an inhibitory leachate or extract is obtained, a preliminary chromatographic separation can be made. Information about the "polarity" of the active allelochemical(s) can be obtained by extracting the leachate with a series of organic solvents ranging from nonpolar to polar, for example, n-hexane, methylene chloride, methanol, and water. Further separations will be dictated by the polarity of the active fraction(s). Hash (1975) and Shaw (1981) discussed strategies for purifying toxic natural products.

Bioassays are used at each step of the purification process to determine which fractions contain allelochemicals and to monitor their activity. While there is no standard bioassay for allelochemical purification, the bioassay procedure used to determine the presence of allelochemical activity in the field may no longer be suitable for allelochemical purification. For example, wilting of whole susceptible plants watered with aggressor plant foliage leachings requires relatively large quantities of leachate and would become cumbersome after several separations. New bioassays must be devised that preserve the original biological definition of inhibition (wilting) and that are sensitive, repeatable, quantitative, fast, and easily performed, and that use tissue of the susceptible species. Thus electrolyte leakage from leaf disks, cell cultures, or protoplasts of susceptible plant leaves may be a more useful bioassay at this stage of the work. Useful discussions on designing bioassays for purification of toxic natural products are included in Yoder (1981), Yopp (1985) and Leather and Einhellig (1985, 1988).

Fate of allelochemicals

Except for volatile inhibitors absorbed directly from the air, the soil mediates all known allelopathic responses. Potential allelochemicals must remain active in the soil to have an allelopathic effect. Studies of the decomposition of lignin, the formation and decomposition of humus, and the fate of pesticides (Guenzi 1974) have provided abundant information on the behavior of natural products of a variety of chemical classes in the soil. The biological activity, persistence, and movement of natural products in soil depends on their interaction with the soil adsorption complex and soil microbial populations. Both of these factors tend to reduce the quantity of natural products in the soil.

The chemical characteristics of the natural product, the nature of the adsorbent surfaces, the species and size of the microbial population, and the soil chemical environment are important factors that determine the fate of natural products in soil (Weed and Weber 1974). The adsorption complex includes both inorganic and organic surfaces. Adsorbed natural products may be biologically active or they may be rendered inactive, depending on the nature of the adsorbing surface, but adsorbed molecules usually are less available to soil microorganisms (Green 1974, Haider and Martin 1975). Soil clay minerals, iron and aluminum hydroxides, and organic material are important sites of natural product adsorption (Dalton et al. 1983, Green 1974, Weed and Weber 1974). Some natural products also may be irreversibly bound in soil humic substances (Wang et al. 1967a, 1967b).

Soil microorganisms have a tremendous capacity to reduce the quantity of natural products in the soil. Microbial metabolism of organic compounds may result in an increase or decrease in toxicity due to the release of organic carbon as carbon dioxide, fixation into microbial biomass, or transformation to other products. Thus, whether an allelopathic effect is observed depends on the relative rates of allelochemical addition and decomposition or fixation in the soil.

Despite the large amount of information available about the behavior of organic compounds in the soil, there are few studies in which the investigator has attempted to determine the fate of allelochemicals in the soil. Several investigators have noted that large quantities of presumed allelochemicals added to the soil were not recovered by leaching (Blum and Rice 1969, Wang et al. 1967b). Interestingly, several researchers who initially reported that phenolic compounds were operative allelochemicals later concluded that their concentration in the soil was below the threshold required for biological activity (Börner 1971, McCalla 1971). Einhellig (1987) suggested that the presence of environmental stresses (for example, deficiencies in water, light, or nutrients) or the presence of pesticide residues modifies the threshold for allelochemical activity. Investigation of the fate of ^{14}C-labeled allelochemicals in soil is an important area for future research. Haider and Martin and their associates have studied the fate of phenolic compounds reported to be allelochemicals in soil systems (Haider and Martin 1967, 1975, Haider et al. 1977, Martin and Haider 1971, 1976, 1979, Martin et al. 1967, 1972, 1974, 1978). Waller and associates (1987) investigated the natural products in soils beneath coffee trees and in soils under no-tillage versus conventional tillage agriculture.

FUTURE RESEARCH DIRECTIONS

Agroforestry in particular respresents an opportunity to obtain immediate benefits from fundamental research on plant-plant interference. Systematic evaluation of crop and woody plant combinations and mulches for interference should provide useful information in designing new agroforestry systems. This information could offer land managers the opportunity to manipulate interference phenomena to reduce costs through such measures as the control of weed species with allelopathic ground covers or mulches that may reduce the need for herbicides (Putnam et al. 1983). Moreover, determining whether interference is competitive, allelopathic, or the result of other influences could affect the land management strategy that is chosen. A knowledge of the mechanism of interference may result in a manager's choosing species that tolerate allelochemical residues or that have compatible light, water, or nutrient requirements. The important message for the land manager is that understanding the basic biology of interference processes will make it possible to maximize economic yield. The important message for the researcher is that the use of appropriate controls and bioassays and good organic chemistry techniques in establishing the involvement of allelopathic natural products is imperative.

Studies of interference are complex but approachable. Most of the deficiencies in work done to date are the result of approaching interference from the point of view of a single discipline, without adequate knowledge of other disciplines. For example, plant biologists usually are not trained in the methods of natural products organic chemistry, and natural products organic chemists are unaware of the opportunity to apply their discipline or are not attracted by the presumed lack of "interesting" chemistry. There is a need for well-documented case studies of interference that consider competition, allelopathy, and other influences. Such studies are best undertaken by teams of scientists who represent such disciplines as plant ecology, plant physiology, soil science, microbiology, plant pathology, and natural products organic chemistry.

For allelopathy to be operative, adequate amounts of allelochemicals must be available in the soil to have a physiological effect on the growth of another plant. Allelochemicals may have direct or indirect effects on susceptible higher plants. Direct effects occur when the susceptible plant absorbs an allelochemical that results in detrimental effects on its use of site or environmental resources in metabolic processes. Indirect effects result when allelochemical intereference with the metabolism of another plant, usually a microorganism, has a detrimental effect on another higher plant. For example, allelopathic interactions with nitrogen-fixing or nitrifying bacteria, mycorrhizal fungi, or pathogenic microorganisms may have a negative effect on a higher plant sharing the same space. Allelochemical modes of action have received relatively little study compared with attempts to determine the involvement of allelochemical activity in interference. In many cases, investigations of allelochemical modes of action have begun from a suggestion in the literature that a particular natural product or class of natural products is responsible for interference. Consequently, much of the information in the literature on allelopathy concerns the effects of phenolic compounds (Horsley 1977b, Rice 1984, Einhellig 1985, 1986). Future research on allelopathic interference must concentrate on determining the physiological modes of action in plants. Defining the symptoms of interference more closely should aid in achieving this objective.

LITERATURE CITED

Barnes, J. P., and A. R. Putnam. 1983. Rye residues contribute weed suppression in no-tillage cropping systems. *Journal of Chemical Ecology* 9: 1045–1057.

Bell, D. T., and D. E. Koeppe. 1972. Noncompetitive effects of giant foxtail on the growth of corn. *Agronomy Journal* 64: 321–325.

Blum, U., and E. L. Rice. 1969. Inhibition of symbiotic nitrogen-fixation by gallic and tannic acid and possible roles in old-field succession. *Bulletin of the Torrey Botanical Club* 96: 531–544.

Börner, H. 1971. German research on allelopathy. In: *Biochemical interactions among plants*. Washington, DC: US National Committee for International Biology Programs, National Academy of Science. pp 52–56.

Boulter, D. T., J. J. Jeremy, and M. Wilding. 1966. Amino acids liberated into the culture medium by pea seedling roots. *Plant and Soil* 24: 121–127.

Bradow, J. M., and W. J. Connick, Jr. 1988a. Volatile methyl ketone seed-germination inhibitors from *Amaranthus palmeri* S. Wats. residues. *Journal of Chemical Ecology* 14: 1617–1631.

Bradow, J. M., and W. J. Connick, Jr. 1988b. Seed germination inhibition by volatile alcohols and other compounds associated with *Amaranthus palmeri* residues. *Journal of Chemical Ecology* 14: 1633–1648.

Cheng, H. H. 1983. Allelochemicals in soils and their impact on soil nitrogen transformations. In: *Allelochemicals and pheromones* (C. H. Chou and G. R. Waller, eds.). Taipai, Taiwan: Institute of Botany, Academia Sinica. pp 209–216.

Chou, C. H. 1983. Allelopathy in agroecosystems in Taiwan. In: *Allelochemicals and pheromones* (C. H. Chou and G. R. Waller, eds.). Taipai, Taiwan: Institute of Botany, Academia Sinica. pp 27–64.

Chou, C. H., and G. R. Waller. 1983. *Allelochemicals and pheromones*. Taipai, Taiwan: Institute of Botany, Academia Sinica. 314 p.

Chou, C. H., and Y. L. Kuo. 1986. Allelopathic research of subtropical vegetation of Taiwan. III. Allelopathic exclusion of understory by *Leucaena leucocephala* (Lam.). *Journal of Chemical Ecology* 12: 1431–1448.

Chou, C. H., S. Y. Hwang, C. I. Peng, Y. C. Wang, F. H. Hsu, and N. J. Chung. 1987. The selective allelopathic interaction of a pasture-forest intercropping in Taiwan. *Plant and Soil* 98: 31–41.

Clayton, M. F., and J. A. Lamberton. 1964. A study of root exudates by the fog-box technique. *Australian Journal of Biological Science* 17: 855–866.

Cochran, V. L., L. F. Elliot, and R. I. Papendick. 1977. The production of phytotoxins from surface crop residues. *Soil Science Society of America Journal* 41: 903–908.

Coll, J. C., B. F. Bowden, and D. M. Tapiolas. 1982. *In situ* isolation of allelochemicals released from soft corals (Coelenterata: octocorallia): A totally submersible sampling apparatus. *Journal of Experimental Marine Biology and Ecology* 60: 293–299.

Cross, J. H., R. C. Byler, R. F. Cassidy, Jr., R. M. Silverstein, R. E. Greenblatt, W. E. Burkholder, A. R. Levinson, and H. Z. Levinson. 1976. Porapak-Q collection of pheromone components and isolation of (Z)- and (E)-14-methyl-8-hexadecenal, sex pheromone components from the females of four species of *Trogoderma* (Coleoptera: dermestidae). *Journal of Chemical Ecology* 2: 457–468.

Dalton, B. R., U. Blum, and S. B. Weed. 1983. Allelopathic substances in ecosystems:: effectiveness of sterile soil components in altering recovery of ferulic acid. *Journal of Chemical Ecology* 9: 1185–1201.

Dekker, J. H., W. F. Meggitt, and A. R. Putnam. 1983. Experimental methodologies to evaluate allelopathic plant interactions: the *Abutilon theophrasti-Glycine max* model. *Journal of Chemical Ecology* 9: 945–981.

del Moral, R., and R. G. Cates. 1971. Allelopathic potential of the dominent vegetation of western Washington. *Ecology* 52: 1030–1037.

del Moral, R., and C. H. Muller. 1969. Fog drip: a mechanism of toxin transport from *Eucalyptus globulus*. *Bulletin of the Torrey Botanical Club* 96: 467–475.

del Moral, R., and C. H. Muller. 1970. Allelopathic effects of *Eucalyptus camaldulensis*. *American Midland Naturalist* 83: 254–282.

Einhellig, F. A. 1985. Allelopathy—A natural protection, allelochemicals. In: *Handbook of natural pesticides: methods, Vol. 1, Theory, practice and detection* (N. B. Mandava, ed.). Boca Raton. Florida: CRC Press. pp 161–200.

Einhellig, F. A. 1986. Mechanisms and modes of action of allelochemicals. In: *The science of allelopathy* (A. R. Putnam and C. S. Tang, eds.). New York: John Wiley and Sons. pp 171–188.

Einhellig, F. A. 1987. Interactions among allelochemicals and other stress factors of the plant environment. In: *Allelochemicals: Role in agriculture and forestry* (G. R. Waller, ed.). American Chemical Society Symposium Series 330. Washington, DC: American Chemical Society. pp 343–357.

Fischer, N. K., and L. Quijano. 1985. Allelopathic agents from common weeds *Amaranthus palmeri, Ambrosia artemisiifolia,* and related weeds. In: *The chemistry of allelopathy biochemical interactions among plants* (A. C. Thompson, ed.), American Chemical Society Symposium Series 268. Washington, DC: American Chemical Society. pp 133–147.

Fuerst, E. P., and A. R. Putnam. 1983. Separating the competitive and allelopathic components of interference: theoretical principles. *Journal of Chemical Ecology* 9: 937–944.

Green, R. E. 1974. Pesticide-clay-water interactions. In: *Pesticides in soil and water* (W. D. Guenzi, ed.). Madison, Wisconsin: Soil Science Society of America. pp 3–37.

Grodzinzky, A. M. 1965. *Allelopathy in the life of plants*. Kiev, USSR: Namkova Dumka (Russian).

Guenzi, W. D. (ed.). 1974. *Pesticides in soil and water*. Madison, Wisconsin: Soil Science Society of America.

Guenzi, W. D., and T. M. McCalla. 1966a. Phenolic acids in oats, wheat, sorghum and corn residues and their phytotoxicity. *Agronomy Journal* 58: 303–304.

Guenzi, W. D., and T. M. McCalla. 1966b. Phytotoxic substances extracted from soil. *Soil Science Society of America Proceedings* 30: 214–216.

Guenzi, W. D., T. M. McCalla, and F. A. Norstadt. 1967. Presence and persistence of phytotoxic substances in wheat, oat, corn and sorghum residues. *Agronomy Journal* 59: 163–165.

Haider, K., and J. P. Martin. 1967. Synthesis and transformation of phenolic compounds by *Epicoccum nigrum* in relation to humic acid formation. *Soil Science Society of America Proceedings* 31: 766–772.

Haider, K., and J. P. Martin. 1975. Decomposition of specifically carbon-14 labeled benzoic and cinnamic acid derivatives in soil. *Soil Science Society of America Proceedings* 39: 657–662.

Haider, K., J. P. Martin, and E. Rietz. 1977. Decomposition in soil of ^{14}C-labeled coumaryl alcohols; free and linked into dehydropolymer and plant lignins and model humic acids. *Soil Science Society of America Journal* 41: 556–562.

Harper, J. L. 1977. Mechanisms of interaction between species. In: *Population biology of plants*. New York: Academic Press. pp 347–381.

Hash, J. H. (ed.) 1975. *Methods in enzymology, Vol. 43, Antibiotics*. New York: Academic Press.

Horsely, S. B. 1977a. Allelopathic inhibition of black cherry by fern, grass, goldenrod, and aster. *Canadian Journal of Forest Research* 7: 205–216.

Horsley, S. B. 1977b. Allelopathic interference among plants. II. Physiological modes of action. In: *Proceedings of the 4th North American forestry and biology workshop* (H. E. Wilcox and A. F. Hamer, eds.). State University of New York, College of Environmental Science and Forestry, Syracuse, New York. pp 93–136.

Horsley, S. B. 1986. Evaluation of hayscented fern interference with black cherry. *American Journal of Botany* 73: 668–669.

Horsley, S. B. 1987. Evaluating the mechanism of hayscented fern interferece. *Proceedings of the Northeastern Weed Science Society* 41: 58.

Horsley, S. B. 1988. Nitrogen species preference of *Prunus serotina* Ehrh. and *Betula alleghaniensis* Britt. seedlings. *American Journal of Botany* 75(6) part 2: 75.

Kaminsky, R. 1980. The determination and extraction of available soil organic compounds. *Soil Science* 130: 118–123.

Kimber, R.W.L. 1973. Phytotoxicity from plant residues. III. The relative effect of toxins and nitrogen immobilization on the germination and growth of wheat. *Plant and Soil* 38: 543–555.

Kuo, Y.L., C.H. Chou, and T. W. Hu. 1983. Allelopathic potential of *Leucaena leucocephala*. In: *Allelochemicals and pheromones* (C. H. Chou and G. R. Waller, eds.). Taipei, Taiwan: Institute of Botany, Academia Sinica. pp 107–119.

Leather, G. R., and F. A. Einhellig. 1985. Mechanisms of allelopathic action in bioassay. In: *The chemistry of allelopathy. Biochemical interactions among plants* (A.C. Thompson, ed.). American Chemical Society Symposium Series 268. Washington, DC: American Chemical Society. pp 197–205.

Leather, G.R., and F.A. Einhellig. 1988. Bioassay of naturally occurring allelochemicals for phytotoxicity. *Journal of Chemical Ecology* 14: 1821–1828.

Liang, J.C., S.S. Sheen, and C.H. Chou. 1983. Competitive allelopathic interaction among some subtropical pastures. In: *Allelochemicals and pheromones* (C. H. Chou and G. R. Waller eds.). Taipai, Taiwan: Institute of Botany, Academia Sinica. pp 121–133.

Lovett, J.V., and K. Jokinen. 1984. A modified stairstep apparatus for studies of allelopathy and other phytotoxic effects. *Journal of Agricultural Science in Finland* 56: 1–7.

Martin, J. P., and K. Haider. 1971. Microbial activity in relation to soil humus formation. *Soil Science* 111: 54–63.

Martin, J. P., and K. Haider. 1976. Decomposition of specifically Carbon-14-labeled ferulic acid: free and linked into model humic acid-type polymers. *Soil Science Society of Americal Journal* 40: 377–380.

Martin, J. P., and K. Haider. 1979. Effects of concentration on decompostition of some ^{14}C-labeled phenolic compounds, benzoic acid, glucose, cellulose, wheat straw and *Chlorella* protein in soil. *Soil Science Society of America Journal* 43: 917–920.

Martin, J.P., K. Haider, and C. Saiz-Jimenez. 1974. Sodium amalgam reductive degradation of fungal and model phenolic polymers, soil humic acids and simple phenolic compounds. *Soil Science Society of America Proceedings* 38: 760–765.

Martin, J.P., K. Haider, and D. Wolf. 1972. Synthesis of phenols and phenolic polymers by *Hendersonula toruloidea* in relation to humic acid formation. *Soil Science Society of America Proceedings* 36: 311–315.

Martin, J.P., A.A. Parsa, and K. Haider. 1978. Influence of intimate association with humic polymers on biodegradation of (^{14}C) labeled organic substrates in soil. *Soil Biology and Biochemistry* 10: 483–486.

Martin, J.P., S.J. Richards, and K. Haider. 1967. Properties and decomposition and binding action in soil of "humic acid" synthesized by *Epicoccum nigrum*. *Soil Science Society of America Proceedings* 31: 657–662.

McCalla, T.M. 1971. Studies on phytotoxic substances from soil microorganisms and crop residues at Lincoln, Nebraska. In: *Biochemical interactions among plants*. Washington, DC: US National Committee for International Biology Programs, National Academy of Science. pp 39–43.

Muller, C.H. 1969. Allelopathy as a factor in ecological process. *Vegetatio* 18: 348–357.

Norstadt, F.A., and T.M. McCalla. 1968. Microbially induced phytotoxicity in stubble-mulched soil. *Soil Science Society of America Proceedings* 32: 241–245.

Putnam, A.R. 1985. Weed allelopathy. In: *Weed physiology, Vol. 1, Reproduction and ecophysiology* (S. O. Duke, ed.). Boca Raton, Florida: CRC Press. pp 131–155.

Putnam, A.R., and C. S. Tang. 1986. *The science of allelopathy*. New York: John Wiley and Sons.

Putnam, A.R., J. DeFrank, and J.P. Barnes. 1983. Exploitation of allelopathy for weed control in annual and perennial cropping systems. *Journal of Chemical Ecology* 9: 1001–1010.

Radosevich, S.R. 1987. Methods to study interactions among crops and weeds. *Weed Technology* 1: 190–198.

Ramos, L., A.L. Anaya, and J. N. de Pascual. 1983. Evaluation of allelopathic potential of dominant herbaceous species in a coffee plantation. *Journal of Chemical Ecology* 9: 1079–1097.

Rice, E.L. 1984. *Allelopathy*. 2nd ed. New York: Academic Press.

Rietveld, W.J. 1982. The significance of allelopathy in black walnut cultural systems. In: *Black walnut for the future*. USDA, Forest Service General Technical Report NC-74. pp 73–86.

Rose, S.L., D.A. Perry, D. Pilz, and M.M. Schoeneberger. 1983. Allelopathic effects of litter on the growth and colonization of mycorrhizal fungi. *Journal of Chemical Ecology* 9: 1153–1162.

Rovira, A.D. 1969. Plant root exudates. *Botanical Reveiw* 35: 35–57.

Shaw, P.D. 1981. Production and isolation. In: *Toxins in plant disease* (R. D. Durbin, ed.). New York: Academic Press. pp 21–44.

Stowe, L.G. 1979. Allelopathy and its influence on the distribution of plants in an Illinois old-field. *Journal of Ecology* 67: 1065–1085.

Suresh, K.K., and R.S.V. Rai. 1987. Studies on the allelopathic effects of some agroforestry tree crops. *International Tree Crops Journal* 4: 109–115.

Tang, C.S., and C.C. Young. 1982. Collection and identification of allelopathic compounds from the undisturbed root system of Bigalta Limpograss (*Hemarthria altissima*). *Plant Physiology* 69: 155–160.

Thompson, A.C. 1985. *The chemistry of allelopathy biochemical interaction among plants*. American Chemical Society Symposium Series 268. Washington, DC: American Chemical Society.

Tubbs, C.H. 1973. Allelopathic relationship between yellow birch and sugar maple seedlings. *Forest Science* 19: 139–145.

Tukey, H. B., Jr. 1970. The leaching of substances from plants. *Annual Review of Plant Physiology* 21: 305–324.

Turner, J.A., and E.L. Rice. 1975. Microbial decomposition of ferulic acid in soil. *Journal of Chemical Ecology* 1: 41–58.

Waller, G.R. 1987. *Allelochemicals: role in agriculture and forestry*. Washington, DC: American Chemical Society.

Waller, G.R., E.G. Krenzer, Jr, J.K. McPherson, and S.R. McGown. 1987. Allelopathic influences on no tillage versus conventional tillage in wheat production. In: *Allelochemicals: role in agriculture and forestry* (G. R. Waller, ed.). American Chemical Society Symposium Series 330. Washington, DC: American Chemical Society. pp 371–383.

Wang, T.S.C., S.Y. Cheng, and H. Tung. 1967a. Extraction and analysis of soil organic acids. *Soil Science* 103: 360–366.

Wang, T.S.C., S.Y. Cheng, and H. Tung. 1967b. Dynamics of soil organic acids. *Soil Science* 104: 138–144.

Watanabe, H., P. Sahunalu and C. Khemnark. 1988. Combinations of trees and crops in the taungya method as applied in Thailand. *Agroforestry Systems* 6: 169–177.

Weed, S.B., and J.B. Weber. 1974. Pesticide-organic matter interactions. In: *Pesticides in soil and water* (W. D. Guenzi, ed.). Madison, Wisconsin: Soil Science Society of America. pp 39–66.

Willis, R. J. 1985. The historical basis of the concept of allelopathy. *Journal of Historical Biology* 18: 71–102.

Wood, P. J. 1988. Agroforestry and decision-making in rural development. *Forest Ecology Management* 24: 191–201.

Yoder, O.C. 1981. Assay. In: *Toxins in plant disease* (R.D. Durbin, ed.). New York: Academic Press. pp 45–78.

Yopp, J.H. 1985. Bioassays for plant hormones and other naturally occurring plant growth regulators. In: *Handbook of natural pesticides: methods, Vol. 1, Theory, practice and detection* (N. B. Mandava, ed.). Boca Raton, Florida: CRC Press. pp 329–477.

PART FOUR

Managing the Soil

CHAPTER 11

Soil Fertility

Anthony Young

Agroforestry depends on ecological and economic interactions between trees and other agricultural components. It has both productive and service functions —the leading products in most areas are some combination of fuelwood, fodder, and fruit; the most important service function is, without doubt, maintaining or improving soil fertility.

An agroforestry *system* is a specific local land-use system, characterized by a unique set of environmental plant management and social and economic factors. An agroforestry *practice* is a distinctive arrangement of plant components in space and time. There are thousands of agroforestry systems but only some 20 distinct practices. These may be classified as rotational, spatial mixed, or spatial zoned based on the components present and the type of association between the tree and nontree components. In *rotational* practices, ecological interactions between trees and crops (including effects on soil fertility) take place over time, whereas in spatial practices they occur in space. In *spatial mixed* practices, the trees are distributed over more or less the whole land-use system, while in *spatial zoned* practices, they are planted in distinct areas, such as rows or blocks. Spatial mixed practices are subdivided into open and dense according to the spacing of the trees.

The distinction between rotational, spatial mixed, and spatial zoned practices forms a rational basis for the planning of agroforestry research (Huxley, 1986a, 1986b). An indication of the probable effects of each practice on soil fertility, based on a previous review, is given in table 11.1.

Land productivity is the capacity of land to support the growth of plants useful to man, including crops, trees, and pastures. It is a property not of soil alone, but also of land, where *land* refers to all environmental factors that affect potential for use, including landforms, climate, hydrology, soils, vegetation, and fauna. *Soil fertility* is therefore the capacity of soil to support the growth of plants on a sustained basis, under given conditions of climate and other properties of land.

Two kinds of fertility problems are commonly encountered in tropical land-use systems: low soil fertility and declining fertility. *Low soil fertility* refers to soils that have inherent problems, such as low water-holding capacity, strong acidity, or low nutrient content caused by nutrient-poor parent materials. *Declining fertility* is soil degradation brought about by the action of man, most often by the prolonged failure to replace what is removed in harvest (commonly described

Table 11.1. Effects of agroforestry practices on soil fertility

Agroforestry systems	Effects on soil fertility
Mainly agrosilvicultural (trees with crops)	
Rotational	
Shifting cultivation	+ or –
Improved tree fallow	+
Taungya	+ or –
Spatial mixed	
Open—Trees on cropland	+
Dense—Plantation crop combinations	+
Multistory tree gardens	+
Spatial zoned	
Hedgerow intercropping (alley cropping, barrier hedges) (also agrosilvopastoral)	+
Boundary planting	+ or 0
Trees on erosion control structures	+
Windbreaks and shelterbelts (also silvopastoral)	+
Biomass transfer	+
Mainly or partly silvopastoral (trees with pastures and livestock)	
Spatial mixed	
Open—Trees on rangeland or pastures	+
Dense—Plantation crops with pastures	+ or 0
Spatial zoned	
Live fences	nr
Fodder banks	nr
Multipurpose forestry (cf. also *taungya*)	
Woodlots with multipurpose management	+
Reclamation forestry leading to multiple use	+
Other components present	
Entomoforestry (trees with insects)	nr
Aquaforestry (trees with fisheries)	nr

Effects on soil fertility: + = positive; 0 = neutral; – = negative; nr = not relevant.

Source: Young 1989a.

as "over-cultivation"). This distinction is important in attempts to treat fertility problems: tackling a decline in fertility depends on working with nature and assisting the reestablishment of natural processes, whereas increasing low fertility requires improving upon nature. Since agroforestry effects changes through a natural medium, the plant, there is intrinsically a greater opportunity to remedy problems of decline in fertility than of low natural fertility, which is more likely to call for such artificial treatments as fertilizers.

While evidence for the effects of agroforestry on soil fertility is both direct and indirect, direct evidence, in the form of trials of agroforestry systems, currently is sparse. Indirect evidence, in the form of reasoning derived from research in agriculture, forestry, and soil science, is therefore analyzed for applicability to agroforestry. This state of affairs will soon change, as results become available from the many hundreds of agroforestry trials recently started or currently planned. Clearly, it is desirable to replace inferential evidence with data from direct research on soil fertility under agroforestry.

The effects of agroforestry on soil moisture, which in many areas is as important to plant growth as soil fertility, are excluded from the present discussion; these aspects are considered in Reifsnyder and Darnhofer 1989.

This chapter is based on a review of the potential of agroforestry for soil conservation (Young 1989a). The original review contains 480 references; only key items are cited here. Previous reviews include Nair (1984) and Sanchez (1987).

RESEARCH QUESTIONS ON SOIL-AGROFORESTRY INTERACTIONS

The basic hypothesis about the effect of agroforestry on soil fertility is this:

> Appropriate agroforestry systems have the potential to control erosion, maintain soil organic matter and physical properties, augment nitrogen fixation, and promote efficient nutrient cycling.

Appropriate here means systems that are fitted to local environmental conditions, are well designed and properly managed, and are applicable within the constraints and meet the needs of farmers or other land users. A system may be technically favorable to fertility, but if it calls for inputs to which farmers do not have access, or fails to meet their needs (such as for food or cash), then it will not sustain soil fertility simply because it is not put into practice!

From this central hypothesis, 10 critical questions for agroforestry research on soil fertility are derived (Young 1989b). In each of the following questions, *can* means "to what degree, under which environmental conditions, and by what means, can," and *maintain* means "help to maintain or improve."

1. Can agroforestry systems control soil erosion?
2. Can agroforestry systems maintain soil organic matter?
3. Can agroforestry systems maintain soil physical properties?
4. Can agroforestry systems augment nitrogen fixation?
5. Can agroforestry systems augment soil nutrient inputs?
6. Can agroforestry systems promote efficient nutrient cycling?
7. Can agroforestry systems reduce soil toxicities?
8. Can agroforestry systems promote desirable soil faunal activity?
9. Can agroforestry systems augment soil water availability to crops?
10. What is the role of root systems in agroforestry?

There are substantial links between these questions, particularly between the effects of erosion on organic matter and nutrients, between effects of organic matter on physical properties, nutrients, and soil fauna, and between effects of roots on organic matter, physical properties, and nutrient cycling. The plant-soil processes relevant to each question, with the status of evidence for each, are summarized in table 11.2.

Table 11.2. Status of evidence for plant-soil processes involved in soil-agroforestry hypotheses

Hypothesis	Processes	Evidence	
		Direct	Indirect
1. Erosion control	Reduce losses: of organic matter nutrients	+	++
	of nutrients	+	++
2. Organic matter	Augment additions: through carbon fixation	++	++
	Reduce losses: by erosion	see 1	
	by slower decomposition	0	+
3. Soil physical properties	Effects of organic matter	see 2	
	Effects of roots	+	++
	Break up compact layers	0	+
	Reduce soil temperature extremes	0	+
4. Nitrogen fixation	Augment additions: through N-fixing trees	++	
5. Nutrient inputs	Augment additions: from the atmosphere	0	+
	from rock minerals	0	+
6. Nutrient cycling	Reduce losses: from erosion	see 1	
	by inter-root transfer	see 10	
	by retrieval	see 6A	
	by synchrony	see 6B	
6A. Nutrient retrieval	Reduce leaching losses	+	++
6B. Synchrony	Synchronize nutrient release with demand	+	+
7. Soil toxicities			
7A. Acidity	Reduce acidity or acidification	+	+
7B. Salinity	Assist reclamation of saline or alkaline soils	+	0
8. Soil fauna	Promote biomass of beneficial types	0	0
9. Soil water	Reduce losses: from runoff	+	0
	from evaporation	+	0
10. Roots	Role in C fixation, physical properties	see 2-6	
	N fixation, nutrient inputs, nutrient cycling		
	Exudation of growth-promoting substances	0	0
	Transfer of assimilate between root systems	0	0
10A. Root competition	Tree-crop competition for nutrients	0	+

++ = substantial evidence; + = some evidence; 0 = no evidence.

These questions are restated as testable hypotheses, each followed by an assessment of the research that has been done to test them and suggestions for future research.

Hypothesis 1 — Soil Erosion

Agroforestry systems have the potential to control erosion, thereby reducing losses of soil organic matter and nutrients.

Discussion

The potential of agroforestry systems to control soil erosion has been reviewed by Young (1989a) and Wiersum (1984 and this volume). The key conclusion is that there is considerable apparent potential to reduce water erosion to levels commonly regarded as acceptable, although many of the arguments for this are based on indirect evidence, and experimental data is at present scarce. There is a larger body of evidence for the potential of windbreaks to control wind erosion in semiarid areas.

A major reason for controlling erosion is to reduce losses of soil organic matter and nutrients. Eroded sediment normally contains higher levels of carbon and nutrients than the soil from which it is derived; the difference, the nutrient enrichment ratio, is commonly 2 or more, sometimes as high as 10. In addition, dissolved nutrients are removed in runoff water. An order-of-magnitude calculation shows that severe erosion, such as 50 t/ha/yr, can remove about 1 tonne of soil carbon, 100 kg of nitrogen, and corresponding amounts of other nutrients. Reducing erosion to 10 t/ha/yr will save 80% of these amounts, equivalent to the application of several bags of fertilizer! The financial saving, if applied to large areas, is considerable (Stocking 1986).

The erosion hypothesis poses a highly important question with respect to soil fertility. The reviews cited show it is very probably true, although a larger body of direct evidence is required for confirmation.

Hypothesis 2—Soil Organic Matter

Under agroforestry systems, soil organic carbon can be maintained at levels that are satisfactory for soil fertility due to the contribution of decomposed residues from the tree component. This contribution may come from above-ground litter and prunings, root residues, or indirectly as farmyard manure where prunings are fed to livestock.

Discussion

Soil organic matter has many roles in maintaining fertility. These include the beneficial effects on soil physical properties, including water-holding capacity; the slow release of nutrients, particularly significant in low-input farming systems; enhancement of the cation exchange capacity, significant where fertilizers are applied; and the provision of a favorable environment for soil faunal activity.

Much remains to be learned about the nature and behavior of soil organic matter, but current knowledge suggests the existence of several fractions, distinct or intergrading. The *plant litter fraction* consists of organic residues that are not yet humified; it can be subdivided into *structural* (lignin-rich) and *metabolic* (carbohydrate-rich) litter. The *soil faunal biomass* is small in amount but has a continuous turnover.

There is a large loss of biomass by bacterial oxidation, the *litter-to-humus conversion loss*, in the transformation of plant residues to humus carbon. This loss is of the order of 85% for above-ground litter and 67% for carbon, varying with climate, soil texture, type of plant residue, and manner of addition to the soil.

Humified soil carbon contains at least two fractions, labile and stable humus. *Labile humus* is that which decomposes at a moderate rate, losing carbon through bacterial oxidation according to the following relation:

$$Ct = Co\,(1 - K)^t \text{ or } Ct = Co.e^{-rt}$$

where

Ct = carbon at time t
Co = initial carbon
K = the humus decomposition constant
t = time in years
e = the exponential constant
r = a rate parameter

For slow rates, r is nearly equal to K. The decomposition constant, K, is of the order of 0.03 to 0.04 for labile humus, that is, labile humus is lost from soil at some 3% to 4% of its mass per year. It is this labile fraction that is largely responsible for nutrient release by mineralization and that is also susceptible to change through soil management.

Almost nothing is known about *stable humus*, other than that it forms part of the soil carbon shown in analysis and that it decomposes extremely slowly, probably with a half-life of more than 100 years. It has been suggested that woody, lignin-rich plant residues may contribute preferentially to its formation and that stable humus makes some distinctive contribution to soil properties, possibly to soil structure, but these suggestions are both unproven (Swift 1987).

It is reasonable to suppose that much topsoil organic matter consists of labile humus, since this is observed to decline under continuous arable cropping and increase under forest fallow. It is, moreover, the fraction that is responsible for release of most nutrients and that is susceptible to alteration through management.

The plant litter needed to replace soil losses can be compared with the potential of trees to produce such amounts of biomass in tables 11.3a and 11.3b.

Table 11.3. Soil requirements and agroforestry potential for maintenance of soil organic matter

(a) Indicative plant biomass requirements

Climatic zone	Topsoil carbon percent (%)	Initial topsoil carbon (kg C/ha)	Oxidation loss (kg C/ha/yr)	Erosion loss (kg C/ha/yr)	Required addition to soil humus (kg C/ha/yr)	Required plant residues added to soil (kg DM/ha/yr) above ground	Roots
Humid	2.0	30,000	1,200	400	1,600	8,400	5,800
Subhumid	1.0	15,000	600	200	800	4,200	2,900
Semiarid	0.5	7,000	300	100	400	2,100	1,400

(b) Biomass production of multipurpose trees

Climatic zone	Land use	No. of records	Above-ground biomass production kg DM/ha/yr (range)
Humid	Tree plantation	5	11,800 - 30,000
	Plantation crop combination	5	4,600 - 22,700
	Hedgerow intercropping	1	4,400
Subhumid	Tree plantation	3	10,000 - 38,200
	Hedgerow intercropping	6	2,400 - 7,400
Semiarid	Tree plantation	1	3,700

Typical values of topsoil carbon percentage are converted to mass of carbon in table 11.3a. The oxidation loss is calculated by assuming a decomposition constant of 0.04, adding a small erosion loss, and summing the amounts to give required additions to soil humus carbon. This is translated into requirements for plant biomass addition, using the conversion losses given above and assuming that roots amount to 40% of above-ground biomass. This gives a required annual addition of above-ground plant biomass, in round figures, of 8,000 kg, 4,000 kg, and 2,000 kg DM/ha/yr (DM = dry matter) for humid (forest), subhumid (savanna), and semiarid environments respectively.

The biomass production of natural vegetation is self-evidently able to meet such requirements: typical values for climatic zones in kg DM/ha/yr are 20,000 for humid, 10,000 for moist subhumid, 5,000 for dry subhumid, and 2,500 for semiarid. Table 11.3b gives data for multipurpose tree species used in agroforestry systems, for tree plantations (pure tree stands), and for two agroforestry practices. For the latter, the ranges span the maintenance requirements of table 11.3a. A feature of spatial zoned systems, such as hedgerow intercropping, is that the tree biomass production is more than that of pure tree stands by proportional area probably because incident light reaches the tree rows laterally as well as from above.

Besides contributing directly to humus, a tree cover, if unpruned, can also reduce humus losses, first, through control of erosion, and second, by reducing the rate of decomposition by reducing topsoil temperatures through canopy and litter shade.

A number of descriptive accounts suggest that soil organic matter is in a steady state under agroforestry systems; this is particularly evident with the large biomass production of home gardens (Soemwarto 1987), and probably with spatial dense plantation crop combinations (see below). However, the only cases to date in which soil properties have been monitored over a period of years appear to be at Ibadan, Nigeria, under hedgerow intercropping (figure 11.1). In the *Leucaena* trial (figure 11.1a), soil carbon was maintained over six years by adding tree prunings and crop residues; it declined on control plots where prunings were removed. Two other tree species (figure 11.1b) maintained soil carbon for two years, and *Cassia siamea*, with the highest biomass production, raised it. Computer modeling has produced promising results, showing soil organic matter equilibria at levels generally accepted as satisfactory, some 40% below equilibria under natural vegetation (Young 1990).

For research into this question, the basic need is to establish the plant-soil carbon cycle for agroforestry systems. As with many branches of agroforestry research, this calls for both trials of complete systems and research into individual elements of them. In this case, the need is to monitor soil carbon changes over time with additions of different types of plant litter—for example, tree litter (herbaceous and/or woody), tree roots, crop residues, crop roots, farmyard manure, and compost—and with different rates of addition. A "kill SOM" plot, in which loss of soil organic matter is monitored under tillage with no plant additions, is used for control. Plots of about 10 m square are sufficient. Apart from measurement of plant additions, the technique is one of repeated sampling and standardized analysis. Results from such microplots can be compared with those on complete agroforestry systems. The main problem is that organic carbon, in an apparently uniform soil, may have a coefficient of variation of 25% to 35%, necessitating 10 or more samples for each treatment to

A.

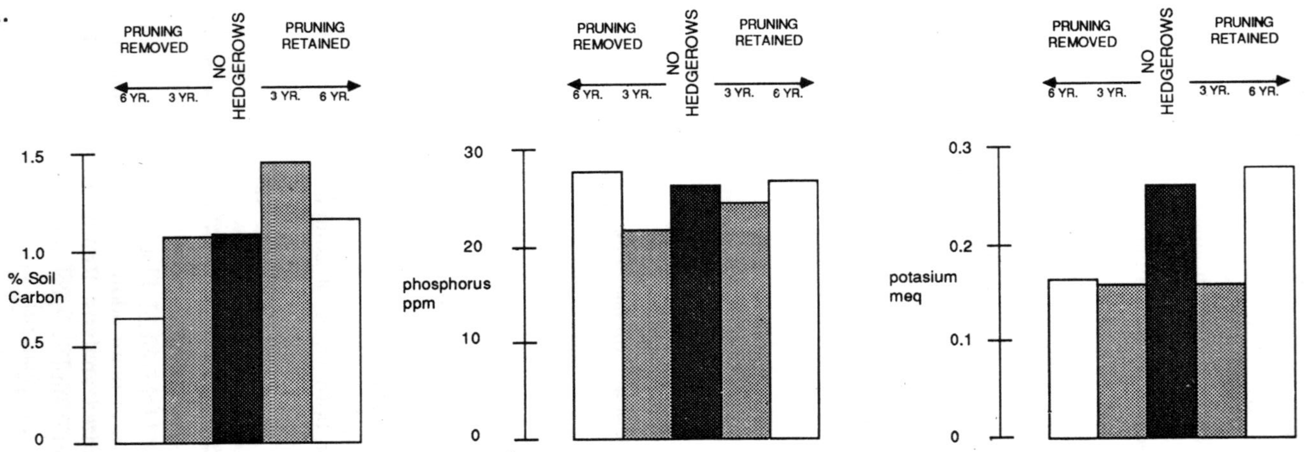

Figure 11.1. Soil changes under hedgerow intercropping at Ibadan, Nigeria. All values are for soil depth 0-15 cm.
A. Changes over 3 and 6 years, with hedgerow prunings removed and retained, *Leucaena*/maize (Kang et al. 1981, 1985).
B. Changes over 2 years, *Gliricidia sepium*, *Flemingia congesta* and *Cassia siamea* with maize (Yamoah et al. 1986c).
C. Bulk density after 2 years, treatments and source as for B.

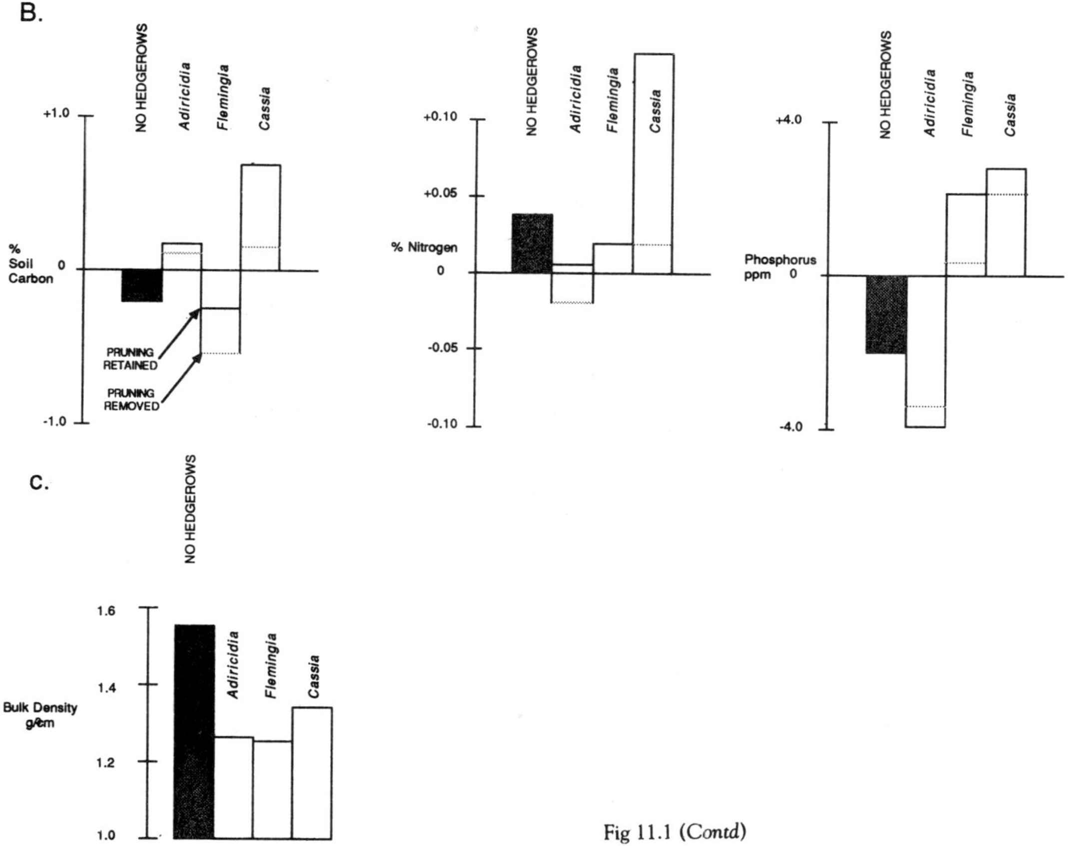

Fig 11.1 (Contd)

establish statistically valid trends. Owing to the time and cost of standard Walkely-Black carbon analysis, loss on ignition for nine hours at 375°C may be used as an alternative (Ball 1964).

Far more research under different climatic conditions and soil types and for different agroforestry systems is needed before the organic matter hypothesis can be validated or rejected. The tree biomass production of a system, and whether all or part of this reaches the soil, are clearly critical elements. The potential benefits from productive land-use systems that sustain organic matter, which is the most basic element of soil fertility, are so great that a substantial research effort directed at this question is a priority.

Hypothesis 3—Soil Physical Properties

Agroforestry systems maintain more favorable soil physical properties than agricultural systems (including soil structure, porosity, and water-holding capacity) through maintenance of organic matter and the effects of roots.

The physical properties of soil include not only organic matter, but also the effects of roots. It is established that physical conditions of soil, independent of nutrient content, can substantially affect fertility (Lal and Greenland 1979). The stability of soil aggregates when saturated after heavy rains is also important, so porosity plays a key role. A well-structured, porous soil combines high water-holding capacity with sufficient permeability to drain away excess water.

Evidence from soil under hedgerow intercropping at Ibadan, Nigeria, shows consistently more favorable physical propeties than soil under agriculture in control plots, even over as short a period as two years (figure 11.1c).

Agroforestry research should include measurements of soil physical properties as a matter of regular practice. Specific research can be incorporated in the microplots suggested above for organic matter.

It is known from forestry plantations that tree roots may penetrate and possibly break up compact soil layers such as stone lines and nodular laterite, and this process can improve both physical properties and nutrient intake from the B/C horizons. An additional physical process, resulting from canopy and litter shading, is the reduction of very high topsoil temperatures that can occur on bare soil. Evidence is lacking from agroforestry systems, but the effect of litter is known from studies of mulching and zero tillage.

Hypothesis 4—Nitrogen Fixation

Nitrogen-fixing trees can substantially supplement, and in some circumstances replace, the use of fertilizers in maintaining the nitrogen economy of agroforestry systems (see Avery, this volume).

Discussion

In tropical soils, the commonly limiting nutrient is nitrogen, to which a growth response is immediatey obtained on previously unfertilized soils. Where fertilizers are unavailable to farmers due to cost or other reasons, improving the nitrogen economy can make a considerable contribution to crop production (10 kg more maize for 1 kg N is a typical ratio).

Partly through the efforts of the Nitrogen Fixing Tree Association, this question has been comparatively well studied, and lists of nitrogen-fixing trees, with their broad climatic adaptations, are available (NIFTAL 1984, Brewbaker 1986, von Carlowitz 1986a, table 11.3). Large numbers of species in the family Leguminosae, particularly the subfamilies Mimosoideae and Papilionoideae, nodulate in association with nitrogen-fixing bacteria of the genera *Rhizobium* and *Bradyrhizobium*, while a small number of nonlegumes, notably *Alnus* and *Casuarina* spp., nodulate in association with *Frankia* bacteria. Many nitrogen-fixing legumes benefit from the improved nutrient uptake brought about by association with vesicular-arbuscular mycorrhizae. Large increases in fixation rates can sometimes follow bacteria and/or mycorrhizal soil inoculation (von Carlowitz 1986b, Dommergues 1987).

Recorded rates of fixation for trees commonly range on an annual basis from 20 kg to 200 kg/ha of N; the highest recorded is 500 kg/ha of N for *Leucaena*. *Sesbania* spp. and *Acacia mearnsil* also have high rates (Dommergues 1987, Young 1989a). Fixation rates are higher in soils with good organic matter and with phosphorus fertilization.

Nitrogen-fixing trees can be incorporated in all types of agroforestry practice: in rotational, as improved tree fallows; in spatial mixed dense, as in plantation crops combined with *Erythrina* or *Inga* spp. (Bornemisza 1982, Roskoski 1982, Roskoski and van Kessel 1985); in spatial mixed open, as in *Acacia albida* systems (Felker 1978); and in spatial zoned, as in hedgerow intercropping (Mulongoy 1986). In hedgerow intercropping, fixed nitrogen is transferred to intercrops (Sanginga et al. 1986), but the effectiveness for the soil nitrogen economy is obviously reduced if prunings are removed for fodder.

Earlier studies of nitrogen fixation rates were based on soil and plant balance estimates, and they contained many assumptions about nonmeasured inputs and outputs. Methods based on acetylene reduction and the use of ^{15}N isotope labeling are now available (Dommergues 1987).

Nitrogen fixation by the tree component represents a clear gain to the nutrient economy in agroforestry systems, with substantial economic value. Its effectiveness is proven, and research into improvement of rates of fixation, through species selection and inoculation, should be continued.

Hypothesis 5—Nutrient Inputs

The tree component in agroforestry systems can increase nutrient inputs by (1) providing favorable conditions for deposition of atmospheric dust and (2) improving nutrient uptake from deeper soil layers.

Discussion

Trees can augment nutrient inputs through more ways than nitrogen fixation. Trees are involved with inputs of all nutrients from the atmosphere and from weathering of primary rock as wet deposition (dissolved in rainfall) and dry deposition (in dust). The amounts are appreciable in relation to natural ecosystems but are small compared to the requirements of crops. Since trees reduce near-surface wind velocities, it is likely that they will augment dry deposition. This process is probably minor but would be straightforward to test.

Nutrients other than nitrogen enter ecosystems from the weathering of primary rock minerals, mainly in the lower B and B/C horizons of soils. It has been suggested that the deeper root systems of trees may be more effective than crops in taking up newly released nutrients. This process could be important, but it is unproven and appears difficult to test directly. Indirect evidence can be derived from nutrient cycling studies (for example, Alpízar et al. 1986, 1988).

Hypothesis 6—Nutrient Cycling

Agroforestry systems can lead to more efficient nutrient cycling, thereby slowing the rate of crop yield decline, or leading to a steady state in low-input systems, or making more effective use of fertilizers in high-input systems.

Under low-input agricultural systems without inorganic fertilizers, crop yields normally decline, leading either to abandonment of the land, as in the various forms of shifting cultivation, or to a condition of low-level equilibrium, with stable, but unsatisfactorily low, yields (for example, below 1 t/ha/yr for cereals). The benefits to low-input systems would be substantial if, under certain conditions, nutrient recycling were so efficient that harvest removal would be compensated by natural inputs. For sustained crop production at high levels, fertilizer inputs are essential, particularly to replace phosphorus. By reducing losses, the amount of recycling should be greater for agroforestry than agricultural systems, with consequent economic benefit through greater efficiency of fertilizer use. The two-pronged nature of these benefits, if proven, would rank among the top potential benefits from agroforestry.

Four specific processes contribute to this hypothesis (see table 11.2), of which two, the major reduction of nutrient losses through erosion control and the more speculative possibility of inter-root transfer, are derived from other hypotheses. The remaining processes, nutrient retrieval and synchrony, form two subhypotheses.

Subhypothesis 6A

Through the retrieval of nutrients in the soil solution by tree roots, agroforestry systems have more closed nutrient cycling than agricultural systems.

Discussion

This hypothesis states that the root systems of trees, including mycorrhizae, take up nutrients from the soil solution more efficiently than those of crops, thereby reducing leaching and increasing the ratio between internal recycling and external losses.

Indirect evidence comes from the low ratios of leaching losses to internal recycling under natural forest ecosystems (for example, Lelong et al. 1984, Moreau 1984, 1985). Even on sandy, permeable soils under high rainfall, natural forests achieve a high level of nutrient recycling (Golley et al. 1975, Sioli 1985). This contrast between forest and agricultural ecosystems is so great that if the tree component in agroforestry could achieve a partial degree of that effect, the reduction of nutrient losses would be substantial.

Direct evidence comes from studies of plantation crop combinations in Central and South America. Coffee and cacao are grown in spatial mixed dense combinations with (misnamed) "shade trees," particularly *Erythrina*, *Inga*, and *Cordia* spp. The leaf litter (from both crops and trees) is returned to the soil, and its nutrient content per hectare per year is of the order of 150 to 300 kg N, 10 to 20 kg P, 75 to 150 kg K, and 100 to 300 kg Ca. When these systems are fertilized, the nutrients recycled in litter can exceed the annual fertilizer input, an impressive tribute to the efficiency of retrieval (Aranguren et al. 1982, Glover and Beer 1986, Alpízar et al. 1986, 1988).

For the spatial zoned system of hedgerow intercropping, nutrients are returned to the soil if hedge prunings are added. The annual amounts per hectare are substantial, typically 100 to 200 kg N, 4 to 20 kg P, and 50 to 150 kg K (Kang and Duguma 1985, Yamoah et al. 1986b, Weerakoon and Gunasekera 1985). The nitrogen may be derived from fixation, but this does not apply to other nutrients or to the considerable amounts of nitrogen contained in the leaves of supposedly non-fixing trees such as *Cassia siamea*. At Maha Illuppallama, Sri Lanka, hedgerows of *Gliricidia* have raised maize yields considerably compared with monoculture control plots; it appears that the nitrogen economy is being sustained, but probable phosphorus deficiency has been encountered after several years (L. Weerakoon, personal communication).

Full nutrient-cycling studies involve a substantial research effort, but the techniques are known. It is straightforward to measure the biomass and, by tissue analysis, the nutrient content of tree and crop leaf, fruit, and wood. It should be noted, however, that deciduous plants translocate most nutrients out of their leaves prior to shedding, so the nutrient content differs between natural litter and prunings (Bernhard-Reversat 1987). It is known which parts of plants are removed in harvest and which are returned to the soil. More difficult, but equally important (see below), is estimating nutrients in root shedding. Nutrients in eroded soil and runoff water can be measured. The most difficult to measure experimentally is leaching, and many studies assign the unaccounted balance to this process. The necessary equipment, the lysimeter, is cumbersome and even more difficult to set up below trees than below crops, but attempts have been made by planting trees on established lysimeters (for example, at Dehra Dun, India).

There is strong indirect evidence for this subhypothesis and clear direct evidence, though from a limited range of agroforestry practices and site conditions. It is potentially a process of great importance, both economically and with respect to efficiency in the use of limited supplies of nutrients. Both basic and specialized research should be undertaken. Basic research, carried out in association with all trials of agroforestry systems, should include measurements of the biomass of all plant parts, both harvested and returned to the soil, including an estimate of root biomass, together with tissue analyses of samples. Specialized research is that in which 11 components of the nutrient cycle are measured, including erosion and leaching losses.

Subhypothesis 6B

Agroforestry systems present opportunities to synchronize the release of nutrients from decomposition of plant residues with requirements for uptake by plants through the choice of tree species, timing of pruning, and manner of incorporation of residues into the soil.

Discussion

One of the basic hypotheses of the Tropical Soil Biology and Fertility program is that synchrony can be achieved by management of land-use systems in general (Swift 1987). The essential facts are that plant residues form a nutrient reserve, protected from loss; once these are mineralized, the nutrients become available for plant uptake, but at the same time are highly subject to leaching. Crop nutrient requirements are strongly phased, and by synchronizing mineralization of residue nutrient with maximum demand, the ratio of uptake to leaching loss is likely to be higher. This contributes to the process of subhypothesis 6A.

The quality of plant residues is linked to this process. In simplified terms, high-quality residues are relatively high in nitrogen and low in lignin and polyphenols, low-quality residues are low in nitrogen and high in lignin and/or polyphenols. Woody residues and straw are of low quality; herbaceous matter is usually of high quality, although some trees, notably many *Eucalyptus* species, contain polyphenols that lower their litter quality. High-quality residues decay rapidly, giving short-term release of nutrients, while low-quality residues decay more slowly, extending the period over which nutrients are protected against leaching.

Manipulation of the timing of nutrient release can be attempted in all land-use systems, but agroforestry presents a particularly wide range of opportunities. The timing of nutrient release is influenced by a number of factors:

Choice of tree species (with high- or medium-quality leaf residues, or a mixture)
Types of residues added to the soil (herbaceous, woody, or both)
Timing of residue addition through pruning
Manner of addition to the soil (surface or buried)
Treatment of residue (raw, composted, or manure)

Such management options occur most clearly in spatial zoned systems where pruning and other management interventions are normal. They are available to a lesser degree in spatial mixed systems, although some of the tree components may be regularly pruned. In rotational agroforestry systems, much of the litter input necessarily comes at the end of the tree fallow, a feature which may account for their lesser efficiency, compared with spatial systems, in terms of land requirements. Synchronization is most important for annual intercrops, although it is also desirable for perennials during early growth and fruiting times.

For annual crops, the period of highest nutrient demand comes during the phase of early growth, and thus in seasonal climates is soon after the commencement of the rains. In hedgerow intercropping, there is the fortunate coincidence that pruning is necessary to reduce shading immediately prior to planting crops. Many species commonly used, such as *Leucaena*, *Gliricidia*, *Flemingia congesta*, and *Cassia siamea*, have a half-life for litter decay of less than 60 days. At Ibadan, Nigeria, crop nutrient uptake surpasses release from litter after 40 to 50 days (Yamoah et al. 1986a, Wilson et al. 1986). Release from roots lies beyond the direct control of management, although root dieback is likely to occur following pruning. Fine woody residue (twigs and branches) provide low-quality litter for more extended nutrient release.

Although the principles involved are clear, quantitative knowledge of this question is scanty. Research involves use of the litter-bag technique in the field and laboratory incubation. Methods are given in Anderson and Ingram (1989). Determination of the phasing of crop requirements, by growth measurements and foliar analysis, is equally necessary.

Hypothesis 7—Soil Toxicities

The cycling of bases in tree litter can assist in (1) ameliorating soil acidity or checking acidification and (2) reclaiming saline or alkaline soils.

Strong acidity (pH < 5.5, exchangeable aluminum present) and alkalinity are problems inherent to soils of low fertility that will be discussed only briefly. One known case in which it can be modified through agroforestry is the "chitemene" form of shifting cultivation found in Zambia and neighboring African countries. Trees from natural savanna are felled, piled up on part of the area from which they have come, and burned. Rises of up to 2.0 pH points, caused by bases in the ash, have been recorded (Stromgaard 1985). However, these results are obtained from bases accumulated over many years of growth and from an area larger than that which was treated. The mass of calcium and magnesium in tree growth is to small by an order of magnitude to appreciably modify natural soil acidity; moreover, these bases have necessarily been abstracted from the soil. There may be some limited role of trees in checking the soil acidification that commonly occurs under continuous agriculture.

Trees have been successfully incorporated in the reclamation of saline and alkaline soils with associated cereal intercropping, for example at Karnal, India (Grewal and Abrol 1986). The part played by the trees, as distinct from drainage and gypsum treatment, has not been isolated.

Hypothesis 8—Soil Fauna

Agroforestry systems have beneficial effects upon, or offer management options to modify, soil fauna, with consequent improvements in soil fertility.

Discussion

Like synchrony, this is a hypothesis originating from the Tropical Soil Biology and Fertility program as applied to agroforestry. Fertile soils have a higher soil faunal biomass and different species composition than soils impoverished by continuous cropping, but this is a two-way relationship. A specific suggestion is that surface mulching encourages fauna with benefits to soil physical properties (Swift 1987, p 46).

No reports compare soil fauna under agroforestry with agriculture, but the prima facie arguments for supposing the existence of differences significant to fertility are sufficient to justify research.

Hypothesis 9—Soil Water

Some agroforestry systems have the potential to augment soil water availability to crops.

Discussion

In drier climates, water availability is often as, if not more, important for crop production than nutrient supplies. For one agroforestry system the benefits are proven: reductions in evaporation of the order of 20% to 30% have been recorded for windbreaks. On sloping land, it is known that one of the benefits from conventional soil conservation measures is reduced runoff and therefore improved infiltration; research is needed into whether hedgerows have similar effects. In general, however, we do not know if trees increase or reduce the water available to associated crops. A possible adverse effect is the danger that the established root systems of perennial trees will rob newly growing annual crops of water. On the other hand, it is possible that trees, whether intimately mixed with crops or planted in rows, will improve the total water supply by reducing evaporation.

Hypothesis 10—The Role of Roots

The role of roots in maintaining soil fertility in agroforestry systems is at least as important as that of above-ground biomass.

Discussion

Roots play a part in nearly all the above hypotheses and processes, particularly in organic matter input, soil physical conditions, nitrogen fixation, and nutrient retrieval and cycling. At the same time, competition between tree and crop roots for nutrients is a potentially adverse feature of agroforestry with respect to fertility and crop production, although no such case has yet been demonstrated.

The root biomass of trees is typically 20% to 30% of total plant biomass, equivalent to 25% to 43% of above-ground biomass. This, however, underestimates net primary production since many finer roots are annually sloughed off into the soil and regrown (the below-ground equivalent of litter fall). The nutrient content of roots, all of which is ultimately released into the soil, is substantial. Rain forests have a lower ratio of roots to shoots than most ecosystems, yet in forests of Sri Lanka and Sarawak, roots were found to contain 12% to 28% of the N, P, K, and Ca in the standing plant biomass (Andriesse et al. 1984, 1987).

Tree and crop roots are a basic, rarely harvested input to organic matter maintenance that is additional to above-ground residues. It is believed that the litter-to-humus conversion loss of roots is lower than that of shoots. In nutrient retrieval, roots are the primary agency. Not all the elements of the plant-soil nutrient cycle follow the path soil-root-shoot-litter-soil, part follows that of soil-root-root residue-soil.

An intriguing result was obtained in trials of hedgerow intercropping at Ibadan, Nigeria, where three treatments were compared: intercropped hedgerows with prunings placed on the soil, the same with prunings removed, and crops only (Yamoah et al. 1986b). The root weight of intercropped maize was higher on the hedgerow plots with prunings removed than for the crop-only control. This feature was attributed to the better physical conditions of the soil brought about by the hedgerow root systems, although microclimatic effects are an alternative explanation.

Root studies are currently neglected in agroforestry research, partly because roots cannot be seen and partly because of the labor involved. Their potential role, particularly for organic matter maintenance, soil physical properties, and nutrient cycling, is so great that this situation should be remedied. A guide to methods is given in Anderson and Ingram (1989).

Two processes not cited in other hypotheses are exudation by roots of growth-promoting substances, and direct transfer of assimilate between root systems, possibly via mycorrhizal bridges (Fitter 1985). Both are unproven but merit investigation through their obvious potential role in mixed tree/crop systems.

Subhypothesis 10A

Competition for nutrients between trees and crops can be reduced by selection of species with different rooting patterns.

Discussion

The possibility of reduced crop production is often cited as the major objection to agroforestry. It is essential that such reduction should be minimized or eliminated. The three major potential causes are shading, competition for moisture, and competition for nutrients, only the last of which falls within the scope of this discussion.

There is clearly a danger that the permanent (albeit regularly renewed) root systems of perennial trees will rob nutrients from the systems of annual crops, the more so where tree roots extend laterally beneath the area planted to crops. While often mentioned, there is little evidence for or against this contention. For *Leucaena*, *Cassia siamea*, and *Prosopis chilensis*, the vertical distribution of fine roots was found not to differ greatly from that of adjacent maize, although the fine root biomass of the trees was greater (Jonsson et al. 1988).

The hypothesis is that such competition will be reduced (and the efficiency of nutrient retrieval increased) if tree and crop root systems occupy different parts of the soil. Since most crops have shallow rooting systems, trees with a predominance of deeper roots are preferred. However, the evidence from Jonsson and associates (1988) in Tanzania (rainfall 870 mm/yr) suggests that the root distribution of five important trees (including *Leucaena*) in unfertilized fields is similar to that of maize, suggesting that competition for below-ground resources is the likely outcome.

In hedgerow intercropping, the ideal pattern is one in which the hedge roots first grow downwards below the topsoil but then spread laterally, allowing them to act as a "safety net" to intercept nutrients that would otherwise be leached. In southern Sumatra, Indonesia, *Peltophorum pterocarpa* has been found to develop such a pattern. Root density, depth, and pattern can be substantially affected by the timing of the shoot pruning (van Noordwijk 1989, van Noordwijk et al. in press).

Research is clearly desirable and can be undertaken at two levels. The simpler approach is to study crop yields near the tree/crop interface, eliminating as far as possible all interactions except nutrient competition. Moisture competition can be removed by irrigation, and microclimatic effects minimized by frequent pruning. Such experiments should be matched by the converse arrangement, minimizing nutrient competition by fertilization and isolating the soil moisture effect.

FUTURE RESEARCH DIRECTIONS

All 10 hypotheses and the subhypotheses require research. The hypothesis on erosion control is particularly important for landscapes dominated by moderate and steep slopes, while the questions of soil toxicities are applicable to areas with special soil problems. The other hypotheses are of general applicability to agroforestry systems. Of greatest importance are the questions on soil organic matter, linked to both physical properties and nutrient release, and on nutrient cycling, based on the subhypotheses of nutrient retrieval and synchrony. Only for nitrogen fixation is the present research effort commensurate with needs.

In many parts of Asia, there is little prospect of using tree prunings for improving soil fertility since they are too valuable as fodder for livestock, and virtually all crop residue and organic matter are consumed by livestock. Therefore, research priorities should concentrate on the contribution of below-ground litter such as root residue to organic matter improvement, and the effects of roots on soil structure. This is a much neglected area of investigation as most studies have emphasized the benefit of prunings as green leaf manure.

The evidence in support of the claim that trees improve the yield of associated trees is largely based on a few examples located on good soils that are in little need of improvement (Sanchez 1987). The challenge is to examine the changes in degraded or marginal soils where agroforestry is considered to be especially applicable. The most appropriate test for any soil-agroforestry hypothesis is to measure changes in soil properties over time on the *same* site. A more popular and common approach is to sample soils of nearby sites at the same time, but the initial condition and soil properties of the sites are unknown. The latter approach should be examined carefully to separate the variability of a site by comparing the particle size distribution, a good indicator of whether the soils were similar when the comparisons started. For example, a comparison of the soil profile of teak plantations up to 120 years of age in Kerala, India, revealed major changes in soil organic matter and bulk density with age of the plantations (Jose and Koshy 1972). However, analysis of the clay and sand content showed considerable variation among the sites, indicating that changes in organic matter and bulk density were a function of soil texture, not age of the teak plantations.

The three questions of research, What? Why? and How? (or What happens? Why does it happen? and How does it happen?), represent three levels of research. Applied to soil research, studies at the What level consist of trials of agroforestry systems, usually with differences in several variables (for example, tree species, spacing, manner of litter addition), with measurements of changes in soil properties. Such monitoring should form a standard part of all agroforestry trials. A minimum set of observations to be undertaken could include the following:

1) Before commencing any trial, take soil samples from the site in a statistically based pattern, and carry out standard soil analyses, including physical properties.

2) Repeat this soil sampling and analysis approximately every three years using identical methods of analysis; where applicable, stratify the sampling pattern for different treatments, for example, beneath hedgerows and within cropped alleys.

3) Measure above-ground biomass production from all plants in the system, partitioned into leaf, fruit, wood, and root. Carry out tissue analyses for the nutrient content of each plant part.

4) Make some attempt to estimate root production, vertical and horizontal distribution, and nutrient content.

5) If the trial is on sloping land, measure runoff and soil loss.

By including this basic set of observations within general-purpose agroforestry system trials, substantial contributions to knowledge of their effectiveness in maintaining soil fertility will be made. For clear, statistically valid evidence on soil changes, a period of five to six years is desirable, but provisional indications of trends can be obtained from a three-year experiment.

The Why level is concerned with special-purpose research in which soil studies are the primary objective. While some aspects call for specialized equipment or knowledge, much can be achieved with the facilities available at most research stations. This level of research is frequently conducted, not on complete agroforestry systems, but on isolated elements such as single hedgerows to study erosion processes, on individual tree/crop interfaces, or on plots of the order of 10 m square to investigate the effects of different litter applications. Comprehensive nutrient cycling studies are possibly the most central and important type of work to be undertaken by major research centers.

The How level of research consists of studies of basic processes, not specific to agroforestry, and may require sophisticated techniques, for example, isotope labeling. This is usually undertaken by universities and other specialized research institutions.

Based on the argument that the objective is to find methods that can be of practical use to farmers, there is a current tendency to concentrate research on trials of complete agroforestry systems. This is false reasoning. By analogy, a car is a product intended for users, but one does not construct cars of every possible size, shape, and specification and test each to see if it works! Research is conducted into properties of materials, applied thermodynamics, and the like, on the basis of which a small number of promising designs are constructed and, as a final stage, road-tested.

Such is the complexity of biological systems that it will never become possible to design an agroforestry system and predict its performance precisely in advance. Nevertheless, field trials with large numbers of variables are costly in terms of effort and expense. There can be a great economy of research effort in first studying individual processes and components, then applying the knowledge so gained to the design of prototype systems that can reasonably be expected to perform well. Owing to the urgency of needs, try-it-and-see research will certainly continue, and no doubt will sometimes come up with systems that work. Concurrent with such trials, however, there should be experiments directed at understanding the processes which occur within the systems. This two-pronged approach offers the best hope to fulfill the considerable potential of agroforestry for maintenance of soil fertility and sustainable land use.

LITERATURE CITED

Alpízar, L., H.W. Fassbender, J. Heuveldop, H. Fölster, and G. Enríquez. 1986 and 1988. Modeling agroforestry systems of cacao (*Theobroma cacao*) with laurel (*Cordia alliodora*) and poro (*Erythrina poeppigiana*) in Costa Rica. I, II, and III. *Agroforestry Systems* 4: 175–190 and 6: 37–62.

Anderson, J.M., and J.S.I. Ingram (eds.) 1989. *Tropical soil biology and fertility: a handbook of methods*. Wallingford, UK: CAB International. 171 p.

Andriesse, J.P., T.T. Koopmans, and R.M. Schelhaas. 1984 and 1987. A monitoring study of nutrient cycles in soils used for shifting cultivation under various climatic conditions in tropical Asia. I, II, and III. *Agroforestry Ecosystems and Environment* 12: 1–16 and 19: 285–332.

Aranguren, J., G. Escalenta, and R. Herrera. 1982. Nitrogen cycle of tropical perennial crops under shade trees. I. Coffee. II. Cacao. *Plant and Soil* 67: 247–269.

Ball, D.F. 1964. Loss-on-ignition as an estimate of organic matter and organic carbon in non-calcareous soils. *Journal of Soil Science* 15: 84–92.

Bernhard-Reversat, F. 1987. Les cycles des éléments minéraux dans un peuplement à *Acacia seyal* et leur modification en plantation d' *Eucalyptus* au Sénégal. *Acta Oecologica* 8: 3–16.

Bornemisza, E. 1982. Nitrogen cycling in coffee plantations. In: *Nitrogen cycling in ecosystems of Latin America and the Caribbean* (G. P. Robertson, R. Herrera, and T. Rosswall, eds.) Dordrecht, The Netherlands: Nijhoff. pp 241–246.

Brewbaker, J.L. 1986. Significant nitrogen fixing trees in agroforestry systems. In: *Agroforestry: realities, possibilities and potentials* (H. L. Gholz, ed.). Dordrecht, The Netherlands: ICRAF and Nijhoff. pp 31–45.

Dommergues, Y.R. 1987. The role of biological nitrogen fixation in agroforestry. In: *Agroforestry: a decade of development* (H. A. Steppler and P. K. R. Nair, eds.). Nairobi: ICRAF. pp 245–271.

Felker, P. 1978. *State of the art:* Acacia albida *as a complementary permanent intercrop with annual crops*. Report to USAID. Riverside, California: University of California. 133 p.

Fitter, A.H. (ed.). 1985. *Ecological interactions in the soil environment*. Oxford: Blackwell. 400 p.

Glover, N., and J. Beer. 1986. Nutrient cycling in two traditional central American agroforestry systems. *Agroforestry Systems* 4: 77–87.

Golley, F.B., J.T. McGinnis, R.G. Clements, G.I. Child, and M.J. Duever. 1975. *Mineral cycling in a tropical moist forest ecosystem*. Athens, Georgia: University of Georgia Press. 248 p.

Grewal, S.S., and I.P. Abrol. 1986. Agroforestry on alkali soils: effect of some management practices on initial growth, biomass accumulation and chemical composition of selected tree species. *Agroforestry Systems* 4: 221–232.

Huxley, P.A. 1986a. Simplifying experimental agroforestry. In: *ICRAF Newsletter* 16, Nairobi: ICRAF. pp 1–2.

Huxley, P.A. 1986b. *Rationalising research on hedgerow intercropping: an overview*. ICRAF Working Paper 40. Nairobi: ICRAF 151 p.

Jonsson, K., L. Fidjeland, J.A. Maghembe, and P. Hogberg. 1988. The vertical distribution of fine roots of five tree species and maize in Morogoro, Tanzania. *Agroforestry Systems* 6: 63–70.

Jose, A.I., and M.M. Koshy. 1972. A study of morphological physical and chemical characteristics of soils as influenced by teak vegetation. *Indian Forester* 98: 338–384.

Kang, B.T., and B. Duguma. 1985. Nitrogen management in alley cropping systems. In: *Nitrogen management in farming systems in humid and subhumid tropics* (B. T. Kang and J. van der Heide, eds.). Haren, The Netherlands: Institute of Soil Fertility. pp 269–284.

Lal, R., and D.J. Greenland. 1979. *Soil physical properties and crop production in the tropics*. Chichester, UK: John Wiley and Sons. 551 p.

Lelong, F., E. Roose, G. Aubert, R. Fauck, and G. Pedro. 1984. Géodynamique actuelle de differents soils à végétation naturelle ou cultivés d' Afrique de l'ouest. *Catena* 11: 343–376.

Moreau, R. 1984, 1985. Etude sur parcalles comparatives de l'évolution des sols ferrallitiques sous différents modes de mise en culture en zones forestiéres et préforestiére de Cote d'Ivoire. *Cahiers ORSTOM Sér. Pédol.* 21: 43–56.

Mulongoy, K. 1986. Nitrogen cycling in alley cropping systems. *IITA Research Briefs* 7(4): 3–5.

Nair, P. K. R. 1984. *Soil productivity aspects of agroforestry*. Nairobi: ICRAF. 85 p.

NIFTAL Project. 1984. *Nitrogen fixing tree master list*. NIFTAL Project, University of Hawaii, Honolulu, HI.

Reifsnyder, W.S., and T.O. Darnhofer. 1989. Meteorology and agroforestry. In: Proceedings of an international workshop on the application of meteorology to agroforestry systems planning and management. Nairobi: ICRAF. 546 p.

Roskoski, J.P. 1982. Nitrogen fixation in a Mexican coffee plantation. In: *Nitrogen cycling in ecosystems of Latin America and the Caribbean* (G. P. Robertson, R. Herrera, and T. Rosswall, eds.). Dordrecht, The Netherlands: Martinus Nijhoff. pp 283–292.

Roskoski, J.P., and C. van Kessel. 1985. Annual, seasonal and diel variation in nitrogen fixing activity by *Inga jinicuil*, a tropical leguminous tree. *Oikos* 44: 306–312.

Sanchez, P.A. 1987. Soil productivity and sustainability in agroforestry systems. In: *Agroforestry: a decade of development* (H. A. Steppler and P. K. R. Nair, eds.). Nairobi: ICRAF. pp 205–223.

Sanginga, N., K. Mulongoy, and A. Ayanaba. 1986. Inoculation of *Leucaena leucocephala* (Lam.) de Wit with Rhizobium and its nitrogen contribution to a subsequent maize crop. *Biological Agriculture and Horticulture* 3: 347–352.

Sioli, H. 1985. The effects of deforestation in Amazonia. *Geographical Journal* 151: 197–203.

Soemwarto, O. 1987. Homegardens: a traditional agroforestry system with a promising future. In: *Agroforestry: a decade of development* (H. A. Steppler and P. K. R. Nair, eds.). Nairobi: ICRAF. pp 157–170.

Stocking, M. 1986. *The cost of soil erosion in Zimbabwe in terms of the loss of three major nutrients.* Consultants' Working Paper 3. Soil Conservation Program, AGLS, FAO, Rome. 164 p.

Stromgaard, P. 1985. Biomass, growth and burning of woodland in a shifting cultivation area of South Central Africa. *Forest Ecology and Management* 12: 163–178.

Swift, M.J. (ed.) 1987. Tropical soil biology and fertility (TSBF): inter-regional research planning workshop. *Biology International Special Issue* 13: 68.

van Noordwijk, M. 1989. Rooting depth in cropping systems in the humid tropics in relation to nutrient use efficiency. In: *Nutrient management for food crop production in tropical farming systems* (J. van der Heide, ed.). Haren, The Netherlands: Institute for Soil Fertility. pp 129–144.

van Noordwijk, M., Kurniatun Hairiah, and Syekhfani Ms. In press. *Peltophorum pterocarpa* (DC) Back (Caesalpiniaceae), a tree with a root distribution suitable for alley cropping on acid soils in the humid tropics. In: *Proceedings of the International society of Root Research Symposium*, 1988, Uppsala, Sumatra.

von Carlowitz, P.G. 1986a. *Multipurpose tree and shrub seed directory*. Nairobi: ICRAF. 265 p.

von Carlowitz, P.G. 1986b. Multipurpose tree yield data—their relevance to agroforestry research and development and the current state of knowledge. *Agroforestry Systems* 4: 291–314.

Weerakoon, W.L., and L. G. Gunasekera. 1985. In situ application of *Leucaena leucocephala* (Lam) de Wit. as a source of green manure in rice. *Sri Lankan Journal of Agricultural Science* 22: 20–27.

Wiersum, K.F. 1984. Surface erosion under various tropical agroforestry systems. In: *Symposium on effects of forest land use on erosion and slope stability* (C. L. O'Loughlin and A. J. Pearce, eds.). Honolulu, Hawaii: Environment and Policy Institute, East-West Center pp 231–239.

Wilson, G.F., B.T. Kang, and K. Mulongoy. 1986. Alley cropping: trees as sources of green-manure and mulch in the tropics. *Biological Agriculture and Horticulture* 3: 251–267.

Yamoah, C.F., A.A. Agboola, and K. Mulongoy. 1986a. Decomposition, nitrogen release and weed control by prunings of selected alley cropping shrubs. *Agroforestry Systems* 4: 239–246.

Yamoah, C.F., A.A. Agbolla, and G.F. Wilson. 1986b. Nutrient competition and maize performance in alley cropping systems. *Agroforestry Systems* 4: 247–254.

Yamoah, C.F., A.A. Agboola, G.F. Wilson, and K. Mulongoy. 1986c. Soil properties as affected by the use of leguminous shrubs for alley cropping with maize. *Agricultural Ecosystems and Environment* 18: 167–177.

Young, A. 1989a. *Agroforestry for soil conservation*. Wallingford, UK: CABI International and ICRAF. 274 p.

Young, A. 1989b. Ten hypotheses for soil-agroforestry research. *Agroforestry Today* 1(1): 13–16.

Young, A. 1990. Agroforestry for the management of soil organic matter. IBSRAM Proceedings. In press.

CHAPTER 12

Soil Erosion and Conservation in Agroforestry Systems

K. F. Wiersum

Soil erosion is one of the most widespread causes of land degradation. In humid areas, it is primarily the result of water action; in arid regions, it is primarily the result of wind. This chapter is confined to water-related erosion—for a discussion of research methods for wind erosion, see Skidmore (1988).

Water erosion is a process by which soil particles are first loosened and broken apart and then carried, rolled, or washed away. It is basically caused by the interaction between rainfall as an erosive agent and soil as the medium that is detached and transported. The potential ability of rainfall to cause erosion, called *erosivity*, depends on such characteristics of rainfall as the energy of the falling raindrops and the intensity, the length, and total number of rainstorms. These characteristics determine the ability of raindrops to detach soil particles and the possible occurrence of surface runoff, a primary means for transporting detached soil particles. The susceptibility of soil to erosion, its *erodibility*, depends on various soil characteristics: aggregate stability, transportability of loosened soil particles, and infiltration rates. The aggregate stability of a soil determines how easily soil particles are detached; the transportability determines how easily these detached soil particles may be carried away; and the infiltration rate influences surface runoff. The degree of erosion also depends on the slope of the land, on vegetation, and on artificial soil conservation practices. Details of these factors are given in all standard textbooks on erosion (Hudson 1981, Morgan 1986).

Erosion may take various forms based on the interaction between the erosive agent and the soil. Erosion caused by water flowing over the soil surface and forming minute channels is *rill erosion*. Detachment and removal of soil more or less evenly between rills by rainsplashes is *interrill* or *sheet erosion*. *Pedestal erosion* occurs when impermeable objects such as rocks, stones, or roots provide cover for small areas of the soil and protect them from erosion in the shape of small columns. In *gully erosion*, the channels formed by erosion are so deep and extensive that the land cannot be used for normal cultivation.

Other specialized types of erosion, such as piping, pinnacle, and streambank erosion, as well as landslides and soil creep, are also caused by water, but as these processes are distinctly different from the other types, they will not be treated here.

Erosion is a normal geological process, but accelerated or man-made erosion is cause for grave concern, especially in tropical regions (El-Swaify et al. 1982). It often results in a decrease of land quality for two reasons:

1) Soil quality deteriorates because of the loss of the most fertile topsoil layers and the decline of physical properties.
2) The soil infiltration rate decreases, and overland flow of water increases, accelerating the erosion process.

The results of this deterioration are far-reaching: Not only do eroded farmlands have lower yields per hectare, but cropping area and accessibility may also decrease due to gully formation. More runoff in the rainy season in the form of sediment-rich overland flow from erosion changes the hydrological regime of rivers. High sediment concentrations in rivers results in undesired sedimentation in downstream reservoirs and irrigation systems, decreasing their efficiency and increasing the flood hazard. These processes finally reduce the living standards of the people, not only where erosion is taking place, but also down-stream. *Soil conservation* measures are directed at preventing such erosion, while *erosion control* measures aim to reduce erosion after it has already begun.

EROSION CONTROL THROUGH AGROFORESTRY

Agroforestry systems are assumed to be superior to other cropping systems with respect to erosion protection because trees and other forms of vegetative cover protect the soil against erosion by the ways they affect both the erosive agent (rainfall) and the medium being eroded (soil).

Role of Trees and Other Vegetation Layers

Trees not only influence rainfall and soil independently, they also affect the level where these factors interact, the soil surface (figure 12.1). These three processes by which trees influence erosion may be summarized as follows (Wiersum 1985, Young 1986).

1) Rainfall interception in tree canopies decreases the quantity of water reaching the soil and alters the spatial distribution of that water by stemflow and throughfall, which includes concentrated drip-points. As a result of the interception, the initial erosive power of rain is broken, but through coalescence of raindrops into larger drops, the erosive power of throughfall drops may exceed the initial erosive power under trees that have a free dropfall height of about 5 m or more.
2) Trees exert a positive influence on soil detachability and infiltration capacity because sustained litter input to the soil causes higher soil humus content. In addition, favorable microclimate conditions under tree canopies positively influence various soil organisms that affect such soil processes as decomposition, humification, and pore for-mation.
3) Tree litter and surface vegetation protect the soil directly against the erosive force of raindrops and surface runoff. By filtering splashed soil particles, surface vegetation and litter also prevent clogging of soil pores, which decreases the infiltration rate and increases surface runoff.

There is ample evidence that the protective influence of trees on surface erosion depends more on their ability to produce a sustained litter cover on the

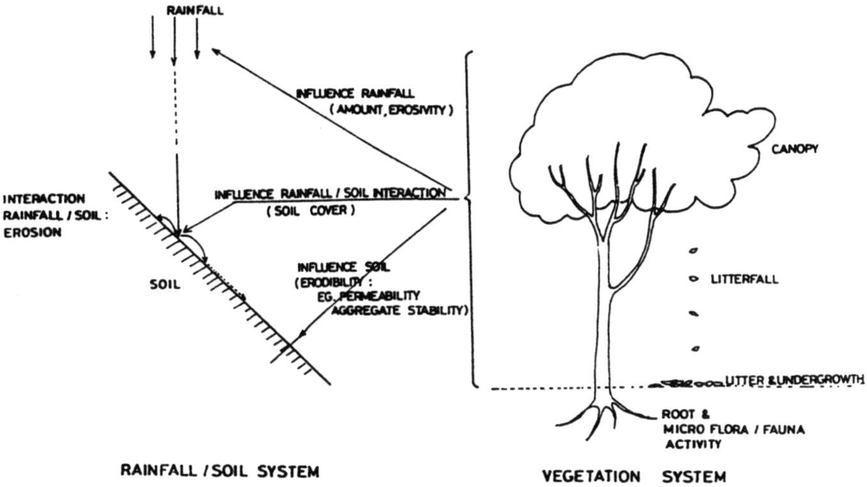

Figure 12.1. Influence of vegetation on various erosion factors.
Adapted from: Wiersum 1984.

soil surface than on their direct influence on rainfall or soil (Wiersum 1984, Young 1986). This surface cover gives optimal interception of the erosive force of the falling waterdrops and reduces the velocity of overland flow. The effect of the tree canopy in intercepting raindrops and changing drop size, distribution, and fall velocity (Brandt 1987) has a variable, but relatively minor, effect when compared with the influence of the surface cover (Wiersum 1985). The decrease in soil erodibility from humus incorporation in the soil develops only over long periods.

The major protective effect of agroforestry on erosion is thus its capacity to establish and maintain a regularly renewed ground cover of litter consisting of crop residue, tree prunings or natural leaf fall, and/or a good soil cover of grasses, herbs, or cover crops. If planted densely along the contour, trees may also serve as barriers to surface runoff, which may add additional protection against erosion. But in general terms, there are greater opportunities to reduce erosion by maintaining soil cover than by creating a vegetative barrier (Young 1986, 1989).

The effect of vegetative cover on erosion depends not only on the proportion of the ground that is covered by the vegetation, but also on the depth of the cover. Figure 12.2 illustrates how erosion depends on the canopy cover, its distance to the ground, and the amount of bare soil that is open to splash from raindrops and/or waterdrops falling from the canopies. Where there is more than 60% ground cover, the presence of trees hardly influences erosion (Stocking 1988), but where the soil surface is less covered, characteristics of the tree canopy may significantly affect erosion.

The amount of soil litter or vegetative mulch and the rate of decomposition also affect erosion. For optimal erosion control, decomposition should be slow. However, enhancement of soil fertility or other reasons may require faster decomposition rates.

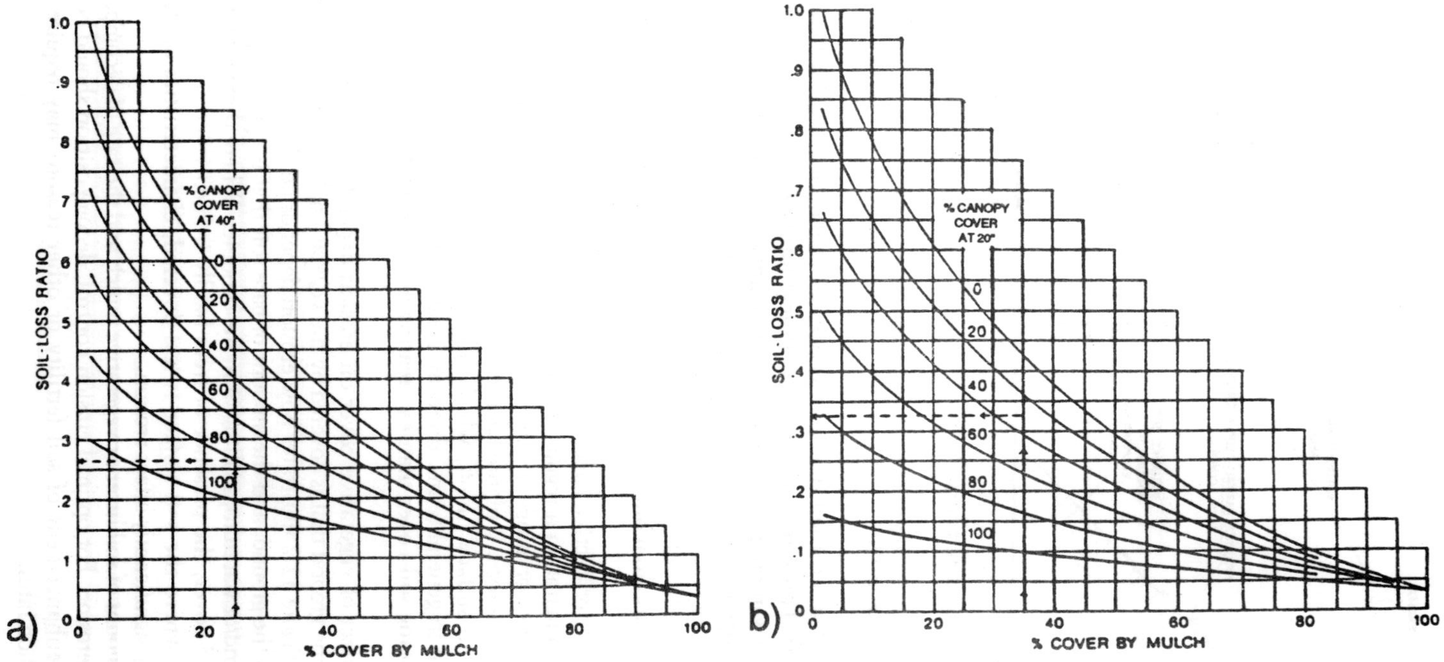

Figure 12.2. Combined effects of mulch and canopy cover on erosion at two distances from canopy to ground: (a) 40 inches and (b) 20 inches. Adapted from: Wischmeier and Smith 1978.

Role of Management

Because of the forest-like characteristics of agroforestry systems, it is often claimed that introducing agroforestry on lands vulnerable to erosion will conserve soils, check erosion, and sustain production (Wiersum 1984). However, it is dangerous to generalize from the one ecosystem to the other about erosion control for two reasons:

1) Many different agroforestry systems are possible on a continuum from almost pure agriculture on one end to almost pure forestry on the other. The kinds and numbers of trees grown, as well as the configuration in which they are cultivated, may vary greatly within and between systems. The same is true of agricultural crops, which may range from short-lived semi-annuals to perennials, and of any livestock included in the various systems.

2) The management aims of agroforestry systems may vary as widely as the species of vegetation. In some agroforestry systems, erosion control is the primary goal; in others it is of secondary importance. In fact, some agroforestry systems may require management practices that favor erosion—for example, soil tillage or frequent biomass harvesting.

Thus the ability of agroforestry systems to control soil erosion cannot be taken for granted. Each specific protective feature requires careful identification. While individual trees cannot be expected to exert the same positive erosion control as undisturbed forest ecosystems, manipulating the tree and vegetative components and possibly including anti-erosion management practices in agroforestry systems can make important contributions to erosion control and soil conservation. The key to success is good management practices (Wiersum 1984).

Present Knowledge

Firm knowledge about the effects of most agroforestry systems on erosion is still sparse. Therefore, it is important to collect field data to substantiate claims on the efficacy of specific agroforestry systems for soil conservation. However, a summary of reported rates of erosion under different agroforestry systems and related forms of land use can be made, such as is found in table 12.1. This table combines experimental data collected in various countries under different soil and rainfall regimes. The range of erosion observed in the different systems indicates several important factors that affect the rate of erosion, most notably the effect of the inclusion of annual crops in sequential agroforestry systems and the presence or absence of a protective plant cover directly on the soil surface. Based on such observations and theoretical considerations, it is possible to estimate the relative value of various types of agroforestry systems for soil conservation. These systems, in order of decreasing importance to erosion control, are summarized below (Young 1986).

1) Barrier hedges may control erosion through the combined effects of checking runoff, providing litter cover, maintaining soil organic matter, and progressively transforming slopes into terraces.

2) Trees planted on either grass barrier strips or on earth structures for erosion control may help stabilize these structures and, in some cases, add to

their protective value. Equally important to the landowner is that these soil conservation works become directly productive due to the tree products.

3) Contour alley cropping may form erosion barriers, especially if pruned branches are used to establish trash ridges anchored by the trees; the contour plantings may also provide a litter cover of green manure on the cropped alleys.

4) Multistory gardens and plantation crop combinations may provide good erosion control, provided that there is a sufficiently dense tree cover, there is a good ground cover of litter, and vegetation or cover crops are maintained. Erosion rates are high when such plantations are clean-weeded or if litter is removed (table 12.1).

5) Silvopastoral systems will only contribute to erosion control if proper pasture management practices are applied, including rotational grazing and control of livestock numbers below the carrying capacity of the grazing lands.

6) Planted tree fallows and *taungya* systems can check erosion only during the period of tree growth, but erosion control as a whole depends on practices during the intermediate cropping periods.

Table 12.1. Rates of erosion in various tropical forest and agroforestry systems (t/ha/yr)

	Minimal	Median	Maximal
Multistoried tree gardens (4/4)*	0.01	0.06	0.14
Natural forests (18/27)	0.03	0.30	6.16
Shifting cultivation fallow period (6/14)	0.05	0.15	7.40
Forest plantations, undisturbed (14/20)	0.02	0.58	6.20
Tree crops with cover crop/mulch (9/17)	0.10	0.75	5.60
Shifting cultivation, cropping period (7/22)	0.40	2.78	70.05
Taungya cultivation (2/6)	0.63	5.23	17.37
Tree crops, clean-weeded (10/17)	1.20	47.60	182.90
Forest plantations, burned/litter removed (7/7)	5.92	53.40	104.80

*(x/y), x = no. of locations, y = no. of treatments/observations.
Adapted from: Wiersum 1984.

In addition to the configuration and density of the tree crops and other woody vegetation and the status of the undergrowth, the proportion of annual crops needing regular soil tillage strongly influences the soil erosion rates in agroforestry systems. Another important factor is the characteristics of the slope. In general terms, agroforestry should not be practiced on slopes exceeding 60%. On slopes of 60% to 80%, agroforestry is generally allowed only if combined with engineering works for soil conservation. If important areas of regularly tilled soils are present under or between the trees, terraces may also be needed on slopes between 20% and 60% (RAPA 1986).

OBJECTIVES OF EROSION RESEARCH IN AGROFORESTRY

As in any research, the first step in studying erosion is to carefully identify the specific objectives of the research. A careful identification is critical because these objectives may determine what type of research is used and how data is collected. These objectives fall into the following broad categories (Young 1986):

1) Assessing actual erosion rates in specific agroforestry systems and establishing whether these systems and their practices are really as effective as indicated by inferential, or limited experimental, evidence.

2) Determining how specific agroforestry practices achieve control of erosion, using this knowledge to refine the design of specific systems, and testing the improved designs.

3) Reconciling designs that are optimal for erosion control with those that meet other design requirements.

4) Testing the response of farmers to controlling erosion by means of agroforestry.

5) Evaluating the overall effectiveness of erosion control through agroforestry on land with an erosion problem, taking into account environmental, economic, and social criteria.

Thus research on the role of agroforestry with respect to erosion may assess not only the rates and effects of erosion in existing or newly designed agroforestry systems, but also the willingness of farmers to accept these systems for erosion control and soil conservation. Acceptability depends both on the effectiveness of the systems in preventing or reducing erosion and on such characteristics as their ability to produce subsistence or commercial products. Some protective characteristics are not exclusive to agroforestry but may also be found with other vegetation systems or erosion control measures. However, several agroforestry systems have an advantage over other forms of soil conservation and erosion control because they combine protective and productive functions, provided that the right species mix, proper plant configurations, and optimal management techniques are chosen.

The remainder of this chapter discusses methods for obtaining information about the characteristics of different agroforestry systems in relation to soil erosion and conservation; it concentrates on the description of well-proven methods for applied erosion research. New approaches to erosion research are being developed that show promise for future application in agroforestry research, but where such methods are still in an experimental stage and have not yet been made operational for applied research, they will be noted only briefly.

RESEARCH METHODS

The techniques for assessing erosion rates and erosion factors in agroforestry come from both agricultural and forestry research. While these methods must sometimes be adjusted to fit specific characteristics of agroforestry systems, such as the large spatial variation in vegetation encountered in several systems, they fall into three general categories:

1) Assessment of rates of erosion through the measurement of soil loss in specific land-use types.

2) Assessment of the effects of specific soil, climate, and vegetation factors and soil conservation practices on erosion.

3) Assessment of the effects of erosion on soil properties and crop productivity.

Regardless of the category, there are two general methods for conducting research: observational and experimental (Hayward 1968, 1969). Data for each of these methods may be either qualitative or quantitative, depending on the purpose of the research and the detail needed. Qualitative assessments, if not carefully planned and executed, may be selective and serve as exemplary data only. This limitation may be overcome by following up with quantitative research that will supply data that can then be used to develop parameter values for erosion prediction models.

Observational Method

The observational method measures erosion rates at a specific location—erosion is measured where the researcher finds it. The observed quantity of eroded material can be accurately measured, but it cannot be explained as a product of values of more specific erosion factors. These values can only be estimated. Because of the many factors influencing erosion, data obtained with such observational methods should be considered as random. Ideally, such observations should be replicated on the basis of an adequate sampling design in order to obtain reliable data, with means and standard deviations, on erosion rates under specific conditions. Provided that enough data is collected, estimates of values for the separate erosion factors may be calculated by multivariate analysis. To do so requires many replications, however, and often it is impossible to carry out the number of measurements needed for statistical accuracy. Consequently, experimental methods of erosion assessment are preferred in many cases.

Experimental Method

In experimental methods of erosion research, the rates of soil loss during a specific period are measured in established plots with controlled treatments of land-use practices, erosion control measures, and slope conditions. In such experiments, the effect of specifically imposed treatments are compared with a standard condition. This enables parameter values to be determined for specific erosion factors from which conclusions may be drawn about the relationships between the investigated variables. For such experiments, a proper experimental design based on randomized block procedures should be used.

In addition to collecting data on the rate of erosion in different land-use systems and on the effect of specific erosion factors, it is also important to collect data on the effects of erosion. These effects are by no means limited to loss of soil depth. Erosion also affects the soil's water-holding capacity and levels of nutrient and organic matter, thus affecting crop productivity. An important factor to consider in erosion research is the intricate interdependencies between soil characteristics and soil erosion. Soil erosion depends not only on characteristics of the soil—which may themselves change as a result of erosion—but also on the soil's level of plant productivity. For example, soils that have suffered high erosion rates tend to grow protective plant cover slowly; therefore erosion rates tend to remain high.

Classification of Vegetation Characteristics

Several characteristics of vegetation determine an agroforestry system's soil erosion and conservation characteristics: the percentage and distribution of ground and canopy cover, the structure of the tree canopy (single or multilayered, width and height), and the structure of other plants (Armstrong and Mitchell 1987, Brandt 1987). Obviously, in agroforestry there may be numerous combinations of these factors depending on the number and nature of crop and tree species in the system. Because of the great variety of crop combinations and configurations in agroforestry, it is often useful to test the effect of general vegetation characteristics for erosion control rather than each possible agroforestry combination separately. This may be done by classifying different agroforestry crops into groups with similar structural characteristics (table 12.2). Also, the percentage of vegetative cover may be measured by well-established botanical techniques, for example, the point intercept method, for which a quadrat sighting frame may conveniently be used (Stocking 1988). For standardized measurements on vegetative cover, increasing height classes are often useful, for example, 0 to 25 cm, 25 to 50 cm, 50 to 100 cm, 1 to 2.5 m, and over 5 m (figure 12.2).

In measuring vegetative cover, special attention should be given to monitoring the presence of soil litter or mulch material. Not only is quantitative data about the amount of such material (in the form of average percentage of cover or dry weight per unit area) importance, but also qualitative data about the time it takes to decompose.

OBSERVATIONAL STUDIES OF EROSION

Observational studies of erosion rates of different agroforestry systems provide useful estimates for comparing the erosion control functions of these systems. The empirical data base on the effects of different agroforestry systems on erosion control under field conditions greatly needs to be improved. Much progress can be made through observational studies on the rate of erosion in different existing and experimental agroforestry systems, although data from such studies should only be considered as approximate. Ideally, observations should be carried out over the whole range of occurrences of a specific agroforestry system in a certain region rather than be limited to supposedly typical locations. Improtant advantages of such observational studies are that they can be carried out on a general reconnaissance level and that they normally take relatively little time and effort. By regularly repeating the observations over a period of time, it is possible not only to obtain one-time estimates of the rate of erosion, but also to establish longer-term trends. This is particularly important if data is collected from newly established agroforestry systems where trees are still immature.

Two Approaches to Observational Studies

The first approach to observational studies on erosion rates measures sediment transport rates past a point in a stream at the outlet of a small watershed. Such measurements integrate the effects of erosion over the total area of a watershed, but they do not provide data about the spatial distribution of erosion within the watershed. This approach will not be further treated here because it requires specialized instruments and (hydrological) knowledge, and in many cases, agroforestry systems do not occur as a uniform land-use type covering a whole watershed. Further information on such sediment sampling techniques is given in many textbooks (Walling 1988, Hadley and Walling 1984, Dunne 1977).

Table 12.2. Classification of erosion conditions according to SSF system

Soil movement	Depth of recent deposites around obstacles, or in microterraces, and/or depth of truncated areas is between 0 and .1 in (0 to 2.5mm).	Depth of recent deposits around obstacles, or in microterraces, and/or depth of truncated areas is between .1 and .2 in. (2 to 5 mm).	Depth of recent deposits around obstacles, or in microterraces, and/or depth of truncated areas is between .2 and .4 in. (5 to 10 mm).	Depth of recent deposits around obstacles, or in microterraces, and/or depth of truncated areas is between .4 and .8 in. (10 to 20 mm).	Depth of recent deposits around obstacles, or in microterraces, and/or depth of truncated areas is over .8 in. (20 mm).
	0 or 3	5	8	11	14
Surface litter	No movement, or if present, less than 2% of the litter has been translocated and redeposited against obstacles.	Between 2% and 10% of the litter has been translocated and redeposited against obstacles.	Between 10% and 25% of the litter has been translocated and redeposited against obstacles.	Between 25% and 50% of the litter has been translocated and redeposited against obstacles.	Between 50% of the litter has been translocated and redeposited against obstacles or removed from the area.
	0 or 3	6	8	11	14
Surface rock fragments	Depth of soil removal around the fragements and/or depth of recent deposits around the fragments is less than .1 in. (2.5 mm).	Depth of soil removal around the fragments and/or depth of recent deposits around the fragments is between .1 and .1 in. (2.5 to 5 mm).	Depth of soil removal around the fragments and/or depth of recent deposits around the fragments is between .2 and .4 in. (5 to 10 mm).	Depth of soil removal around the fragments and/or depth of recent deposits around the fragments is between .4 and .8 in. (10 to 20 mm).	Depth of soil removal around the fragments and/or depth of recent deposits around the fragments is over .8 in. (20 mm).
	0 or 2	5	8	11	14
Pedestalling	Pedestals are mostly less than .1 in. (2.5 mm) high and/or less frequent than 2 pedestals per 100 sq ft.	Pedestals are mostly .1 to .3 in. (2.5 to 8 mm) high and/or have a frequency of 2 to 5 pedestals per 100sq ft.	Pedestals are mostly .3 to .6 in. (8 to 15 mm) high and/or have a frequency of 5 to 7 pedestals per 100sq ft.	Pedestals are mostly .6 to 1 in. (15 to 25 mm) high and/or have a frequency of 7 to 10 pedestals per 100sq ft.	Pedestals are mostly over 1 in. (25 mm) high and/or have a frequency of over 10 pedestals per 100sq ft.
	0 or 3	6	9	11	14
Flow patterns	None, or if present, less than 2% of the surface area shows	Between 2% and 10% of the surface area shows	Between 10% and 25% of the surface area shows	Between 25% and 50% of the surface area shows	Over 50% of the surface area shows evidence of

	evidence of recent translocation and deposition of soil and litter. 0 or 3	evidence of recent translocation and deposition of soil and litter. 6	evidence of recent translocation and deposition of soil and litter. 6	evidence of recent translocation and deposition of soil and litter. 6	recent translocation and deposition of soil and litter. 15
Rills	Rills, if present, are mostly less than .5 in. (13 mm) deep, and generally at infrequent intervals greater than 10 ft. 0 or 3	Rills are mostly .5 to 1 in. (13 to 25 mm) deep, and generally at infrequent intervals greater than 10 ft. 6	Rills are mostly 1 to 1.5 in. (25 to 38 mm) deep, and generally at 10 ft intervals. 9	Rills are mostly 1.5 to 3 in. (38 to 76 mm) deep, and at intervals of 5 to 10 ft. 12	Rills are mostly 3 to 6 in. (76 to 152 mm) deep, and at intervals of less than 5 ft. 14
Gullies	No gullies, or if present, less than 2% of the channel bed and walls show active erosion (are not vegetated), gullies make up less than 2% of the total area. 0 or 3	Between 2% and 5% of the channel bed and walls show active erosion (are not vegetated), or gullies make up between 2% and 5% of the total area. 6	Between 5% and 10% of the channel bed and walls show active erosion (are not vegetated), or gullies make up between 5% and 10% of the total area. 9	Between 10% and 50% of the channel bed and walls show active erosion (are not vegetated) or gullies make up between 10% and 50% of the total area. 12	Over 50% of the channel bed and walls show active erosion (are not vegetated) along their length, or gullies make up over 50% of the total area. 15

Adapted from: Clark 1980.

The second approach involves the direct qualitative or quantitative measurement of erosion at a number of sampling sites of one land-use type. These measurements provide information on spatial distribution and local variation of the rates of erosion under the system being studied. If a sufficient number of observations are made, it is also possible by multivariate statistical methods to obtain estimates on the effect of certain key variables within the land-use system on erosion.

Qualitative observations on erosion

Qualitative measurements of erosion are most suited for use in reconnaissance surveys of different systems of land use to obtain preliminary and comparative data. The following features may be used as qualitative indicators of erosion in the field (Dunne 1977, Clark 1980, see also figure 12.3):

1) Pedestals or columns of soil protected at the top by stones, clumps of grasses, or debris (litter, wood).
2) Exposed roots of trees and shrubs.
3) Erosion pavement, which is a layer of rock fragments on the ground surface, similar in kind, size, and shape to rock fragments embedded in the soil beneath the surface, formed when the finer particles of surface soil are washed away leaving behind the larger fragments.
4) Rills and/or gullies.
5) Deposits of soil in ditches, hillside depressions, or field boundaries.
6) Depth of horizon, or other profile characteristics, that can be compared with similar profile characteristics of known undisturbed profiles, particularly variations in surface soil color relative to color differences known to occur through the profile.

By combining these indicators, several land-use classes with different rates of erosion can be identified (Morgan 1986). An example of a classification system is the soil surface factor (SSF) classification system developed by Clark (1980). It was designed to determine ecological range conditions by correlating soil properties and ground cover; estimate rates of soil loss; measure apparent ecological trends by using erosion condition data as an index; and monitor changes in erosion activity over time as a result of alternative land management practices.

Composite SSF values are calculated from evaluations of seven soil surface indicators of possible erosion (table 12.3). Although this system was developed for rangeland conditions, it also seems appropriate for collecting qualitative data on erosion in agroforestry systems.

Quantitative observations on erosion

Quantitative measurements of erosion may be made either by relating changes in soil surface to a permanent reference point or by measuring erosion in a sediment collection trough. These measurements offer more refined information than qualitative observations, but they require repeated measurements.

Changes in Soil Surface.

There are two methods for measuring this change: detailed leveling of permanent transects or direct measurement of soil loss around pins and rods

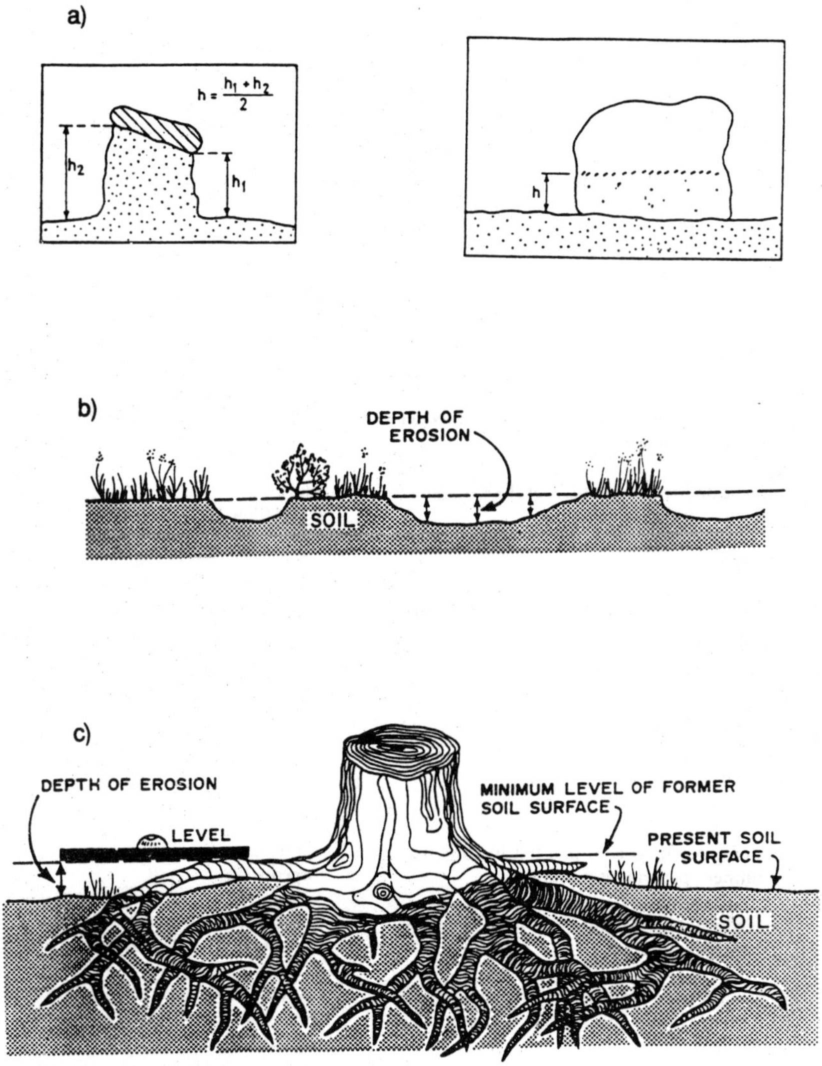

Figure 12.3. Methods for measurement of field indicators of erosion.
(a) Pedestals and exposed stones.
(b) Sheet erosion between vegetative remnants.
(c) Exposed roots.
Adapted from: Dunne 1977.

Table 12.3. Classification of crops based on similarities in soil protective characteristics

Description	Examples
A. Row crops	
1. tall, upright crops generally grown on unridged lands	annuals: maize, sorghum, sunflowers; perennials: napier fodder, sugarcane
2. leguminous, annuals; short, bunch, and procumbent varieties	beans: soya, velvet, jack, dolichos, and French; groundnuts, cowpeas
3. tall, upright crops grown on ridged lands	tobacco varieties, group 1 crops on ridges
4. woody, bushy row crops with individual growth and leaf development	cotton varieties
B. Broadcast crops	
1. tall, upright crops broadcast for fodder	see A 1
2. short, leguminous crops broadcast for fodder and green manure	see A 2
3. medium height plants for fodder, green manure and weed fallow	sun hemp, weed fallow
C. Orchards/plantaions	
1. individual trees and bushes planted in a regular pattern	coffee, citrus, deciduous fruit
2. hedged crops	tea
3. thick stands of natural and exotic trees with little to no grass cover	forestry
D. Grasslands	
1. stoloniferous grasses planted in rows from runners; permanent pastures	star, Kikuyu, torpedo
2. seed established grasses usually broadcast; bunch varieties	love grass, *Sabi panicum*, Katambora Rhodes, Giant Rhodes
3. species composition closely related to the natural regions soil types and condition of the veld	Natural veld grasses, usually mixed species predominantly bunch grasses, both annual and perennial

Adapted from: Elwell and Stocking 1976.

(figure 12.4). Repeated level surveys along the contour between two well-established benchmarks have the advantage that data on inter-rill, rill, and gully erosion may be collected. The use of pins provides data only on sheet and rill erosion. The most accurate measurements with pins are made if a washer is placed on the soil surface around the pin, providing a firm surface from which to measure and preventing soil splash against the pins. Erosion pins are cheap and easy to install, but sufficient numbers (at least 25 per location placed in a grid

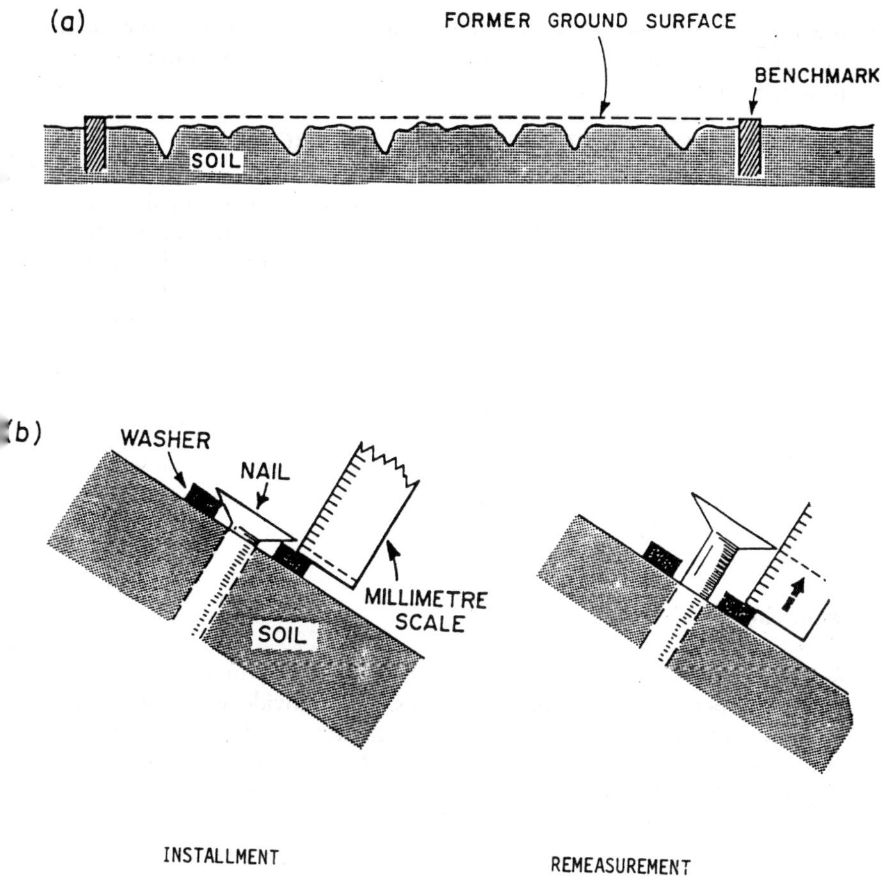

Figure 12.4. Measuring rates of erosion by (a) leveling between two benchmarks and (b) use of erosion pins.
Adapted from: Dunne 1977.

design) must be used to obtain reliable data. Pins should not be used on soil likely to crack or swell. Although such measurements are relatively easy and cheap to make, they are not very accurate: the minimum change in soil level that can be measured is 1 mm.

Sediment Collection Troughs.

Troughs can provide data that is at least 100 times as detailed as that provided by surface change measurements. Troughs of various sizes, installed along the contour, catch erosion material from clearly demarca-ted plots. Unbounded plots are occasionally used, but then the area of sediment production is defined by a topographic survey. This method is not very accurate.

One of the most widely used methods for obtaining quantitative measurements of erosion is the establishment of erosion plots. However, for observational erosion studies, their use is limited because they are more elaborate and costly to establish and monitor than the surface change methods and because they require intense measurements. It is often impossible to establish and monitor the number of plots that are required for a sample. Their use can best be restricted to experimental studies. Unfortunately, observation studies do use erosion plots for a few selected measurements, but data obtained in this way is more anecdotal than scientific.

EXPERIMENTAL STUDIES OF EROSION

Quantitative values for the effect of separate erosion factors (rainfall erosivity, soil erodibility, and vegetation effects) can be derived from experimental studies designed to measure the dependence of soil loss on each factor. Standardized erosion plots are very suitable for such research. Further details on the effect of trees or other agroforestry components on rainfall erosivity and soil erodibility may be collected in auxiliary tests.

Experimental Erosion Measurement Techniques

The best way to obtain information on the effect of specific components of agroforestry systems on erosion and to compare alternative designs is to establish experimental erosion plot studies in which comparative data on different treatments are collected.

Erosion plots

Erosion plots are well suited to experimental studies to determine the effects and interrelations between various erosion factors. They provide quantitative values for the effect of different erosion factors by measuring erosion rates of different land-use treatments under standardized field conditions. The layout and methods for data collection of erosion plots are discussed in all standard textbooks on erosion (Hudson 1981, Morgan 1986), to which readers are referred for full details (see also Mutchler et al. 1988).

Erosion plots basically consist of three parts:

a bounded plot with a sediment collection trough at its lowest end
a sediment measuring tank connected to the trough
a rainfall recorder

The size of erosion plots varies depending on the purpose of the study. Small plots of a few square meters are usually restricted to detailed studies on basic erosion processes—for example, special aspects of sheet erosion such as raindrop detachment, splash transport, and surface sealing with its effect on runoff. Such small plot experiments should complement experiments on larger plots where the complete process of rill and inter-rill erosion is represented. Medium plots of approximately 22 m × 10 m—the minimum width for obtaining representative data on rill erosion—have been widely used for standardized erosion measurements under various land-use treatments to develop comparative data on erosion factors that may be incorporated in predictive models (see below). A disadvantage of such medium-size plots is that not all erosion processes are represented (for example, gully erosion) and that certain erosion control measures (such as terraces with controlled waterways) cannot be adequately represented. Large plots of roughly two to four hectares (unit-source watersheds) provide the results of all erosion and conservation processes in a single measurement. Such large plots, however, cannot be easily replicated.

Depending on the plot size, large amounts of runoff and sediment may occur that cannot be accommodated in the collection troughs. Therefore, each trough usually is connected to one or more sediment measuring tanks. Where these tanks cannot catch all the runoff and sediment, some kind of divider system is installed to collect only a predetermined portion. Normally, the amounts of sediment collected are measured each day, allowing the calculation of daily rates of soil loss from the plots that can then be correlated with the daily rainfall conditions.

To monitor incoming rainfall, each erosion plot should be equipped with a rainfall gauge. Automatic rainfall recorders are most suitable, as the data of these recorders may be used to calculate rainfall erosivity values by using well-established empirical formulas (El-Swaify et al. 1982, Lal 1988). (Rainfall simulators are used in measuring erosion from small plots, but such equipment is not useful in agroforestry research because of the high spatial diversity in the agroforestry systems.)

Although the erosion plot research methods have been developed specifically for agricultural research, they are also suitable for agroforestry research. For instance, the medium-size erosion plots may be used to test the effect of various crop and tree combinations and configurations or different management practices such as mulching, selective soil tillage, or use of cover crops and contour ridges. In planning such studies, the following specific aspects of agroforestry research should be taken into account:

Spatial Configuration

In many agroforestry systems, trees are more widely spaced than agricultural crops. Plots large enough to include a representative part of such complex land-use systems must often be wider than plots in agricultural research. The sediment collection systems must be well adapted to plots of such large size. There are no norms for deciding how large plots should be for measuring erosion in different agorforestry systems. The size of the plots generally increases as tree spacings increase and as the spatial diversity of the agroforestry system increases. As a minimum, it seems desirable to include at least two to three rows of trees in each plot. In very complex agroforestry systems, it may be better to use small unit-source watersheds instead of erosion plots and to measure various details of the erosion process separately with auxiliary measurement in order to assess the role of specific elements (such as trees) of the system (see below).

Duration of Experiments

Trees may take many years to develop to their full size and to exert their full effect on erosion. Consequently, one-year measurements do not give accurate data on the soil conservation value of newly established experimental agroforestry designs.

Variability of Conditions

Where erosion plots are established in existing agroforestry systems, it may not be possible to standardize all plot conditions, for example, differences in microrelief and distribution of trees. Therefore, the same degree of accuracy cannot be obtained as in agricultural erosion research. Care should be taken in planning treatments that the expected effects of these treatments are sufficiently large not to be disguised by variability of erosion plot conditions.

Site Selection

Due to interdependencies between various erosion factors, data on the effect of agroforestry systems on recently opened or well-managed fertile soils may not be applicable to eroded and impoverished soils. Agroforestry systems that aim to control erosion on degraded lands should have trials on sites that adequately represent such conditions.

Auxiliary measurements

If complex agroforestry systems require large erosion plots or even unit-source watersheds for adequate data collection on rates of erosion, it is often impossible (or too costly) to compare various treatments in replicated trials. In such cases, it may be useful to carry out auxiliary measurements within the plot in order to assess the influence of specific agroforestry components on erosion. For instance, simple observational measurements (discussed above) may be used to obtain estimates about the rate of erosion in different parts of the agroforestry system. Auxiliary measurements may also provide detailed information about the influence of specific agroforestry components on various elements of the erosion process. While erosion plot experiments basically measure input (rainfall) and output (sediments), they do not provide information about the precise effects of soil conservation and erosion control measures on the various elements of the erosion process such as rainfall erosivity and soil erodibility.

Rainfall Erosivity

In order to be able to quantify the effects of rainfall interception by vegetation on rainfall erosivity, the drop-size distribution under different canopy types should be compared to drop-size distribution in the open. Also, interception losses due to evaporation and stemflow and the velocity of the waterdrops should be assessed. An enormous data set and much computational work to establish such detailed comparisons will be needed. This can be avoided by simplifying the process of data collection by concentrating only on measuring the actual erosive energy of rainfall and throughfall. Examples of various methods for erosivity measurements are given by Hudson (1981) and Lal (1988), the most widely used technique being the splashcup technique (Kinnell 1974).

Soil Erodibility

Many different tests have been developed to measure soil erodibility under different parts of a complex agroforestry system (Lal 1988, Morgan 1986, El-Swaify et al. 1982, Bryan 1976). Some of the tests are dynamic, others are static. Dynamic tests simulate the processes operating during erosion, for example, simulated rainfall tests that may be performed either in the field or on soil samples in the laboratory, or wet sieving tests that measure properties effecting dispersion and water transmission. Static tests measure certain constant soil properties that are empirically known to relate to erosion susceptibility. These tests are normally carried out in the laboratory.

These different tests measure different aspects of soil susceptibility to erosion, such as the aggregate stability of the soil or its infiltration characteristics. The aggregate stability of a soil determines the ease of detachment and dispersion of soil aggregates, while the infiltration rate determines the likelihood of overland flow. Tests have been developed that incorporate both properties, but there does not exist any standard test that is simple to measure, reliable in operation, and capable of universal application (Bryan 1976). A careful choice should therefore be made of the most appropriate test to be used, particularly as many static tests are based on empirical data from temperate, not tropical, regions. Generally, therefore, dynamic tests are to be preferred. Best results are obtained by carrying out several tests and correlating their results (Ambar and Wiersum 1980).

EROSION PREDICTION MODELS

Erosion prediction models have been developed to indicate which land-use management systems maintain erosion within permissible limits and to estimate how erosion control measures may be modified to obtain optimal results. The most commonly used equations have been developed and verified on the basis of large amounts of systematically collected soil loss data in the United States. These equations have yet to be fully tested for applicability in tropical regions in general and in agroforestry in particular. For agroforestry research, special attention should be given to verifying or modifying the equations for calculating the effect of crop cover and management factors. Nonetheless, provided that the equations are cautiously applied, they may be used to obtain indicative data for selecting appropriate agroforestry designs.

Possible Uses of Erosion Prediction Models

Many years of systematic research on erosion in agricultural systems, especially in the United States, have yielded much quantitative data on the effect of different erosion factors. Various erosion prediction models have been developed that indicate the relations between these parameters and ways in which to estimate them. Such erosion prediction models may be used to estimate erosion rates in situations where no locally specific field data is available. This is achieved by making a systematic inventory of site-specific conditions and by evaluating their effects on erosion through extrapolation of information from other areas. The models may also serve as tools for selecting suitable conservation measures for sites with specific conditions of rainfall, soil, and slope (Foster 1988). It must be remembered, however, that each erosion prediction model has been developed for a particular purpose and each model should be used only for that purpose (Wischmeier 1976). Although erosion prediction models may be used to extrapolate experimental data beyond the range and area of data used to derive the models, care should be taken that this is not done indiscriminately, especially in cases where little base-line data is available for local verification of the model.

Until the present, most erosion prediction models have been developed for monocropped agricultural systems, which differ markedly from the much more diverse multistoried agroforestry systems. Therefore, before using any erosion prediction model in agroforestry research, its suitability for use should be carefully assessed. Much work still remains to be done to verify and/or modify the existing erosion prediction equations for use in agroforestry on the basis of results of long-term experimental studies. If cautiously applied, however, these models may by used to estimate erosion rates under specific agroforestry parctices and to assess how erosion factors may be manipulated to decrease soil loss. Thus they may serve to make provisional estimates about the effects of alternative agroforestry designs and to judge the effectiveness of specific management practices in controlling erosion.

The most widely used erosion prediction model (Wischmeier and Smith 1978) is the universal soil loss equation (USLE), which was developed in the United States for predicting long-term annual soil losses from sheet and rill erosion on monocropped agricultural fields. This empirical model requires relatively little data and computational effort. It recognizes that the total annual soil loss (A) from a specific site depends on a rainfall erosivity factor (R), a soil susceptibility factor (K), slope length and steepness factors (L, S) and crop and land management factors (C, P), or

$$A = RKLSCP$$

All values relate to mean annual values, and the formula cannot be used for predicting erosion from individual storms. In many years of erosion research in the United States, specific parameter values have been obtained or empirical formulas have been developed to calculate expected erosion rates under various forms of land use. Recently, efforts have been undertaken to modify the USLE and to develop erosion prediction models that may be applied on a watershed basis or for individual rainstorms (Knisel 1980, Laflen et al. 1985, Hartley 1987). These approaches are of varying mathematical complexity because, for example, some replace the simple rainfall energy factor, R, with a more complex runoff factor. They are still undergoing development and testing and are not yet ready for standard operational use in applied erosion research.

Because of its successful application in the United States, the USLE has also been introduced to tropical regions (Foster et al. 1982). Several studies have verified the general applicability of the model outside the United States, and gradually increasing numbers of specific parameter values for tropical regions are becoming available (El-Swaify et al. 1982). Application of this model and details about the calculations involved are given in many standard textbooks on erosion (Hudson 1981, El-Swaify et al. 1982, Morgan 1986).

A simplified erosion prediction model for tropical regions, similar to the USLE, has been developed in and for southern Africa (Elwell and Stocking 1982). This soil loss estimation method (SLEMSA) has simplified data requirements, but it has as yet been little used and tested in areas other than where it was developed.

Assessing Vegetative Cover and Management Effects

The effect of vegetative cover and management practices which are of special relevance to assessing erosion factors in agroforestry, can be calculated using techniques developed for estimating the USLE cover and management factor in undisturbed areas, including forests. Wischmeier (1975) listed three subfactors that determine the composite values for the cover management factor:

canopy height and cover
ground surface cover
below-surface effects

Dissmeyer and Foster (1980, 1981) added additional subfactors, and they adjusted and further developed the subfactor relations for forests. This method takes into account the following subfactors with respect to vegetative cover and forest management:

amount of bare soil
canopy height and density
soil reconsolidation factor
high organic matter content of forest soils
presence of fine root mat
residual binding effect on newly reclaimed forest lands
on-site sediment storage in local depressions or behind debris
irregular soil surface as a result of step formation caused by local sediment deposition
contour tillage during forest site preparation

Many of these subfactors also influence the erosion rate in agroforestry systems. Although there is little experience with this approach in agroforestry, it looks promising for applying existing knowledge about erosion rates to a wide range of agroforestry conditions and thus in choosing from among alternative agroforestry designs. This approach can also help in planning experimental studies by focusing data collection on the most relevant local subfactors determining erosion.

FUTURE RESEARCH

In future studies on the effects of erosion factors in agroforestry, priority should be given to evaluating the precise effects of vegetative cover. Erosion control in agroforestry is most effectively achieved by a cover of surface litter, consisting of crop residues, tree prunings or natural leaf fall, and/or a good soil cover of grasses, herbs, or cover crops. Therefore, in testing agroforestry systems for erosion control, a primary concern should be judging how well surface ground cover is established and maintained. In addition, it may be worthwhile to test the need to include structural erosion control measures, and to obtain data about the precise role of trees with respect to erosion in the agroforestry system.

LITERATURE CITED

Ambar, S., and K. F. Wiersum. 1980. Comparison of different erodibility indices under various soil and land use conditions in West Java. *Indonesian Journal of Geography* 10(39): 1–15.
Armstrong, C. L., and J. K. Mitchell. 1987. Transformation of rainfall by plant canopy. *Transactions of the American Society of Agricultural Engineers* 30(3): 688–696.
Brandt, J. 1987. The effect of different types of forest management on the transformation of rainfall energy by the canopy in relation to erosion. In: *Forest hydrology and watershed management*. International Association of Hydrological Science Publication 167: 213–222.
Bryan, R. B. 1976. Considerations on soil erodibility indices and sheet erosion. *Catena* 3: 99–111.
Clark, R. d. 1980. *Erosion condition classification system*. Bureau of Land Management Technical Note 346. Denver, Colorado: US Department of the Interior. 47 p.
Dissmeyer, G. E., and G. R. Foster. 1980. *A guide for predicting sheet and rill erosion on forest land*. UASD report SA-TP-11. Atlanta, Georgia: Forest Service Southeastern Area. 40 p.
Dissmeyer, G. E., and G. R. Foster. 1981. Estimating the cover management factor (C) in the universal soil loss equation for forest conditions. *Journal of Soil and Water Conservation* 36(4): 235–240.
Dunne, T. 1977. Evaluation of erosion conditions and trends. In: *Guidelines for watershed management* 1. FAO Conservation Guide. Rome: FAO.
El-Swaify, S. A., E. W. Dangler, and C. L. Armstrong. 1982. *Soil Erosion by water in the tropics*. Honolulu. College of Tropical Agriculture and Human Resources, University of Hawaii. 173 p.

Elwell, H. A., and M. A. Stocking. 1976. Vegetal cover to estimate soil erosion hazard in Rhodesia. *Geoderma* 15: 61–70.
Elwell, H. A., and M. A. Stocking. 1982. Developing a simple yet practical method of soil loss estimation. *Tropical Agriculture* (Trinidad) 59(1): 43–48.
Foster, G. R. 1988. Modeling soil erosion and sediment yield. In: *Soil erosion research methods* (R. Lal, ed.). Ankeny, Iowa: Soil and Water Conservation Society. pp 97–117.
Foster, G. R., W. C. Moldenhauer, and W. H. Mischmeier. 1982. Transferability of US technology for prediction and control of erosion in the tropics. In: *Soil erosion and conservation in the tropics*. Madison, Wisconsin: American Society of Agonomy and the Soil Conservation Society of America. pp 135–149.
Hadley, R. F., and D. E. Walling. 1984. *Erosion and sediment yield, some methods of measurement and modelling*. Norwich, UK: Geobook.
Hartley, D. M. 1987. Simplified process model for water and sediment yield from single storms. *Transactions of the American Society of Agricultural Engineers* 30(3): 710–723.
Hayward, J. A. 1968. *The measurement of soil loss from fractional acre plots*. Lincoln Papers in Water Resources 5. Canterbury, NZ: New Zealand Agricultural Engineering. 47 p.
Hayward, J. A. 1969. *The use of fractional acre plots to predict soil loss from a mountain catchment*. Lilncoln Papers in Water Resources 7. Canterbury, NZ: New Zealand Agricultural Engineering. 92 p.
Hudson, N. 1981. *Soil conservation*. 2nd ed. London: Batsford. 323 p.
Kinnell, P. I. A. 1974. Splash erosion: some observations on the splashcup technique. *Soil Science Society of America Proceedings* 37: 617–621.
Knisel, W. G. (ed.). 1980. *CREAMS, a field-scale model for chemicals, runoff and erosion from agricultural managment systems*. Conservation Research Report 26. Washington, DC: USDA.
Laflen, J. M., G. R. Foster, and C. A. Onstad. 1985. Simulation of individua-storm loss for modeling the impact of soil erosion on crop productivity. In: *Soil erosion and conservation* (S. A. El-Swaify, W. C. Moldenhauer, and A. Lo, eds.). Ankeny, Iowa: Soil Conservation Society of America. pp 295–296.
Lal, R. 1988. Erodibility and erosivity. In: *Soil erosion research methods* (R. Lal, ed.). Ankeny, Iowa: Soil and Water Conservation Society. pp 141–160.
Morgan. R. P. C. 1986. Soil erosion and conservation. Harlow, UK: Longman. 298 p.
Mutchler, C. K., C. E. Murphree, and K. C. McGregor. 1988. Laboratory and field plots for soil erosion studies. In: *Soil erosion research methods* (R. Lal, ed.). Ankeny, Iowa: Soil and Water Conservation Society. pp 9–36.
RAPA. 1986. *Land use, watersheds and planning in the Asia-Pacific region*. RAPA Report 1986/3. Bangkok: The East-West Center and FAO Regional Office for Asia and the Pacific, 230 p.
Skidmore, E. L. 1988. Wind erosion. In: *Soil erosion research methods* (R. Lal, ed.). Ankeny Iowa: Soil and Water Conservation Society. pp 203–233.
Stocking, M. A. 1988. Assessing vegetative cover and management effects. In: *Soil erosion research methods* (R. Lal, ed.). Ankeny, Iowa: Soil and Water Conservation Society. pp 163–185.
Walling, D. E. 1988. Measuring sediment yield from river basins. In: *Soil erosion research methods* (R. Lal, ed.). Ankeny, Iowa: Soil and Water Conservation Society. pp 39–73.
Wiersum, K. F. 1984. Surface erosion under various tropical agroforestry systems. In: *Symposium on effects of forest land use on erosion and slope stability* (C. L. O'Loughlin and A. J. Pearce, eds.). Honolulu, Hawaii: Environment and Policy Institute, East-West Center. pp 231–239.
Wiersum, K. F. 1985. Effects of various vegetation layers in an *Acacia auriculiformis* forest plantation on surface erosion in Java, Indonesia. In: *Soil erosion and conservation* (S. A. El-Swaify, W. C. Moldenhauer, and A. Lo, eds.). Ankeny, Iowa: Soil Conservation Society of America. pp 79–88.
Wischmeier, W. H. 1975. Estimating the soil loss equation's cover and management factor for undisturbed areas. In: *Present and prospective technology for predicting sediment yields and sources*. Washington, DC: Agricultural Research Service, USAD. pp 118–124.
Wischmeier, W. H. 1976. Use and misue of the universal soil loss equation. *Journal of Soil and Water Conservation* 31(1): 5–9.
Wischmeier, W. H., and D. D. Smith. 1978. *Predicting rainfall erosion losses. A guide to conservation planning*. Agricultural Handbook 537. Washington, DC: USDA. 58 p.
Young, A. 1986. *The potential of agroforestry for soil conservation. I. Erosion control*. ICRAF Working Paper 42. Nairobi. ICRAF, 68 p.
Young, A. 1989. Agroforestry in the control of soil erosion by water. *Agroforestry Abstracts* 1(2/3): 39–48.

PART FIVE

Integrating Livestock

CHAPTER 13

Livestock in Agroforestry: A Farming Systems Approach

L. Reynolds

In mixed smallholder farming systems, livestock generally play a subsidiary role to crops because the farmer's first objective is to ensure adequate food for the family over the coming year. Where animals are present, they may be integrated into the farming system as draught animals; they may consume crop residues to produce milk or meat; or they may be only scavengers around the village. Any intervention to improve livestock production, therefore, must consider both how animals integrate into the farming system as a whole and how suggested improvements would affect the rest of the farming activities.

The farming systems approach to agroforestry research and development described in this chapter is such an integrated approach to intervention. The objective is to provide answers to identified constraints as economically and effectively as possible through on-station and on-farm studies that evaluate technical, biological and socioeconomic data. A farming system is not simply a collection of crops and animals to which one can apply this or that input and expect immediate results. Rather, it is a complicated, interwoven mesh of soils, plants, animals, implements, workers, other inputs, and environmental influences, the strands of which are held and manipulated by the farmer who attempts to produce output from the inputs and technologies available (CGIAR 1973). Farming systems research (FSR) focuses on these interdependencies in order to generate and test improved agricultural technologies and deliver them to farmers in useful and accepatable forms.

Put more simply, FSR is used to find a solution for a given problem, rather than to search for a problem for which it has a predetermined answer.

Certain weaknesses, however, have surfaced in FSR programs in the past, and the feeling among some donors is that this method for increasing farm productivity does not deliver what it promises. This chapter seeks to show that, on the contrary, FSR can indeed be a useful, productive tool for determining the needs of farmers and testing the sustainability of new technologies.

The particular focus of this chapter is on livestock research as a component of agroforestry systems. It examines the mainstages of an FSR program and briefly discusses two types of agroforestry intervention. Constraints and weaknesses faced by the FSR system are enumerated, and future research topics are suggested.

UNDERSTANDING THE FARMING SYSTEM

Preliminary Surveys

Any agricultural study needs background information on climate, topography, and soils and on the socioeconomic environment within which production takes place. Climatic information on data should include rainfall levels and distribution, temperature ranges, and possibly wind speeds and direction, relative humidity, evapotranspiration rates, and solar radiation, all of which may be available at regional levels from secondary sources.

Evaluation of the major land types in an area should include a description of the major land forms (for example, hillside or valley bottom), an estimation of the proportion of the area covered by that land form, the major soil types, and an indication of any significant features that may affect its agricultural potential. Such information will benefit a wider audience than a purely regional or national group if an internationally accepted system such as FAO's land classification system (1976) or USDA's soil taxonomy (1976) is followed. Again, such data should be available from secondary sources, but some field work may also be needed. Information about the natural vegetation on the major land forms is useful because plants growing on uncultivated land can be an important forage source for small-scale farmers.

Estimates are needed of cultivated, uncultivated, and fallow land, of crops grown and their rotation, and of the types and numbers of livestock species. An outline of the management systems used is necessary.

Socioeconomic evaluations require data on (1) current population densities and the rate of change, family size, and labor availability and cost; (2) infrastructure for transport, communication, education, and health services; (3) level of education; (4) availability and prices of inputs for crops and livestock; (5) demand for agricultural products and the availability of adequate marketing facilities; and (6) tribal, religious, or cultural divisions where they have a significant effect. Farmer's goals and motivations must be understood, since any proposed change will be accepted only if it is consistent with those aspirations.

Government policy, as it affects input supply and marketing, should also be noted. Where production of a particular agricultural commodity for export is encouraged by government, subsidized inputs such as fertilizer, pesticides, or veterinary medication may be available to farmers. Market prices also reflect policy, especially where official commodity boards have monopoly purchasing rights.

The key to conducting the above analysis (and subsequent research) is a multidisciplinary team working in an interdisciplinary framework. At the minimum, the team should contain an animal scientist, an agronomist, and a socioeconomist. It could take years to collect information on all of these various aspects of the farming system, but this approach should be vigorously rejected. The base-line study is not an end in itself, but an important tool for deciding priorities for the next stages of on-station and on-farm testing. The best method for collection of base-line data is the rapid survey. The results of such a survey will not be completely accurate since it must rely often on farmers' recall, but it will be sufficiently valid to allow the selection of a representative site, or sites, where in-depth productivity and socioeconomic data for crops and/or livestock can be collected.

In-depth Research

On the basis of rapid survey data, it is possible to decide on the type of intervention that might be appropriate. The next step is to make a literature search to ensure that time and resources are not spent reinventing the wheel. The area for research should be selected to ensure that results obtained are applicable to other areas, have potential for improved production, be accessible, and have adequate infrastructure to allow research to proceed without unnecessary difficulties.

Zandstra and colleagues (1981) and Mutsaers and colleagues (1986) provide details of data collection for cropping aspects of the farming system. Guidelines for livestock systems are still being formulated, but workshop proceedings edited by Nordblom and others (1985) and Kearl (1986) are helpful.

Once work on-station has begun, concurrent on-farm work can also begin. A representative site is needed for the collection of in-depth information on ownership, management, productivity, and uses of animals and animal products. The time required for this study depends in part on gathering reproductive statistics, which in turn depends on the species of animal involved: cattle may give birth every two years, while sheep or goats give birth at intervals of eight or nine months. Productivity, measured as the weight of offspring weaned/dam/year, depends on parturition intervals, litter size, survival to weaning, and weaning weight. The incidence of diseases must be recorded and their economic impact estimated. The source and availability of feed, whether from grazing or browsing, natural pasture, fallow land, crop residues, cut-and-carry feed, purchased feed, or household waste, should be determined periodically throughout the year. Labor requirements are important because there may be shortages of labor at times of peak demand for cropping activities. Cultural restrictions also may limit the available work force for particular jobs.

Livestock data

When livestock data is collected over a period of time, it is essential to follow individual animals; identification with eartags is therefore necessary. Initial data on the age (determined by dentition), numbers, sex, and species of animals present should be recorded. For an agroforestry project, ruminant livestock are the only logical choice. If the choice of species is open, the relative importance of each species, any major constraints to increasing their numbers, and a realistic appraisal of their potential for improvement should be considered. Monthly visits to the farm are adequate to record births, deaths, sales, and other reasons for entry and exit from the herd. Productivity data should be kept simple. It is not difficult to record monthly weights for small ruminants with a hanging balance and a sling, and cattle weights can be estimated with a weighing tape that converts heart girth measurements to liveweight. Condition score is another useful, simple measurement that can be taken in the field (Nicholson and Butterworth 1986).

During the monthly weighing visits, data can also be collected on health from visual signs and from blood and fecal samples, which can be screened in a simple laboratory for parasites. The objective is to form a picture of disease patterns, and then to concentrate on the two or three that appear to have the most serious effect on productivity. For smallholder farmers, there is little chance that sophisticated veterinary care will be available, and the research group should aim for low-input preventive treatment wherever possible.

Market data

The price and numbers of animals on offer and sold is valuable information. It may be possible to aim production to meet a peak of demand, such as a festival, when prices will be high. There may be a market preference for a certain type of animal (for example, rams for Moslem festivals) that can guide production towards a definite goal. If secondary data is not available, weekly visits to the market are needed. For sites close to an urban center, there may be both rural and urban markets to consider. Supply, demand, and marketing channels for milk and milk products must be determined where relevant to the objectives of the investigation. The economic viability of milk production projects at the small farm level will depend on either local demand or adequate transportation to an urban market. Downstream economics may well be more important than the biological probability of successfully producing milk.

Government policy

The time between conception of a research idea and the impact of a successful extension project being noticed by the farming community will probably approach 8 to 10 years. If the government has a long-term plan for the region, it is important that the objectives of the project are compatible with the likely outcome of the government's plan. For example, will people be moving in or out under transmigration programs? Are feed sources, such as oilseed cakes, exported for foreign exchange or are they available on the home market? Are prices controlled, or are they responsive to external pressures?

Data analysis

Having collected all this information, how does the research team interpret it, decide what are the major constraints to livestock production, and more importantly, determine what line of research to follow? Will agroforestry offer an appropriate answer to some or all of the problems? An understanding is necessary of the advantages and problems of different forms of agroforestry interventions and of the probable effects , in broad terms, of a single component on the rest of the farm activities. ICRAF has developed a design and diagnosis procedure specifically for agroforestry based on rapid survey techniques that can help to guide the investigation (Raintree 1986). The rapid rural appraisal method (Khon Kaen University 1987) takes a similar approach. These methods are not specifically designed for livestock problems but are illustrative of relevant methodologies. The end product of the diagnosis is a set of functional specifications that indicate what the system needs and, in a general way, how these needs can best be satisfied. This is accomplished by narrowing down the range of technical choices to those that are hypothetically capable of meeting the specifications. After a specific technology has been selected, the details of tree species, spacing, and management practices for a given set of circumstances on the ground must still be worked out. Adaptive research follows.

ON-STATION TESTING

On-station testing is an essential stage in the development of FSR technology, enabling the technical and biological aspects of an intervention to be studied under controlled conditions. Any proposed system must be demonstrably successful in the hands of researchers to stand a chance of working in a less controlled on-farm environment. For an agroforestry project, the selection of suitable tree species is the first step, but the researcher must know how the trees will be used in the system so that criteria for selection can be chosen. If the only objective is to produce livestock feed, selection will be based on tree productivity, nutrient digestibility, and palatability; but if the trees are to be integrated with crops, their growth and rooting habits, compatibility with companion food crops, and effects on crop yields under different management systems must be investigated. Some of the information is applicable to only one of the two examples of agroforestry systems that follow in this chapter, alley cropping and plantation systems, but many points are relevant to both. Non-livestock systems are only mentioned briefly here; the reader should refer to other chapters for more details on these systems.

Species Selection

Both exotic and indigenous species should be considered for selection. Preliminary selection of likely exotic species can generally be made from literature sources, but less information is available for indigenous species. A survey at the village level will show what browse species are being used by farmers, but methods of establishment will require attention before the indigenous species can be tested alongside the exotics. Ease of establishment is an important characteristic for farmers, and some species can be disregarded at this stage. Short rows of a few trees are enough to allow selection for growth habit (bushy or upright) and productivity. Productivity can be measured on a per hectare or per tree basis—the method by which the trees are to be incorporated into the farming system determines, which should be used.

Feeding Value

The feeding value of the tree foliage can be screened at an early stage while productivity data is being collected. Simple chemical analysis of young and old leaves for dry matter, crude protein, and fiber can be augmented by a measure of polyphenolics if facilities are available. Levels of tannins and other polyphenolics, which form strong complexes with proteins and inhibit protein digestion in ruminants, are negatively correlated with digestibility. ILCA (1986) found that *Sesbania sesban* had more effect on growth in sheep than *Acacia cyanophylla* and *A. seyal* because, although it had a similar proximate analysis, it had lower levels of polyphenolics. However, some protection of dietary protein from degradation by rumen microorganisms can be advantageous. The inclusion of small quantities of *Leucaena leucocephala* in the diet of dairy cattle improved milk yields even when the basal grass diet contained 18% crude protein (Flores et al. 1979). In growth trials with goats in Indonesia, van Eys and colleagues (1986) found that supplementation with *Leucaena* or *Gliricidia sepium* improved the animals' growth rate and the efficiency of utilization of dietary protein due to the larger proportion of protein "protected" by polyphenols in the tree legume forage, which is hence not degraded by microorganisms in the rumen. Further information can be obtained from relatively small quantities of forage through in vitro digestibility and *in vivo* degradability measurements. Palatability can also be judged using fresh and conserved forage offered in a cafeteria system.

Pasture grasses and herbaceous legumes also require screening and testing. An advantage with grasses and herbaceous legumes is their rapid growth—useful measurements can be taken in the first year. Trials should look at the productivity (dry matter and crude protein) of sole and mixed plots. The possibility of including tree legumes in the mixture should not be ignored. In cut-and-carry systems, large amounts of nutrients are removed from the plot, and a comparison of fertilizer regimes may be necessary to maintain productivity after the first year.

Feeding Trials

Full feeding trials require large amounts of forage and should not be attempted until the original list of possible species or combinations has been reduced to three or four at the most. Feeding trials with ruminants use sheep or goats as the experimental species for the sake of economy, and the results can be extrapolated to larger domestic animals. The number of animals available determine the type of trial. Six animals allow useful statistical analysis of digestibility data. Generally, adult males of roughly the same age and condition are used, since sex, and physiological state affect the results. For productivity trials, 10 animals per treatment are adequate if individual feeding, which reduces variability, is practiced. Again, a homogenous group of animals is required. If, however, the animals are to be fed as a group, 15 animals per treatment may be needed. The number depends on the expected difference between treatments and the variability in the parameter under study (Cochran and Cox 1957). If, for example, we expect the true difference as a result of the treatments not to exceed 15% of the mean, and the coefficient of variation in on-station trials for that parameter is 10%, we will need nine replicates to have a 90% chance of obtaining a significant result.

It is likely that some animals will not complete the trial period, and a large number will be required at the start. In breeding trials, the conception rate may be around 90%, and the mortality rate 10% to 25% between birth and weaning. In growth trials, an allowance for a 10% mortality rate should also be made. If survival rates are higher than expected, the additional animals will not affect the analysis, but if too few complete a trial, the chance of identifying significant differences is reduced.

Methods of Feeding

The next stage depends on whether the forage is to be used as a cut-and-carry feed or grazed and browsed directly by the animals. In the former system, productivity can be measured under different cutting regimes that are determined by the period of major nutritional constraint. For example, cutting can be aimed at providing fresh forage year-round or at providing dry or early wet season feed. Production will be higher in the wet season, so conserving surplus material for dry season use may be studied, bearing in mind that the techniques must be simple and cheap if they are to be used by smallholder farmers.

Grazing in an alley system is generally useful only on medium- or large-scale farms, but can be an integral part of a plantation system. Forage production studies are often slow to include animals, but the whole purpose of growing forage is for animal consumption. No matter how productive a forage system, designed for grazing, is under handcutting, it will have no practical application in the field if it cannot withstand grazing (and overgrazing).

In order to limit many of the possible directions for a grazing study, the ex ante analysis should lay down specific guidelines on how the pasture is to be used. Grazing experiments are costly and not easily changed once they start, and grazing results from one species may not be applicable to another. Fertilization of pasture may not be a valid option for smallholders; the inclusion of legumes may be more appropriate. Similarly, because grazing control may be limited by the cost of fencing, continuous grazing may be the only option. Continuous grazing at different stocking rates is therefore the primary system to be explored in on-station trials. Conventional experimental designs as described by Cochran and Cox (1957) are adequate. A minimum herd size of three animals is desirable, and if stocking rates of, say, 0.5, 0.75 and 1.0 animals per hectare are to be compared for two pastures, a total of 27 ha are needed for each replication. If space is limited, it may be better to compare six treatments without replication, rather than two treatments with three replications. Special care is then needed in laying out treatments to minimize the physical effects of the site.

Measuring Pasture Production

The basic pasture measurements to be taken are dry matter yield and botanical composition. Random samples are taken with a 1 m^2 quadrat for dry matter estimation, cutting herbage above ground level. Sampling at the beginning and end of the growing season provides descriptive data, but it is not sufficient to estimate growth and utilization of forage. Cages to protect a small area are essential for estimates of growth under continuous grazing. Botanical composition is best described in terms of the proportions of species on a weight basis, and this can be performed on the herbage sample collected for dry matter yield determination. Hand sorting should be carried out before the material dries up and becomes friable. Where there are only three or four components, visual estimation may be possible. Changes in the botanical composition over time give an indication of the persistence under grazing of the species being used. Shaw and Bryan (1976) give details for pasture research.

Animal performance can be measured in terms of live-weight gain, milk production, and reproductive performance. Comparisons are easier if a controlled breeding period and homogenous groups of animals are used.

ON-FARM TESTING

It is not necessary, or even desirable, to wait until all the on-station trials have been completed before moving to on-farm studies. The on-station trials grow out of the identification of constraints in on-farm surveys and the selection of possible technological innovations to overcome those problems. It is thus appropriate that on-station work in progress receives information back from the in-depth farm studies so that the original concepts may be modified if necessary. Whenever possible, researchers should be actively involved with both on-station and on-farm work. This reduces the chance of systems being developed on-station that turn out to be totally irrelevant to the needs and resources of farmers. The involvement of extension agents in on-farm trials also ensures that researchers are aware of the views of a cross-section of farmers.

Types of Trials

On-farm research is usually classified on the basis of the degree of farmer and researcher involvement. In cropping research, three types of trials are used:

researcher managed, researcher executed
researcher managed, farmer executed
farmer managed, farmer executed

The first type is the same as an on-station trial, but held at a different location. The second type incorporates the ability of the farmer to follow directions given by the researcher, while the third type permits the farmer more freedom to adapt the techniques to suit his or her circumstances. The first two types of trial provide technical and biological data, but the third type also indicates whether the intervention can fit into the farming system without major disruption, and whether it is acceptable to the community. As the control of a researcher over the trial decreases, the variability in the results will increase, and hence, the number of replicates needed to show significant differences between treatments will rise shaarply. What would otherwise be a very lengthy process can be shortened by running trials concurrently.

Methodology

The methodology for on-farm trials with livestock is less well developed than for crops, but many of the same principles apply. First, representative sites may be chosen on the basis of results of the rapid survey. Second, farmers should be selected from the different sociological (male, married females, widows, tenants, landowners, rich, poor) and environmental groupings (soil, land form, vegetative cover) in the area. Choices must be made on which of these groups to include, since collecting data from on-farm trials is very expensive due to the time and labor required. Third, it is advantageous if the first trials established in a village can be used as demonstration plots to introduce the technology to farmers for the final stage. Fourth, research involving trees is lengthy, a fact that both researchers and funding agencies must bear in mind before passing judgment on the project.

Farmer selection and community participation

Individual farmers can be approached to provide land for the first on-farm tirals, but ILCA's experience has shown that in the long term a community approach is perferable. Extension methods are borrowed to interest the community in the trial. An initial approach is made to the village chief, elders, and leaders to explain the purpose of the trial, to extract their opinions, and to elicit their support. Next, a village meeting is held where the community members are given the chance to air their problems and express their wishes for improvements and the researchers to explain why and how they might be able to help. Following the meeting, the names of interested farmers can be collected and visits organized to select those who would be suitable for inclusion. It may be advisable for good public relations to include some individuals who would not necessarily be chosen on purely technical grounds. The presence of the research team in the village is dependent on the good will of the community, and on-farm research should be the result of collaboration between farmers and researchers.

It is important to spell out clearly at the beginning what assistance in cash or kind, if any, will be given to collaborators in order to avoid misunderstanding and disappointment later. It may be difficult to obtain new recruits later if only the early collaborators receive assistance; latecomers will expect the same treatment as the first group. One solution to this problem is to undertake the later stages of on-farm research in separate villages. This also permits discarding any techniques found to be inappropriate at an early stage without prejudicing the methodologies used with new farmers.

Trial design

Given the complexities of on-farm work, it is essential to keep trial design as simple as possible. Measuring a few paramenters well is a better strategy than looking at many badly. Farm size may be small, particularly for alley-cropping trials, or farmers may be willing to allow researchers to use only a portion of the farm, so a single farm may have to be taken as one replicate. The first trials look at the agronomic aspects of the intervention because, until the trees have been established and are producing forage, the livestock component will be marginal. The research team should be sufficiently confident in the technology to move to farmer-managed and farmer-executed trials at an early stage on-farm, with assistance from the local extension service. (If such confidence is lacking, the system should not be taken to the farmers.) This is the only type of trial that can show the socioeconomic suitability of the intervention and how, for example, the demands for labor fit into the existing farming system.

Variability of data

For the collection of livestock data, the variability between farms will probably be greater than between treatments, and on a single farm, herd size or management practices may make it difficult to ensure separation of animals in treatment groups. Also, if a farmer sees one group doing better than another, he may want to spread the benefits without the researcher being aware of it. Or the researcher may unconsciously pass his expectations to the farmer who will then give more attention to a particular group of animals, so that a management effect is recorded rather than a true treatment effect.

ILCA test results from various ecological zones have shown that the coefficient of variation for growth traits is around 30% and for reproduction parameters, 35% (Sumberg and Mack 1985, Wilson and Durkin 1983). A sample size of at least 70 animals per treatment is needed for productivity data. Allowing for animals that will not complete the trial for reasons unconnected with the treatments (sold to raise cash, slaughtered for festivals) and for expected mortalities, the starting group should be around 120 animals.

Farm visits

The number of farmers involved will obviously depend on average herd size, but it could mean that 40 farms per treatment have to be regularly visited. Each farm must be visited at least once a month so that entries and exits and live weights can be recorded. Entries (births, purchases, loans in) and exist (deaths, sales, slaughters for meat and ceremonies, loans out) provide information on productivity, the reasons for keeping animals, and the cash income from livestock when combined with market price data. It is unlikely that a true answer will be given to a direct question on income, so circuitous means of verification are needed. Records of dam tag number and birth dates allow conception rates, parturition interval, and any seasonal effect on reproduction to be calculated. This last point may be important, since low conception rates in a particular season may be the result of poor nutrition, and this could be rectified by strategic supplementation from browse.

Measurement of intake

Measurements of food intake of free-roaming village animals is very difficult, and it may only be feasible to record quantities of supplementary feed provided by the farmer. Supplements may consist of browse and grass cut from fallow land, browse from planted trees, crop residues or household wastes. Farmers should be requested to hold back supplements on the days of regularly scheduled visits until after the visit. If visits can be arranged on consecutive days, any food remaining from the first day can be recorded to calculate both amounts offered and consumed. Initially, farmers may make extra efforts to provide feed on those days to meet their perceptions of the desires of the researcher, but this soon wears off and the true feeding patterns can be measured. Samples of the feeds can be collected for dry matter determination, but it may be difficult in practice to obtain sufficiently accurate measurements of fresh weights at the time of collection for meaningful dry matter calculations.

Disease

Disease is usually a major constraint on livestock production, and until that is overcome, the beneficial effects of improved nutrition may be limited. However, improved feeding will increase resistance to disease. Records of animal health can be collected at the monthly weighings, and blood and fecal samples collected for the detection of internal parasites. Post-mortems are valuable to determine cause of mortalities, but it is rare for carcasses to be made available to the researchers for post-mortem examinations.

Replication

In plantation systems, persistence and productivity of the herbage under farmer management is critical. Experimental designs are limited by land availability, and a single farm should be taken as one replicate. Interactions between the pasture, animals and the tree crop will require attention to technical, biological, and economic data, but the sociological aspects must also be considered.

Extension

Although not strictly a research topic, the involvement of extension staff during on-farm research activities helps ensure the success of the next stage when the developed package is offered to a broader cross-section of farmers. Often the weakest link in the development chain is that between research and extension. Early involvement of extension services may help to strengthen that link. The research team will be required to provide training for extension staff at first, but as those originally trained gain experience, they should be able to run courses themselves with only limited involvement of the researchers.

TWO AGROFORESTRY INTERVENTIONS FOR LIVESTOCK PRODUCERS

Two contrasting agroforestry schemes are considered in this section, alley farming and plantation systems. Alley farming incorporates trees into traditional farming practices that have generally considered trees as a hindrance to crop production. Research on alley farming is aimed at developing technologies to maintain or improve soil fertility and crop yields, simulataneously producing forage in a way that attracts farmers to tree planting. In plantation systems, farmers already concentrate on the tree crops, and the objective of research and development for livestock is to make better use of space under the trees without negatively affecting tree productivity.

Alley Farming

Continuous cropping on most tropical soils results in declining soil fertility and crop yields. One traditional response has been a bush fallow system to restore fertility, so that for every 2 ha under cultivation, 6 to 10 ha are left fallow. During the fallow period, natural regrowth of shrubs and grasses and subsequent leaf drop provides nutrients for the soil. Animals may feed directly on the plot, or farmers may cut feed to carry back to confined animals. In many areas, however, increasing human population density has caused a shortage of farming land, and fallow periods have been reduced.

Alley farming, also called hedgerow intercropping, integrates the benefits of the fallow period directly into the cropping period. Food crops are planted in the alleys between rows of trees, usually leguminous. The trees are pruned at regular intervals, at about 0.6 m above ground level, and the prunings used for mulch or for animal feed. The trees are not allowed to grow to full height but are maintained as shrubs to prevent shading of companion food crops. The foliage of leguminous multipurpose trees, which can contain over 20% crude protein, provides nitrogen-rich mulch to enhance soil fertility and an on-farm source of high-quality supplementary feed for ruminant livestock. On sloping ground, when planted along the contour, the tree rows act as barriers to prevent soil erosion and can encourage the formation of terracing. Alley farming is a flexible technology that benefits crop and livestock activities (Reynolds and Atta-Krah 1989, Kang et al. 1989, Reynolds and Adediran 1988) and can, through a modification of tree management techniques, provide fuelwood for the household. It is a system that can be adapted to meet the particular priorities of individual farmers.

Two tree species, *Leucaena leucocephala* and *Gliricidia sepium*, have been most widely used for alley farming. Only *Leucaena* is planted for alley farming in Indonesia, but in West Africa alternate rows of *Leucaena* and *Gliricidia* are more common. The use of two or more species is recommended to reduce the possibility of a pest or disease completely destroying production as the psyllid pest (*Heteropsylla cubana*) did to *Leucaena* in much of Asia and the Pacific.

Alley farming at present is recommended for areas with annual rainfall over 1200 mm with a bimodal distribution and a soil pH of over 5.2, where farms are small (around 2 ha) and cultivated by hand or with limited mechanization. Low-input agriculture should be the norm, with maize or cassava as the major food crops. It has proved suitable for both male and female farmers, tenants and landowners. Small ruminant livestock, both free-roaming and confined, are widely owned in areas where alley farming has proved acceptable. These recommended areas for alley farming reflect the conditions in the locations where it has received most research attention, but as testing is extended to other areas and to other tree species, it is likely that alley farming will prove suitable to a wider farming community.

Mulching with foliage from the leguminous trees will improve nutrient levels in the soil (table 13.1) and increase crop yields by up to roughly 40%

Table 13.1. Effects of six years of alley farming maize and cowpea with *Leucaena* and applying nitrogen on chemical properties of surface soil (0-15 cm) of a Psammentic ustorthent.

Treatment (kg N/ha)	*Leucaena* prunings	pH–H_2O	Org. C (%)	Exchangeable cations K	Ca (me/100 g)	Mg
0	removed	6.0	0.65	0.19	2.90	0.35
0	retained	6.0	1.07	0.28	3.45	0.50
80	retained	5.8	1.19	0.26	2.80	0.45
LSD (.05)		0.2	0.14	0.05	0.55	0.11

Adapted from: Kang et al. 1985.

(figure 13.1), over and above any benefit that might be obtained from a basal application of inorganic fertilizer. Incorporation of mulch into the soil gives double the response of a surface application. The inclusion of a fallow period in the system also improves crop production and increases tree productivity after a break from continuous pruning (table 13.2). During the fallow period, the plots can generate fodder and fuelwood.

Supplementary feeding of sheep and goats with *Leucaena* and *Gliricidia* improves growth and survival rates in young stock (table 13.3), increases food intake (figure 13.2), and raises overall productivity of the dams as measured by weight of offspring weaned/dam/year (figure 13.3).

Removal of prunings for animal feed obviously means that less material is available for mulch, and this will diminish the effect on crop output. It is necessary to quantify the effects of this tradeoff in order to determine the economics of the system. In the humid zone, alley farming is more profitable than the conventional crop-fallow cycle, even though more labor is required to

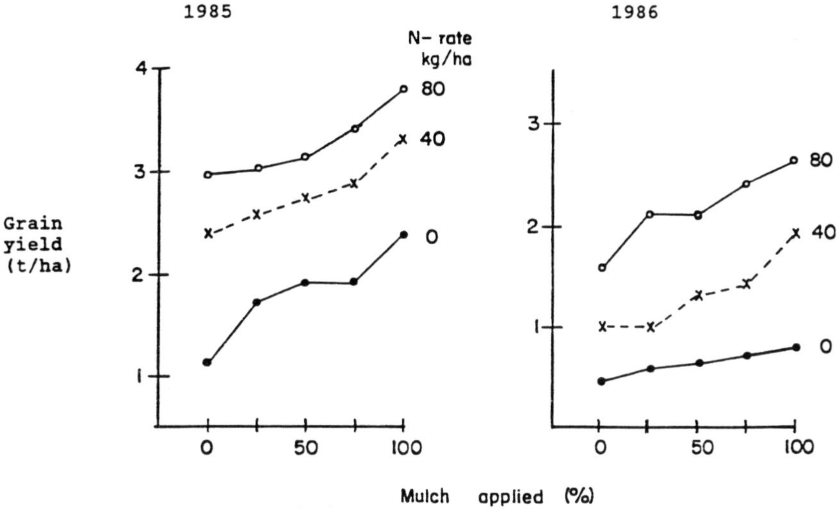

Figure 13.1. Maize grain yield from a *Gliricidia* alley farm, as affected by removal of prunings and inorganic nitrogen application, over two years.
Adapted from: Atta-Krah and Sumberg 1987.

Table 13.2 Dry grain yield (tonnes/ha) of maize and cowpea

Treatment	Maize			Cowpea
	First season	Second season	Total	Second season
Continuous alley cropping (Yr. 5)	2.46	1.47	3.93 (—)	0.24
Alley cropping after grazed fallow*	2.99	1.82	4.81 (22)	0.25
Alley cropping after grazed fallow**	4.40	1.84	6.14 (56)	0.18
LSD (5.0)	0.56	0.50	0.85	0.04
CV (%)	13.30	22.70	13.40	16.10

*3rd consecutive year of post-fallow cropping
**1st year of post-fallow cropping
() Figures in parentheses represent percentage yield increase over continuous alley cropping.
Source: Atta-Krah unpublished.

prune the trees regularly and spread the mulch. The relative benefits to crops and livestock will depend on current market prices. Sumberg and colleagues (1987) found that the inclusion of animals in the system was profitable if net output of livestock products rose 25% to 30% as a result of supplementation.

Table 13.3. Effects of supplementary *Leucaena* and *Gliricidia* browse on growth and survival rates of small ruminants in West Africa

Species	Browse intake (g DM/day)		Growth rate (g/day)		Survival to
	Dam[a]	Offspring[b]	Weaning[c]	24 weeks	24 weeks
Goats	143	39	17.4	14.0	0.36
	254	83	28.7	20.1	0.46
	554	160	25.9	20.9	0.82
	719	246	31.9	28.3	0.94
Sheep	0	0	39.0	25.4	0.50
	120	34	46.7	30.7	0.62
	239	77	57.2	34.0	0.70
	441	136	66.3	44.5	0.89
	741	250	84.0	50.3	1.00

[a] During the final two months of pregnancy up to weaning.
[b] From weaning to 24 weeks.
[c] Weaning at 12 weeks for lambs, and 16 weeks for kids.
Source: ILCA 1988.

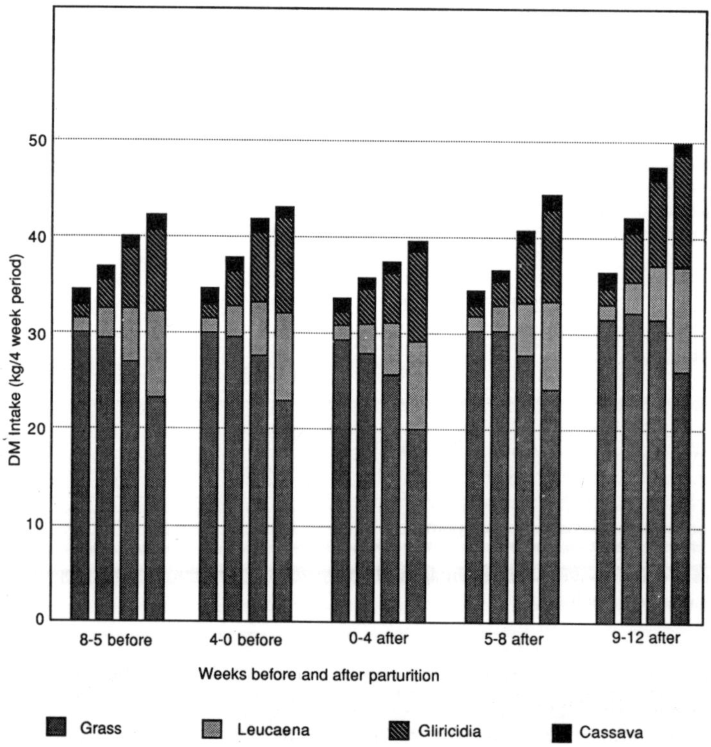

Figure 13.2. Dry matter intake before and after parturition of adult West African dwarf sheep offered different levels of *Leucaena* and *Gliricidia* browse.
Source: Reynolds and Adediran 1988.

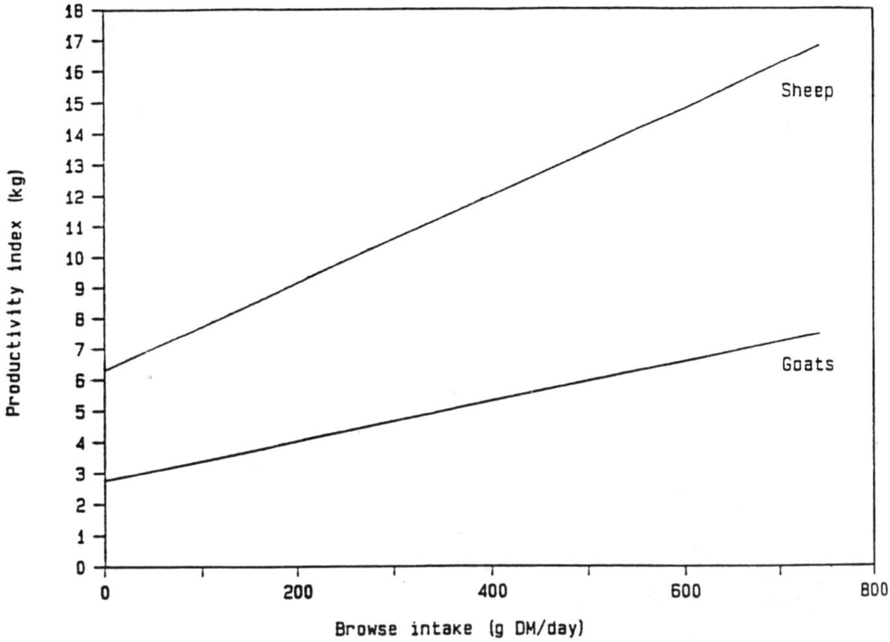

Figure 13.3. The effect of supplementation with *Leucaena* and *Gliricidia* browse on the productivity index (kg offspring weaned/dam/year) of West African dwarf sheep and goats. Source: ILCA 1988.

In the semiarid tropics, however, water competition between trees and adjacent food crops may be the dominant factor. Walker (1987) argued that in India it would be more economic to grow cereals and trees as sole crops rather than as intercrops because sole crops are better able to avoid water stress. Also, when compared with Africa, the use of inorganic fertilizer may be more attractive to the Indian farmer than sacrificing scarce land to trees in an alley farm to produce mulch. Fertilizer prices are more reasonable, and distribution systems are better than in Africa, resulting in wider availability. Fodder shortage during the latter part of the dry season, on the other hand, increases the attractiveness of legume forage to livestock owners (Singh et al. 1989). However, *Leucaena* and *Gliricidia* may not be ideally suited to the semiarid tropics; other tree species, spacing, and management systems should be investigated for different environments.

A modification of the alley-farming system involves planting grass between tree rows, rather than human food crops. Cut-and-carry forage can supplement feed for more intensive livestock operations, such as stall-fed beef or dairy units, or can feed small ruminants. Where soil moisture is not a constraint, productivity of grass adjacent to tree rows improves from increased soil nitrogen levels. In areas of Nigeria that have a rainfall of 1,250 mm, an inter-tree row spacing of 4 m planted with four rows of *Panicum maximum* and 2.5 m alleys with two rows of grass annually produced 20 tonnes of dry matter (DM) per hectare of each crop. Narrower alleys, with more trees per hectare, gave a higher yield of crude protein. Yields are not sustainable, however, at these levels without the application of nutrients from either manure or fertilizer.

Another alternative for livestock producers is a tree-only plot. Tree management is more flexible because shading of companion food crops need not be considered. Forage yields will depend on tree spacing and cutting frequency, as well as tree species and environmental conditions. An inter-row spacing of 0.5 m and a cutting cycle of 12 weeks gave an annual yield of 41 tonnes DM/ha from *Leucaena* in Nigeria. However, without the return of nutrients as manure or inorganic fertilizer to the plot, this level of off-take is not sustainable. Production also decreased as spacing widened and cutting frequency increased (figure 13.4).

Biological Factors

Biological problems that have arisen on-farm have been related to tree establishment and management. On farms where soil fertility was poor (low N and P), the establishment of the trees was patchy. The application of small

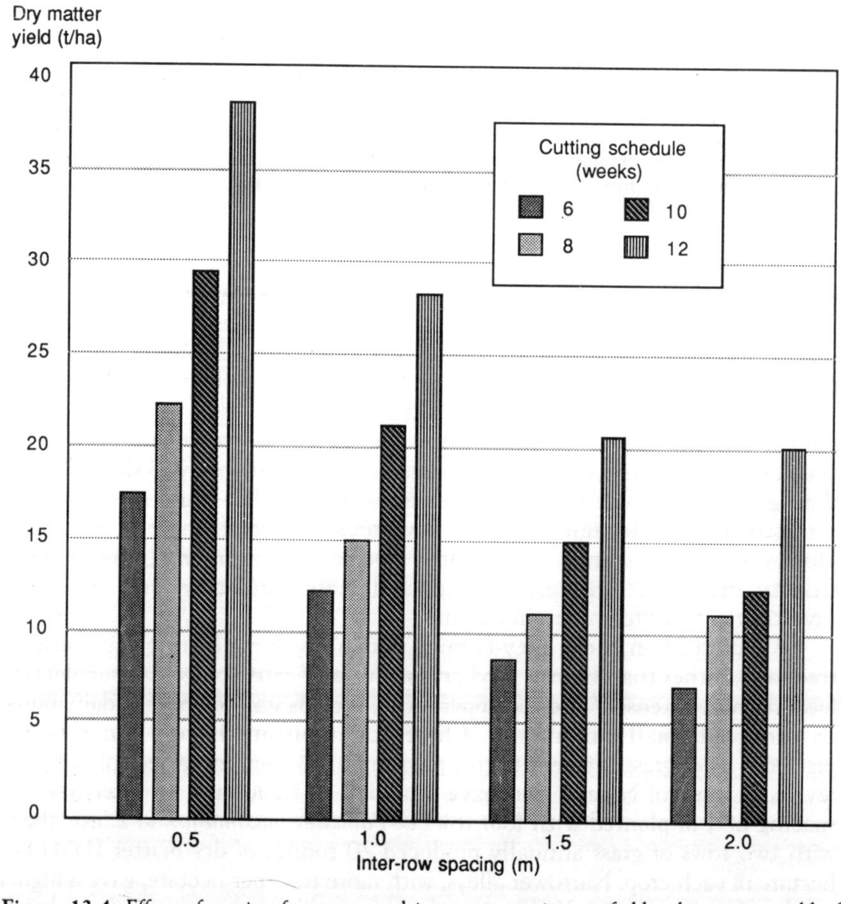

Figure 13.4. Effect of cutting frequency and inter-row spacing on fodder dry matter yield of *Leucaena* under intensive cultivation in southern Nigeria.
Source: ILCA 1987.

amount of inorganic fertilizer during establishment could provide an answer. This is the critical period in which most extension efforts should be made because, if the trees fail to survive, the farmer has no chance of obtaining enough prunings to make an impact. Also, if the species of tree legume is new to the area, it is unlikely that the required strains of rhizobia will be present. Nodulation will occur eventually with nonspecific rhizobia, but these will be less effective than adapted strains.

Plantation Systems

In plantation systems, the main objective is to obtain a commercial tree product (for example, rubber, palm oil, coconut, cashew). Trees are, therefore, the focus of the system, and any activity to grow food crops or produce animals is secondary. Forage production and the role of livestock in plantation systems has been the subject of a number of reviews (Thomas 1978, Plucknett 1979, Reynolds 1988, Shelton et al. 1987). There are over 5 million hectares of coconut plantation, over 2 million hectares of rubber, and about 1 million hectares of oil palm in Asia and the South Pacific, largely in the hands of smallholders. The most promising results from introduction of livestock on natural pasture have been achieved under coconut, with the cost of weeding reduced, coconut production increased, and additional income derived from cattle. The potential for integration with rubber is lower because young trees can be damaged by animals, and cattle will knock latex-tapping cups off mature trees.

The commercial exploitation of natural pastures under coconut will be limited by the low productivity and carrying capacity of the pastures; establishment of improved grass and legume species may be needed. The primary factor limiting pasture productivity under trees is shade, which not only reduces growth but also modifies the effect of inter-species competition. As with other tree crops, the degree of light interception varies with age of the coconut, hence pasture thrives best under young trees (less than five years) or old trees (more than 20 years). Tropical C_4 grasses show a greater decrease in relative growth rate under shade than C_3 legumes, and there are some reports of an increase in legume content at the expense of grass in mixed swards over time (Shelton et al. 1987). It is established practice in many commercial plantations of rubber and oil palm to cultivate fast-growing leguminous cover crops, thereby reducing soil erosion, fixing nitrogen, improving nutrient cycling, and reducing weed growth (Nair 1984). Smallholder plantations are more likely to be interplanted with food crops, such as cassava, yam, and maize, rather than with cover crops. Woody plantation crop systems export relatively low amounts of nutrients out of the system, but the presence of an intercrop will increase the degree of nutrient cycling because of increased nutrient interception by plant-rooting systems.

Since the first objective is production of the tree crop, it is important to check that the introduction of a forage system is beneficial, or at worst neutral, to the tree crop. There may be competition between trees and forage crops for water and nutrients, but the extent and effect will depend upon spacing. Soil characteristics may be modified, both by the pasture itself and by grazing animals. These effects need quantification under controlled conditions. A six-year study in Sri Lanka showed that forage yield declined in a rubber plantation between two and seven years after tree establishment as shading increased, and it showed that tree girth development was adversely affected by *Brachiaria brizantha*, an effect partially mitigated by the inclusion of herbaceous legumes (Waidyanatha et al. 1984). The effect on tree growth was less pronounced with other grass species. A negative effect of pasture on crop yield is also expected in coconut plantations where rainfull is marginal and competition for moisture becomes critical (Plucknett 1979).

Management issues for pastures under tree crops are largely the same as for unshaded pasture. Overgrazing leads to the disappearance of shade sensitive improved grasses and allows unpalatable weeds and less productive naturalized species to take over. Very few high-yielding grass species show shade tolerance, and they fail to persist under sustained grazing. An exception is *Stenotaphrum secundatum*, which has proved successful in Vanuatu (Macfarlane and Shelton 1986, cited in Shelton et al. 1987). Among legumes, *Desmodium intortum*, *D. canum*, and *Leucaena* are not adversely affected by shade. There is some evidence, however, that shade has a detrimental effect on the nutritive value of forage and that animal productivity is lower than when the animals are fed similar forage grown in full light, but more research is needed to clarify these points.

Livestock productivity data have been summarized in table 13.4. In a comprehensive comparison of natural and exotic grasses in the Solomon Islands,

Table 13.4. Livestock productivity under plantation tree crops

Tree species	Location	Livestock species	Annual LWG	System
Coconut	Solomon Islands	Cattle	292 kg/head	175 trees/ha, 60% light transmission, *Axonopus compressus*, *Mimosa pudica* and *Centrosema pubescens*.
Coconut	Western Samoa	Cattle	180 kg/ha	Well-spaced trees, > 50% light transmission, *Axonopus affinis* and *Mimosa pudica*.
Coconut	Western Samoa	Cattle	350 kg/ha	Well-spaced trees, > 50% light transmission, *Brachiaria miliiformis*, *B. mutica*, and *Ischaemum indicum* with *Centrosema pubescens* and *Pueraerea phaseoloides*.
Coconut	Vanuatu	Cattle	175 kg/ha	Old trees, > 50% light transmission, *Stenotaphrum secundatum*.
Coconut	Bali	Cattle	550 kg/ha	Old, well-spaced trees.
Oil palm	Malaysia	Cattle	100-117 kg/ha	Five- to seven-year-old trees
Oil palm	Malaysia	Cattle	210 kg/ha	One- to three-year-old trees
Rubber	Malaysia	Sheep	255-346 kg/ha	Three- to five -year-old trees
Rubber	Malaysia	Sheep	17 kg/head	

LWG = Live-weight gain
Adapted from: Shelton et al. 1987.

Watson and Whiteman (1981) concluded that where there is an adequate cover of naturalized grasses and legumes under coconut, cultivation and planting of exotic species was not recommended since no differences in animal response or copra yield had been observed. In the humid South Pacific, stocking rates as low as 1.5 cattle per hectare have led to the loss of introduced grasses that are grazed in preference to less palatable, but more hardy, indigenous species (Shelton et al. 1987). In Malaysia, six to eight sheep per hectare are recommended for immature rubber and three to five for mature areas. The present recommendations for stocking rates have been derived from experience and can be taken as useful guides, but controlled trials are now needed.

KEY SOCIOLOGICAL ISSUES

Land Tenure

Rights to land, especially for tenant farmers who have no long-term security of tenure, may weigh against adoption of practices that take time to produce results. Farmers are interested in the short-term benefit from improved crop yields and, in some cases, from better livestock productivity. The longer-term effects on soil fertility and sustainability enter few farmers' calculations. With alley farming, a farmer must wait at least one year, and more likely two, after planting before significant amounts of tree foliage are available. This is a greater disincentive to tenants than to landowners. Land ownership by families or communities, rather than by individuals, may also create problems since the farmer, in effect, is a tenant. Land may be reallocated by the community or the family after fallow periods, and there is no guarantee that an individual will be able to return to the same piece of land. If there are changes in the number of adults needing land in the group, the area will be redivided into different-sized portions. Despite these factors, tenants have been well represented among adopters of alley farming.

Gender

Gender is another contentious issue that cannot be divorced from land tenure. In many societies, the rights of women to land are limited. Women may undertake more of the labor, but they often do so on land owned by men. Crops may belong to the man, but the income from sale of livestock produce may belong to his spouse. Rights to crop or livestock produce may reside with different members of the household, giving rise to conflicts of interest.

Gender may also be a problem during the introduction and extension of a new technology. Most extension workers are male, but there may be cultural restrictions on their ability to approach women in the village. Second-hand information gained by women farmers from their spouses will not be as effective as that obtained directly from the extension worker. A solution to the problem is to employ both male and female extension staff. Since it is important during the on-farm testing phase of the project to involve a representative group of farmers, men and women, young and old, this problem has relevance outside the extension phase.

WEAKNESSES AND PROBLEMS OF FSR

Multidisciplinary Research

By its very nature, agroforestry research must be multidisciplinary and take a holistic approach to solving complex land-use issues and human problems. It should judge agroforestry interventions in terms of productivity, profitability, acceptability, and sustainability. And it must be tested on the farm with the farmer. However, Herdt (1987) pointed out that the majority of papers published under the banner of FSR are mono-disciplinary, are not carried out on farmers' fields, and have little or no farmer interaction. Why?

Multidisciplinary research is hard to organize, and difficult to conduct. On-farm researchers end up with dirt on their feet, the results are difficult to analyze statistically, and there are still relatively few journals willing to publish the results. The requirements of a neat and tidy paper containing hard data and falling into disciplinary compartments may be incompatible with true FSR. In Herdt's own words. "farming systems research must be problem driven, not paper driven."

There are many disincentives to interdisciplinary studies. They are difficult to organize, fund, and manage. Disciplinary training promotes a narrow focus, and scientists tend to ignore problems that lie outside their own area of expertise. The result among the majority of agricultural scientists is incomplete problem identification. Each scientist views problems from the perspective of his or her own discipline, but approaches from all the involved disciplines should be considered and accepted or rejected on the basis of the needs, objectives, and resources of the farmer.

Communication

Farming systems research will not work without good channels of communication among team members as well as among farmers, extension staff, and the FSR team. Good communication is most likely to develop where team members have compatible personalities, confidence in their own disciplinary strength, and respect for the role of other disciplines. The ability to listen and a willingness to accept other viewpoints are as important as verbal fluency (Norman 1987).

Leadership

A strong, effective leader is another key element in the success of FSR. The leader should have a sound disciplinary background and a broad understanding of the problems being addressed. All members must feel that their views matter and be handled tactfully if their suggestions are not always acceptable. The team effort must stay focused on the ultimate objective but retain sufficient flexibility to allow the unexpected and possibly critical problem to be addressed during the project. The leader must deal with the donor community, collaborating research institutions and individuals, and extension bodies, and he must be responsible for administration, personnel, and financial matters.

Motivation

Unfamiliarity with FSR among senior administrators and professionals may lead to fewer rewards and incentives and dimmer promotion prospects. Other problems include a lack of trust between disciplines, feelings of superiority or inferiority among disciplines, and unequal commitments by the institutions or individuals participating. Tree work requires a longer time horizon than arable studies, but the pressure for quick results may create strong tensions between team members and bias research toward short-term goals (Gold and Tombaugh 1987).

Location-specific Results

A number of the problems relating to adaptability of results that were experienced by International Agricultural Research Centres working with FSR were aired at a meeting held at ICRISAT in 1986 (IARCs 1987). To some extent, FSR is location-specific because each area has a unique combination of biological and sociological conditions. However, problems selected for study in FSR should be of regional interest so that the principles for one area can be applied to others with local modifications. In commodity research, this problem is circumvented for biological scientists by organizing multilocational trials, but social acceptability may remain unexplored. As stated earlier, biologically successful packages have been ignored by farmers time and time again, and programs have been drawn up to push technologies that will eventually prove inappropriate because socioeconomic aspects have not been considered. The strength of FSR should be that it produces a technology acceptable to farmers, and to define the recommendation domain. Adaptive research, taking that package outside the original environment, and possibly extending the recommendation domain, will be required for any technology, irrespective of how it was derived.

Focus of Study

Farming systems research generally focuses on the household as the appropriate unit to study. However, particular members of the household may require attention. The case of women and the adoption of alley farming is a good example (Francis and Atta-Krah 1988).

National Policy

Agricultural technology packages may have profound implications for society as a whole: an example is the individualization of land holding with the adoption of alley farming. Communication with policy-makers is important to ensure that the objectives of the project are not out of line with national aspirations and that national support is forthcoming. Communicating with policy-makers can also allow the output from FSR to have an impact on national policy formulation.

Conservatism

The alleged conservatism of FSR has also been raised as an issue since successful

FSR programs are said to be constrained by what farmers already know or can perceive (Swindale 1987). It is also said that FSR assumes that the farmer knows best, which is patently incorrect in some cases (for example, soil conservation) and which ignores the conflict of interest between the individual and society (additional animals on communal grazing land incur no extra expense to the individual owner, even when the area is already overstocked). This criticism may apply in the early stages of an FSR project, when the system's description must focus on existing circumstances, but if an improved technology package is developed, it means by definition that the ways of the farmer are not best. This misconception could arise because FSR assumes that a technology which conflicts with social structure is less likely to be adopted, even though it may show large biological benefits. Also, because rural communities are conservative, FSR limits its expectations of change in behavior patterns until on-farm experience demonstrates the value of an intervention. It is not possible to predict with certainty what variations from the recommendation domain are compatible with successful uptake. It will, therefore, be necessary to test a package in greater depth in some areas than in others before extension services can take it over.

Donor Commitment

Some of the initial enthusiasm for FSR in the donor community has faded because the expected burst of new, exciting technologies, ready for extension, have failed to materialize. Although the expectations may have been unrealistic, it is the firm belief of the author that there is no better method of determining the needs of the farmer and of testing the suitability of new technology than by working alongside the farmer, on the farm. Part of this problem may be due to lack of long-term commitment by donors. Farming systems research takes time (studies of sustainability and fallow regimes and livestock productivity trials need long-term projects), but surely it is better to be patient than to rush in and impose unsuitable projects on rural communities, as has happened so many times in the past, Leaving these communities to pick up the pieces at great human and monetary expense.

FUTURE RESEARCH NEEDS

Agroforestry is still in its infancy, and although a long shopping list of further research needs could easily be given, this section will concentrate on a few key areas.

In the realm of basic research, more information is needed on nutrient competition between tree and crop. Work is under way on the development of models to allow competition effects in one ecozone to be predicted from known results in another, and from one tree and crop mixture to another. Sustainability is a key issue, but apart from combinations of *Leucaena* and maize and *Leucaena*, *Gliricidia*, and maize, very little information is available. The extent to which inclusion of leguminous trees will reduce the need for external inputs still needs clarification for most crop mixtures. It has sometimes been claimed that alley farming will make fallowing unnecessary, but this requires verification in practice.

More information is needed on the role of rhizobia and mycorrhizas in nitrogen fixation and on the extent to which leguminous trees fix atmospheric N, as opposed to extraction from the soil through their root network.

Applied research on tree selection for different ecological conditions, different uses, and a variety of farming systems is urgently needed. Reliance on a single species can be disastrous and alternatives must be found. To date, very few of the many possible browse trees have been investigated in any depth.

Research on livestock components should include examining the effects of limited amounts of dietary protein protected by polyphenolics on overall nitrogen utilization and animal productivity. A related question is whether browse from deep-rooted trees provides micronutrients not available from shallow-rooted grasses.

Labor availability may be a constraint, and the development of mechanization will be appropriate to small farmers in some countries. Rooting studies may be needed—some tree species have shallower root systems than others, which could hinder rotation. Tree pruning by hand is also time consuming, and this is another area where mechanization may be beneficial. Cut-and-carry livestock feeding is also labor intensive, and although results so far indicate that grazing small ruminants will debark *Leucaena* and *Gliricidia* trees, studies are needed on grazed or tethered cattle in mixed tree and grass systems. Farmers may decide that they are willing to sacrifice some tree production for the convenience of allowing the animals to feed themselves. Verbal reports have been received that this happened in Central America. Effects of different grazing pressures and of grazing management methods on tree, grass, and livestock production are required.

Alley farming as practiced in the humid zone will clearly not work in semiarid regions, where the demand for animal feed is so great that forage trees surely have a role to play. Most research tests mono-cropped maize in alley farming, but smallholder farmers use mixed cropping. Combinations of food crops have not received any attention in alley farming.

Some work is under way for livestock under coconuts on shade-tolerant grasses and herbaceous and tree legumes. Selected best-bet species and combinations need to be tested on-station under livestock grazing and their persistence studied under farmer management. Economic analyses are also needed to determine the opportunity cost to the farmer of planting forage rather than a food crop under the trees.

Research is needed in plantation systems on the forage yields of natural and improved pastures under trees at different light intensities, in comparison to nontree areas, and on the ability of the more promising combinations to withstand grazing and the resultant effects on livestock productivity. The present recommended stocking rates for plantation systems have been derived from experience and can be taken as useful guides, but controlled trials are needed. Information is also needed, particularly from smallholder systems, on economic aspects of livestock integration.

Modification of the spatial arrangements of trees to allow a longer period for forage production before the tree canopy closes has been suggested for oil palm plantations. But ex ante economic evaluations followed by field trials, if the system appears financially rewarding, will be needed. A minimum of seven years and a long-term monetary commitment are needed for this kind of trial. While data on livestock productivity in general is needed, a few areas of research can be suggested. It has been shown that the nutrient levels of grasses under shade is different from that of the same grasses grown in the open, which raises questions about the effects on the digestibility of forage and on animal performance. Another question is whether shade from the tree canopy and the resulting difference in microclimate have any effect on the levels of external or internal parasites.

Understanding the impact of land tenure on adoption is critically important, particularly where land rights are not individualized. The right of tenant farmers, for example, to plant trees and to reap the long-term benefits is worthy of study in all tenure systems. Gender issues, and the related consideration of how different responsibilities of household members influence the perceptions of a technology and therefore the likelihood of its adoption, also need to be better understood.

A weak link in many programs is that between research and extension. It is not enough for researchers to hand a technology to the extension service and leave them to take it to farmers. Studies are needed on the most effective means of making the handover, on the extent to which extension staff should be involved during the research stage, and on where researchers can best assist adoption.

CONCLUSION

Agroforestry systems are a suitable means of improving livestock production for smallholder farmers, but a sound understanding of the existing farming systems, and the social and physical environment in which they are placed, is needed before judgment can be made. Testing the chosen method of improvement requires a combination of on-station and on-farm trials that evaluate biological, technical, economic, and social aspects of the package. Two agroforestry systems have been described here, alley farming and plantation systems, illustrating contrasting possibilities. The need for a multidisciplinary team approach to the development of any improved methodology is stressed.

LITERATURE CITED

Atta-Krah, A. N., and J. E. Sumberg. 1987. Studies with *Gliricidia sepium* for crop-livestock production systems in West Africa. In: Gliricidia sepium (*Jacq.*) *Walp. management and improvement* (D. Withington, N. Glover, and J. L. Brewbaker, eds.). Hawaii: Nitrogen Fixing Tree Association, pp 31–43.

Cochran, W. G., and G. M. Cox. 1957. *Experimental decisions*. New York: John Wiley and Sons. 611 p.

CGIAR (Consultative Group on International Agricultural Research). 1978. *Farming systems research at the international agricultural research centers*. Rome: CGIAR. 66 p.

Flores, J. F., T. H. Stobbs, and D. J. Minson. 1979. The influence of the legume *Leucaena leucocephala* and formal-casein on the production and composition of milk from grazing cows. *Journal of Agricultural Science* 92: 351–357.

FAO. 1976. *A framework for land evaluation*. Soils Bulletin 32. Rome: FAO. 87 p.

Francis, P. A., and A. N. Atta-Krah. 1988. Incorporating gender concerns into on-farm research: the household and alley farming in Southwest Nigeria. In: *Methodologies handbook on intrahousehold dynamics and farming systems research and extension* (H. Feldstein and J. Jiggins, eds.). In press.

Gold, M. A., and L. W. Tombaugh. 1987. Understanding interdisciplinary research and its application to agroforestry systems research. In: *How systems work. Proceedings of Farming Systems Research Symposium*, 1987. Fayetteville, Arkansas: Winrock International and University of Arkansas. pp 13–26.

Herdt, R. W. 1987. Whither farming systems. In: *How systems work. Proceedings of Farming Systems Research Symposium*, 1987. Fayetteville, Arkansas: Winrock International and University of Arkansas. pp 3–70.

IARCs (International Agricultural Research Centres). 1987. *Proceedings of the Workshop on Farming Systems Research*. ICRISAT: Patancheru, India. 153 p.

ILCA (International Livestock Centre for Africa). 19867. ILCA Annual Report 1985/1986. Addis Ababa: ILCA. 88 p.

ILCA (International Livestock Centre for Africa). 1987. ILCA Annual Report 1986/1987. Addis Ababa: ILCA. 82 p.

ILCA (International Livestock Centre for Africa). 1988. ILCA Annual Report 1987/1988. Addis Ababa: ILCA. 103 p.

Kang, B. T., H. Grimme, and T. L. Lawson. 1985. Alley cropping Sequentially cropped maize and cowpea with *Leucaena* on a sandy soil in southern Nigeria. *Plant and Soil* 85: 267–276.

Kang, B. T., L. Reynolds, and A. N. Atta-Krah. 1989. Alley farming. *Advances in Agronomy* 43: 315–359.

Kearl, S. (ed.). 1986. *Livestock in mixed farming systems: research methodolgies and priorities*. Farming Systems Support Project, Network Report 2. Gainesvile, Florida: ILCA and University of Florida. 220 p.

Khon Kaen University. 1987. *Proceedings of the 1985 conference on rapid rural appraisal*. Rural systems research and farming systems research project. Khon Kaen, Thailand.

Mutsaers, H. J. W., N. M. Fisher, W. O. Vogel, and M. C. Palada. 1986. *A field guide for on-farm research*. Ibadan: IITA. 197 p.

Nair, P. K. R. 1984. *Soil productivity aspects of agroforestry*. Nairobi: ICRAF. 85 p.

Nicholson, M. J., and M. H. Butterworth. 1986. *A guide to condition scoring of Zebu cattle*. Addis Ababa: ILCA. 29 p.

Nordblom, T. L., A. K. H. Ahmed, and G. R. Potts (eds.). 1985. *Research methodology for livestock on-farm trials*. Ottawa: IDRC. 313 p.

Norman, D. W. 1987. Communication and information systems in farming systems work: an overview. In: *How systems work. Proceedings of Farming Systems Research Symposium*, 1987. Fayetteville, Arkansas: Winrock International and University of Arkansas. pp 287–304.

Plucknett, D. L. 1979. *Managing pastures and cattle under coconuts*. Westview Tropical Agriculture Series 3. Boulder, Colorado: Westiview Press, Inc. 364 p.

Raintree, J. B. 1986. *An introduction to agroforestry diagnosis and design*. Nairobi: ICRAF. 55 p.

Reynolds, L., and S. O. Adediran. 1988. The effects of browse supplementation on the productivity of West African dwarf sheep over two reproductive cycles. In: *Goat production in the humid tropics* (O. B. Smith and H. G. Bosman, eds.). Wageningen, The Netherlands: Pudoc. pp 83–91.

Reynolds, L., and A. N. Atta-Krah. 1989. Alley farming with livestock. In: *Alley farming in the humid and sub-humid tropics* (B. T. Kang and L. Reynolds, eds.). Ottawa: IDRC. pp 27–34.

Reynolds, S. G. 1988. *Pasture and cattle under coconuts*. FAO Plant Production and Protection Paper 91. Rome: FAO. 321 p.

Shaw, N. H., and W. W. Bryan. 1976. *Tropical pasture research: principles and methods*. Farnham Royal, UK: Commonwealth Agricultural Bureau. 454 p.

Shelton, H. M., L. R. Humphreys, and C. Batello. 1987. Pastures in the plantations of Asia and the Pacific: performance and prospect. *Tropical Grasslands* 21: 159–168.

Singh, R. P., R. J. van den Beldt, D. Hocking, and G. R. Konwar. 1989. Alley farming in the semiarid regions of India. In: *Alley farming in the humid and sub-humid tropics* (B. T. Kang and L. Reynolds, eds.). Ottawa: IDRC. pp 108–122.

Sumberg, J. E., and S. D. Mack. 1985. Village production of West African dwarf goats and sheep in Southwest Nigeria. *Tropical Agriculture* (Trinidad) 62: 35–40.

Sumberg, J. E., J. McIntire, C. Okali, and A. N. Atta-Krah. 1987. Economic analysis of alley farming with small ruminants. *ILCA Bulletin* 28: 2–6.

Swindale, L. D. 1987. Farming systems and the International Agricultural Research Centers: an interpretive summary. In: *Proceedings of the Workshop on Farming Systems Research, IARCs*. Patancheru: ICRISAT. India. pp 132–139.

Thomas, D. 1978. Pastures and livestock under tree crops in the humid tropics. *Tropical Agriculture* (Trinidad) 55: 39–44.

USDA. 1976. *Soil taxonomy of the national soil survey*. Washington, DC: Soil Conservation Service, US Department of Agriculture.

van Eys, J. E., I. W. Mathius, P. Pongsapan, and W. L. Johnson. 1986. Foliage of the tree legumes gliricidia, leucaena and sesbania as supplement to napier grass diets for growing goats. *Journal of Agricultural Science* 107: 227–233.

Waidyanatha, U. P. de S., D. S. Wijesinghe, and R. Strauss. 1984. Zero grazed pasture under immature *Hevea* rubber: productivity of some grasses and grass-legume mixtures and their competition with *Hevea*. *Tropical Grasslands* 18: 21–26.

Walker, T. S. 1987. *Economic prospects for agroforestry interventions in India's SAT: implications for research resource allocation at ICRISAT*. Resource Management Program, Economics Group, Progress Report 79. Patancheru, India: ICRISAT. 53 p.

Watson, S. E., and P. C. Whiteman. 1981. Animal production from naturalized and sown pastures at three stocking rates under coconuts in the Solomon Islands. *Journal of Agricultural Science* 97: 669–676.

Wilson, R. T., and J. W. Durkin. 1983. Livestock production in Central Mali: weight at first conception and ages at first and second parturitions in traditionally managed goats and sheep. *Journal of Agricultural Science* 100: 625–628.

Zandstra, H. G., E. C. Price, J. A. Litsinger, and R. A. Morris. 1981. *A methodology for on-farm cropping systems research*., Los Baños, Philippines: 147 p.

CHAPTER 14

Overcoming Problems of Fodder Quality in Agroforestry Systems

R. J. Jones
J. B. Lowry

Agricultural systems that combine forestry and livestock production and those that mix crop production, forestry, and livestock are extremely varied. They occur in humid, subhumid, and semiarid areas of the developing world (Gholz 1987). Unfortunately, the animal component in many of these systems has not been adequately evaluated; even the systems themselves, of which animal production is an integral part, are inadequately described and not well understood (Tejwani 1987).

The importance of ruminants in these systems, however, is widely acknowledged (Jones 1988). They may reduce the risks associated with cropping; they provide traction, transport, and fuel and fertilizer (via dung); they are a food source; they satisfy cultural needs and establish prestige; and they generate income (McDowell 1980). Nevertheless, inadequate nutrition is a major constraint to generating more income from ruminants in these systems because often only maintenance levels of energy intake is available for much of the year (Jones 1988). Providing additional energy and protein above maintenance is vital to improving animal productivity. This may be achieved by supplying additional high-quality feed, by using supplements to improve the efficiency of existing feeds, and by removing or reducing the adverse effects of certain secondary plant compounds.

This chapter suggests some options for improving the efficiency of agroforestry species for use in animal production.

CHARACTERISTICS OF FODDER TREES AND UNDERSTORY VEGETATION AND ASSESSMENT OF THEIR FEED QUALITY

All plants contain an array of proteins, lipids, carbohydrates, and other compounds essential to their metabolism. They have well-defined functions in the plant and are utilized by the animal by known pathways. In addition to these primary metabolites, plants contain compounds that are clearly not involved in their internal metabolism. These secondary constituents occur in a very wide range of chemical structures, are found in some plant taxa and not in others, may be extremely rare or widespread, and may be present in traces or be

50% of the biomass. Until quite recently, the function of secondary compounds was regarded as obscure or uncertain. However, it is now generally accepted that they have an ecological role: they are the means by which the plant limits or deters predation (Rosenthal and Janzen 1979).

Secondary compounds in forage plants thus tend to deter feeding through aversion or reduce intake through toxicity. (However, not all herbivores are affected equally. Among ruminants, active hydrolytic and reductive processes in the rumen may break down otherwise adverse compounds.) Furthermore, these compounds give the plant the chemical armory to withstand insect predation in its environment. Secondary compounds may thus be specifically or generally active against invertebrates, higher animals, or plant pathogens.

Secondary compounds, though rarely a problem with temperate forages, are all-important in the development of forages in tropical agroforestry. This is because first, woody plants tend to have higher levels of these compounds than grasses and forbs. Second, tropical plants in general have greater concentrations of them than temperate plants. This generalization follows from the ecological function: because of the biological diversity of the tropics, a given plant will be in contact with a greater array of fungi, bacteria, insects, and higher animals, often at a higher level of activity, than in a temperate environment. Thus, secondary compounds should tend to occur in greater levels and variety in plants in the tropics. There is ample evidence from individual species to support this generalization, but the only systematic comparison is for the most bioactive group, the alkaloids (Levin and York 1978). Third, for plants in a nutrient-poor environment, loss of tissue to herbivores is more serious than for those in an optimum environment that supports rapid growth. As a result, plants adapted to poor soils or low light tend to have higher levels of secondary compounds (Coley et al. 1985). This is potentially relevant where agroforestry systems are used on less favorable land.

Potential Feed Resources in Agroforestry Systems

Because of the great variability in level and occurrence of secondary compounds, their effect on the feed quality of any particular species varies widely. It is useful, however, to consider some biological generalizations that are relevant to the plants likely to be used as feed resources in agroforestry systems.

Fast-growing trees grown in plantations

These trees often have been evaluated and become known purely for wood biomass production. Any forage value is regarded as incidental and is sometimes only revealed by a strong local demand. Thus, *Maesopsis semenii* and *Albizia falcataria* are both well known as fast-growing hardwoods—nowhere does the literature indicate forage potential—but in Java, both are used extensively in cut-and-carry feeding of village ruminants. Many existing plantation trees, and even more of the large number of "new" ones that are being evaluated, are fast-growing pioneer species (rather than slow-growing climax forest species). In nature, these tend to produce seed plentifully, establish readily, and grow rapidly under favorable conditions. Their leaves are generally less protected from predators by secondary chemical defenses than are those of climax species that must persist for a long time in the forest twilight or as emergents. New growth of pioneer species, found in canopy gaps or on landslides and other disturbed sites, provides much of the feed for foresr browsers. Thus there is a better-than-average chance that trees of interest for biomass production will also provide useful forage. Nonetheless, plenty of chemical defenses do exist and will be discussed later.

Crop trees

Included in this category are trees grown primarily for fruit or other products and, possibly, those grown for high-quality wood. They often originate from primary forests, are relatively slow-growing, and as indicated above, should have leaves well protected against herbivores. They form a considerable part of the mixed garden culture of Southeast Asia. Harvesting the produce may make a considerable amount of leaf material available as a potential resource. In the case of the clove tree, the leaf brought down during harvesting of the cloves (which certainly is an unpromising feed) is steam-distilled for essential oil production. In general, little feed value is attached to many of the fruit trees. A notable exception is jackfruit (*Artocarpus heterophylla*) leaf, which is everywhere highly valued as feed for village goats.

True agroforestry trees

A few species, such as *Acacia albida* and *Leucaena leucocephala*, have multiple roles and recognized value to animal production.

Shade-tolerant understory plants

In rubber plantations, these may range from sown shade-tolerent grasses to a mixture of volunteer shrubs and ferns; in hardwood plantations these plants may form a substanital shrub layer with distinctive species. In the mixed garden system, all plants found along shaded banks and paths are used in cut-and-carry feeding. In most cases, ferns make up a high proportion of these plants, and both their nutritive value and their toxicity need to be better known.

Client Problem

In both the wet and the seasonally dry tropics, village animal production is constrained by feed quality (Little et al. 1988). A major factor is the loss of feed quality and quantity during the dry season. Tropical grasses, mostly of the C_4 type, have a characteristic foliar anatomy that gives a higher proportion of vascular fibrous tissue than is found in temperate (C_3) grasses. Protein content and fiber digestibility tend to decrease drastically as the grass matures (Wilson and Minson 1980). Thus, the quality and quantity of grassland become limiting factors.

Crop residues may be a major source of feed. Of these, rice straw is the most abundant but is of notoriously low quality. Tree leaves of high nutritional quality take on additional value as supplements to enable better utilization of such feeds.

Hypothesis

Certain tree leaves are of sufficiently high feed quality to increase animal production when used as supplements to low-quality, fibrous feeds.

Methods of Measurement

The most useful species will be those with high protein and high digestibility. In order to assess the potential, laboratory measurements traditionally include protein, fat, fiber, and an in vitro digestibility measurement. They are well described in many sources (for example, Pigden et al. 1980), and we note here only qualifications or concerns that might be relevant when evaluating feeds for agroforestry.

Protein

Kjeldhal nitrogen in usually converted to crude protein by multiplying by the factor 6.25. This may result in overestimation of true protein in some plant material due to nitrogen being bound in cell wall polymers and to the presence of nitrogen-containing secondary compounds (such as the nonprotein amino acid mimosine in *Leucaena*). However, nitrogen determination is still the best way for estimating and comparing protein content.

Fat

Ether extraction removes not only triglycerides but also cuticular waxes that may be of little nutritive value. However, the lipid content of leaf material is not usually very high or of primary nutritional interest. Any values above about 5% may be worth investigating further.

Fiber

Although publications continue to report on crude fiber, it should be noted that this measurement has been superseded by the sequential detergent methods (Goering and Van Soest 1970, Van Soest and Robertson 1980) that do at least provide an indication of various cell wall constituents. The crude fiber determination does not measure any biologically identifiable fraction. It may have had some value in ranking feeds for monogastric animals but has much less value for ruminants. It may be useful comparing current value with earlier values on the same material, but it is useless for evaluating a new feed.

Research workers in agroforestry may find that the analytical facility closest at hand still offers crude fiber determinations. In such situations, researchers should try to have the detergent system adopted, set it up themselves, or send the samples farther afield—but they should not bother with crude fiber determinations.

A summary of the fractionation by the detergent systems is as follows:

Neutral Detergent Extraction

This should separate cell contents from cell walls. The cell contents, being soluble, are by definition 100% digestible but will not necessarily be of nutritive value. It may include a substantial fraction of soluble secondary compounds, for example, oxalates and phenolics. The cell walls, nominally made up of cellulose, hemicellulose, lignin, and insoluble ash, are the fraction that must be subject to fermentative digestion.

Acid Detergent Extraction

This should make soluble the more easily fermented hemicellulose, leaving primarily lignocellulose fiber. Despite the fact that not all hemicelluloses are readily fermentable and a significant proportion of cellulose can be fermented, the result can be of interest and leads to the next step.

Permanganate Oxidation

When acid detergent fiber is subjected to permanganate oxidation, lignin is degraded and made soluble. Cellulose is recovered as a residue, and lignin is obtained from the difference. The values for lignin obtained in this way may be of considerable interest as they often correlate negatively with in vivo digestibility. In general, the lignin to cellulose ratio is restricted. In a crude sense, lignin can be seen as an encrusting polymer preventing access of enzymes or bacteria.

Ashing

Ashing of acid detergent fiber produces acid insoluble ash, mainly silica, which is often present in high levels in tropical grasses. Cell wall silica may be a constraint to digestibility.

Digestibility determination

Evaluating feeds must ultimately be done via the animal, but the time, labor, expense, and amount of material required have prompted the development of laboratory methods. Most in vitro methods use pepsin-cellulase or rumen fluid-pepsin. If the plant contains microbial inhibitors, results from the rumen fluid method may be abonormally low. This need not reflect the in vivo situation, for in the real rumen these compounds may be broken down or pass out of the rumen with the liquid phase.

Intraruminal nylon bag

This in vivo technique is much less costly than actual feeding trials (Ffoulkes 1986). While it allows estimation of digestibility, acutely toxic or strongly antimicrobial activity will not be detected at all because the compounds in the small nylon bag samples are rapidly diluted in the large rumen volume.

Drying treatments

Traditionally, fiber or in vitro digestibility determinations have been carried out on material dried at 60°C or 100°C, the actual temperature being regarded as of little consequence. While this is true for temperate forages, it is now clear that heating some tropical plants may reduce digestibility (Mahyuddin et al. 1988). Where possible, material should be freeze-dried.

Secondary compounds—Phenolics

The most widely occurring secondary compounds, and those found in greatest amount, are the phenolics. Because these are often equated with tannins, the first point to make is that although all tannins are phenolics, not all phenolics are tannins.

Tannins are those phenolics with the ability to precipitate protein from aqueous solutions, and this imposes certain restrictions on the nature of the phenolic molecule. The actual property of protein precipitation can obviously affect intake and digestibility, for example, through precipitation of salivary mucoprotein or of digestive enzymes. Ruminants appear to tolerate moderate levels of tannin (up to 5% dry weight), but adverse effects appeared above about 10% with sheep in New Zealand (Barry and Duncan 1984). However, it seems likely that tropical ruminants may tolerate much higher levels. Simple phenolic compounds, although they do not generally interact with protein, may inhibit specific enzymes or have other effects. They become significant at the higher levels that can occur (up to 40% leaf dry weight) because, as they are readily absorbed, they must be metabolized and excreted. This imposes metabolic demands on the animal that must be offset by the feed value of the plant. Another possibility is that dietary tannins may reduce sodium retention, a problem in situations where sodium is already deficient for good animal nutrition (Freeland et al. 1985). Although the nutritional implications are still not uderstood, the variety of known effects suggests that it would be advisable to include some analysis of phenolic content in any nutritional evaluation.

Tannin analysis

As the biological effect of tannin is protein precipitation, the only satisfactory analysis is one that measures this activity. The first such method for leaf extracts was that of Bate-Smith (1973), using hemoglobin as a protein. Concentrations of hemoglobin could be easily measured colorimetrically. The test is simple to carry out but requires some dedication from the operator who is expected to provide the hemoglobin, personally, for each run. A variant with stabilized bovine blood has been described (Schultz et al. 1981). A more recent method involves precipitation of tannins with pepsin, separation of the precipitate, and analysis of the phenolics regenerated from the precipitate (Hagerman and Butler 1978).

The only readily available standard for such determinations is tannic acid from commercial sources, which gives results as "tannic acid equivalent." However, it should be noted that not only may the chemical nature of the tannin in a plant sample be different (condensed tannin versus hydrolyzable tannin), but the activity may be different (protein precipitated per unit weight of tannin).

For determination of phenolic compounds in general, a variety of methods are available (Harborne 1973). Perhaps the most convenient is the use of Folin-Dennis or Folin-Ciaccolteau reagent (Ribereau-Gayon 1972). Most methods involve absorbance measurements and require a known compound as a standard. Although tannic acid is often used, it probably gives more intense color development than less highly hydroxylated phenolics and therefore causes them to be underestimated. A more appropriate standard is the flavonol glycoside rutin because it is widespread (found in more than 50% of dicotyledonous species) and commercially available.

It would be desirable if tannins or total phenols could be measured on a direct dry weight basis rather than against a probably inappropriate standard. The Ytterbium precipitation method (Reed et al. 1985) appears to offer just such a method using relatively simple laboratory facilities. However, it should be noted that this method, although it probably precipitates tannins along with some other phenolics, certainly does not give total phenols. We have found a number of common plant phenolics, including rutin itself, to be quite unaffected by the Ytterbium acetate reagent (Lowry and Sumpter in press).

For a plant to act as a useful supplement, the desirable attributes are high protein (> 20%), high degestibility (> 65%), low neutral detergent fiber (> 30%), and low phenolics (> 5%). Poorer quality material can be fed, of course, but not as a supplement.

Critical Review

Currently, information on the composition and nutritional value of species relevant to agroforestry is sparse and scattered. Probably much is unpublished. The inappropriate crude fiber analysis is still being reported.

Although many species still require reliable identification, the importance of this is not recognized by some researchers. There is a tendency for a species name, from whatever source, to be accepted uncritically. Vernacular names are useful but too easily assumed to relate directly to scientific names. The term *roadside grass*, for example, is frequently used as though it were a botanical entity, when in fact it often refers to mixtures of forbs, ferns, and shrubs. The importance of lodging herbarium specimens of material under study is rarely recognized.

There is considerable variation in the conduct and reporting of feeding experiments. In particular, evaluation of very low-sodium feeds will be affected by whether or not salt is provided.

MIXING SPECIES TO OFFSET NUTRITIONAL LIMITATIONS OF INDIVIDUAL SPECIES

As noted earlier, plant toxins and antinutrients are found only in particular taxa. This disjunct distribution may be put to positive use because such compounds often limit the intake of the plant concerned with no adverse effects to the animal below consumption of a certain level.

In general, because of their diverse chemical nature, we would expect different plant compounds to act on the animal independently of each other. Synergistic toxic effects are known but are relatively rare. Thus, we can envisage a diet made up of a mixture of different plants, none of which could be fed as the sole diet. In such a mixture, each antinutrient is diluted below the harmful threshold by addition of other noninteracting material.

This feeding strategy would be expected to occur naturally, most clearly among the arboreal foliovores, and indeed, it does apply to three species of rainforest possums in tropical Australia (Goudberg 1988).

Client Problem

Farmers in tropical areas usually have access to woody species that are highly productive but unutilized or underutilized because they are unpalatable or toxic to livestock. Often these species have high protein content and are digestible. Feeding this material would be one way of increasing the effective feed resources.

Hypothesis

Chemical and nutritional characteristics of individual tree species permit an appropriate blend of species that can be advantageously fed in greater quantity than any of the species individually.

Methods of Measurement

Before attempting to formulate or test various mixtures, it is necessary to know (1) the levels at which each species can be used without adverse effects and (2) the nature of the antinutrients in the species of interest. The former requires feeding experiments that precede the testing of mixtures, and the latter determines which species may be mixed. (There is little point in combining species that have similar compounds, and the prevalence of the more common compounds, such as condensed tannins, may limit the choices available.)

The approach to measurement will depend on the way leaf is to be utilized. Leaf may be fed directly to ruminant livestock, as presumed in most of this chapter. Evaluation of mixtures is carried out in the same manner as for a single species, but may require restricted intake to ensure all components are eaten. Leaf may also be included in rations for intensive poultry and swine production. There is already a significant trade in *Leucaena* leaf meal for this purpose (Anonymous 1984) and in uncontrolled species mixtures (Lowry et al. 1984). The evaluation of this use of leaf meal requires access to a specialized facility. However, there are well-established procedues for nutritional studies on growing broiler chickens and the larger animals that can be used to obtain a more satisfactory statistical treatment.

Critical Review

The hypothesis was tested by using leaf meal in the diet of growing broiler chickens (Lowry and Tangendjaja unpublished). The mixture used was made up of cassava, *Leucaena*, *Calliandra*, *Gliricidia sepium*, and *Sesbania grandiflora*. In order to obtain measurable responses, the percentage of leaf meal was higher than would be found in commercial use. This produced growth depressions compared with an isonitrogenous control diet without leaf meal. At 7% of the diet, birds receiving the mixed leaf meal had 12% lower body weight than the control group. However, the average for birds receiving the pure species leaf meal was 27% below the control group. Thus the mixed leaf meal gave a better result than any single species of which it was composed. This indicates the theory is correct and its application should be considered in particular agrofrestry systems.

OVERCOMING SOME LIMITATIONS OF FODDER BY MINERAL SUPPLEMENTATION

The mineral composition of forage and tree species used for livestock varies greatly from site to site and within sites. Of particular significance is the ability of some species to accumulate specific nutrients relative to others growing under the same conditions. In the case of sodium (Na), Hacker (1974) found that varieties of setaria growing side by side may vary be a factor of 30 or more (0.05% to 1.80%). In 28 fodder shrubs, values for nitrogen (N), phosphorus (P), and sulphur (S) in the leaves ranged from 3.71% to 1.36%, 0.59% to 0.12%, and 0.34% to 0.07% respectively (Bamualim 1981). Tropical legumes are typically low in Na, with 62% of the samples analyzed having values below 0.05% (Norton 1982). The concentration of trace elements in agroforestry species is not well documented, though the low levels of iodine in *Leucaena* may limit animal production (Jones 1979). Rarely are animals given only one forage, so there are opportunities when mixtures are fed for a low level of a particular mineral in one forage to be balanced by a high level in another species. However, such complementary associations usually happen by chance and may only operate for part of the year.

Phosphorus has often been seen as the most important limiting mineral nutrient in the tropics (McDowell et al. 1987). However, the importance of sodium and the possibility of obtaining significant gains from low inputs of this mineral deserve attention.

Client Problem

For tropical forages, particularly shrub and tree species, there is relatively little data on which to assess the mineral deficiencies that could limit animal production. The variety of diets in tropical countries, where differences in the herbaceous and tree species fed vary seasonally as well as from one area to another, impose additional difficulties in identifying a particular problem.

Some of the low growth rates of village ruminants in tropical areas, well documented by Mahadevan (1982) and Thahar and Petheram (1983), are undoubtedly due to mineral deficiencies (Lowry et al. 1983, Little et al. 1988). From results in Papua New Guinea and Indonesia, sodium supplementation or sodium plus supplementation with other nutrients has been shown to improve animal production (Leche 1977, Yates 1982, Juarini et al. 1983). This may indicate a widespread deficiency in the wet tropics. In West Timor, *Leucaena* fed to tethered or penned ruminants is often supplemented with banana pseudostems to provide water and, supposedly, minerals including salt. However, data from the analysis of the banana pseudostems clearly shows that the Na levels are less than those in *Leucaena* (0.01%) although the Potassium (K) levels in banana are very high. In some coastal areas of Timor, seawater is fed to ruminants because rain water collected from the roofs is thought to be unsuitable for the stock. It seems clear that in these Na-deficient areas the local people have overcome the major mineral deficiency by using seawater.

Multiple deficiencies can also occur. In Australia, when leaves from the drought-resistant leguminous tree *Acacia aneura* (mulga) are fed to sheep, wool growth and live-weight gain improved with supplements of Na, S, Ca, and urea (Gartner and Niven 1978, Elliott and McMeniman 1987). Proximate analysis of the leaves would not have predicted the responses to S, Ca, and urea since the leaves are above accepted critical values for these minerals. This unexpected response is undoubtedly due to interference in the availability of the minerals due to the presence of both oxalates (as calcium oxalate) and tannins, which bind proteins and necessitate the excretion of sulphur as sulphate (Gartner and Hurwood 1976). In addition to increasing feed intake, digestibility improved when P, molasses (containing Ca and S), P + molasses, or P + molasses + urea supplemented the feed (McMeniman 1976).

There are doubtless many situations where poor animal production in agroforestry systems is associated with undefined and uncorrected mineral deficiencies.

Hypothesis

Sodium supplementation of leguminous and other browse species feeds for livestock can improve feed intake and utilization (particularly in the wet tropics).

Methods of Measurement

It is important to determine as a base line the level of Na in feeds used for livestock. Rations for growing or lactating animals that contain less than 0.05% Na on a dry matter basis can be regarded as deficient and the animals therefore potentially responsive to Na.

Confirmation of the deficiency and an assessment of the magnitude of the response to supplementation could then be obtained in studies of penned feeding, of tethered animals at pasture, or of free-grazing animals.

Under the most controlled conditions in pens, three treatments are necessary: (1) an unsupplemented control feed, (2) a supplemented feed with intake regulated to that of the controls, and (3) a supplemented feed with free intake. These treatments enable the effect of the supplement to be measured on both voluntary feed intake and on feed utilization. There need to be four to six replications of each treatment, with animals being fed individually in separate pens. Pens would need to be separated to avoid adjacent animals licking each other or drinking each other's urine to obtain Na. The salt (initially NaCl, not sea salt) could be supplied as a block, or granulated in a separate box next to the feed, preferably at head level of the animal to avoid fouling. Salt intake can then be measured by weighing the salt contained at daily or weekly intervals. Adding the salt to fresh feed could cause the plant tissue to blacken and become unattractive to stock. This may then adversely affect intake, especially if the feed is held at ambient temperatures for several hours. This is not the recommended way to provide the salt supplement.

Drinking water should be analysed for Na to ensure that controls are not obtaining a supplement from this source.

Feed intake should be measured daily on a dry matter basis by measuring feed offered and refused. The animals on regulated feed receive the weight of food eaten by the control animals on the previous day. Animals should be weighed weekly over a minimum period of six weeks for reliable information on live-weight change. If the animals used are not obtained from the area where Na deficiency is suspected, a much longer feeding period may be necessary to overcome the effect of Na storage in the tissues of animals previously on a high Na diet. Tethered animals or animsls allowed free grazing during the day could be fed the Na supplement at night when housed. The treatment groups would need to be kept separate for this purpose.

Critical Review

The lack of detailed mineral analyses for the numerous feeds offered to animals in the tropics is a major limiting factor in assessing the likely responses to any form of mineral supplement. Inherent differences in mineral content among species have generally not been measured. Superimposed on any such differences are variations due to annual seasons and soil fertility. The widespread Na deficiency described above suggests that considerable improvement in animal production could be readily achieved by the simple practice of providing salt to animals in salt-deficient areas. The experiment described above could be extended to determine whether minerals other than salt are likely to be deficient. This would require the addition of one or more treatments containing a mixture of minerals (major and trace elements). The magnitude of any difference between the + salt treatment and the + mineral mix treatment would then indicate the severity of deficiencies other than sodium.

The increasing use of sophisticated equipment for mineral analysis in and around the tropics has greatly increased the capacity for plant analysis. A coordinated approach is now required similar to that described by McDowell et al. (1987) for ranches in South America where soil, plant, and animal tissue analyses are combined to identify specific mineral deficiencies. Follow-up work entails feeding studies to confirm the presumed deficiencies established by the analytical data, to devise relevant supplementation strategies, or to select plant genotypes with adequate levels of the deficient nutrients.

DETOXIFYING PROBLEM COMPOUNDS AND INCREASING DIGESTIBILITY WITH BACTERIA

As mentioned earlier, fodder trees and other vegetation in agroforestry systems can contain secondary plant compounds that may adversely affect feeding value or even poison stock. Many leguminous species contain toxins that directly or indirectly may be toxic to ruminants as well as monogastric species (Hegarty 1982, Barry and Blaney 1978). However, the rumen has a tremendous capacity, via microbial populations, to detoxify problem compounds. For example, β-nitro propanoic acid is toxic to chicks but not to ruminants (Hegarty and Pound 1970). Similarly, *Leucaena*, which contains mimosine, is highly toxic to monogastric species but has no immediate toxic effects on ruminants. However, in some countries like Australia, prolonged ingestion of diets high in mimosine results in a toxicity characterized by enlarged thyroids, low-serum thyroxine, ulcerated esophagus and reticulo-rumen, lethargy, alopecia, and reduced appetite. Newborn calves from dams fed *Leucaena* may be severely goitrous and die within a few days (Jones 1985).

This toxicity results from absorption of 3 hydroxy 4-1-H pyridone (DHP), a ruminal metabolite of mimosine now known to be a potent goitrogen (Hegarty et al. 1979). In addition to its cumulative effect on the thyroid, DHP (the 2, 3 isomer) also has a dramatic effect on feed intake when infused into the rumen of sheep in the absence of DHP-degrading bacteria (Bamualim 1984). The 3, 4 isomer also appears to depress intake, for when DHP-degrading bacteria were introduced to Australian goats that had been fed a diet of *Leucaena*, their feed intake doubled (Jones and Lowry 1984). This is a clear example of a secondary plant compound having an effect on feed intake as well as resulting in other toxic effects. Two questions then arise: Do other secondary plant compounds have similar effects? Can they be identified? The successful introduction of DHP-degrading bacteria into Australia has resulted in dramatic improvement in the value of *Leucaena* as a feed for ruminants (Jones and Megarrity 1986, Quirk et al. 1988). The technology for transfer is simple and effective and provides permanent benefits with no recurring costs for maintance of the benefit.

Client Problem

The client problem is not easy to define because there is little information on the plant compounds likely to have adverse effects on livestock in agroforestry systems. According to a survey of urine samples from ruminants in countries where *Leucaena* is fed (Jones and Megarrity unpublished), Africa, China, Japan, and Fiji do not appear to have DHP-degrading bacteria. Therefore, ruminants in these countries could be expected to experience *Leucaena* toxicity. Confirmation for Africa and China was obtained from long-term feeding studies (Lambourne et al. 1987, Wang et al. 1987a). In both Ethiopia and China, DHP-degrading bacteria from Australia were introduced in trials by R. J. Jones and successfully established in the rumen. The anticipated result did indeed happen—exceretion of DHP in the urine stopped and animal production improved (ILCA 1986, Wang et al. 1987b).

Other secondary plant compounds known to adversely affect feed value are tannins and fluoroacetate. Although low levels of condensed tannins can enhance the value of feed, high levels are harmful (Barry and Blaney 1987). It is not known if rumen bacteria can metabolize these condensed tannins, although it was shown that animals conditioned by feeding on tannin-containing plants were more productive than control animals (Barry and Blaney 1987). The possibility that specific bacteria had increased in the rumens of the conditioned group was not explored.

The valuable leguminous shrub and tree genus *Acacia* may contain tannins, oxalates, and fluoroacetate. The last compound is responsible for stock poisonings following ingestion of *Acacia georginae* foliage in Queensland and the Northern Territory of Australia. The legume genera *Gastrolobium* and *Oxylobium* also contain fluoroacetate and can be toxic (Barry and Blaney (1987). Since this toxin appears not to accumulate beneath the tree canopies, soil microorganisms must occur to break them down. It is possible that there are bacteria capable of defluorinating fluoroacetate under anaerobic conditions. If so, these could be identified, multiplied, and introduced into ruminants. Alternatively, transferring the ability to detoxify fluoroacetate to a suitable rumen bacterium by gene-cloning techniques may be possible in the future.

Hypothesis

Transfer of DHP-degrading bacteria to ruminants devoid of these bacteria significantly increases live-weight gain due to the reduction of the goitrogenic and appetite suppressant DHP.

Methods of Measurement

Resident ruminants should be checked to determine if they are capable of degrading DHP. The presence of enlarged thyoids in offspring of ruminants fed *Leucaena* is one indicator that a problem exists. A more direct approach is to check for DHP or mimosine in the urine of animals being fed *Leucaena*. A simple method is to use a ferric chloride solution to test the urine.

The test solution is prepared by dissolving 0.6 g $FeCl_3 6H_2O$ in 1 liter of water and then adding 2 ml of concentrated HCL. If 0.2 ml (a few drops) of urine is added to 2 ml of this solution and a purple coloration results, free mimosine or 3, 4 DHP is present. A blue coloration results if the metabolite in the urine is mainly 2, 3 DHP. A proportion of the mimosine and its metabolites is often bound as glucuronide and must be hydrolyzed to produce the reactive free forms (Hegarty et al. 1964). In animals regularly consuming *Leucaena*, however, there is usually an adequate amount of free mimosine or DHP to produce the color reaction that shows the necessary mimosine and DHP-degrading bacteria are not present. An alternative approach is to use colorimetry or high-performance liquid chromatography (HPLC) analysis to confirm the presence or absence of mimosine or DHP (Megarrity 1978 and 1981, Lowry et al. 1983).

If it is established that DHP-degrading bacteria are not present, it needs to be shown that use of the relevant bacteria will solve the problem.

The experimentation required to assess the value of introduced bacteria is fairly simple, though some details need to be followed exactly. First, two groups of three to six animals (sheep, goats, or cattle) need to be penned individually in separate buildings. Although the bacteria involved in the degradation are strict anaerobes, they can spread in aerosols when animals cough and through feces. It is important, therefore, that the two groups be housed separately. In China, two groups of animal attendants were used to run the separate groups. These never mixed, thus reducing the possibility of accidental transfer of bacteria to the control group. While such strict segregation may not be essential, procedures to minimize the possibility of spread should include a change of overalls or rubber boots when changing from one group of animals to another and always feeding or making collections from the control group before the treated group.

Both groups of animals should be fed the same *Leucaena* diets (> 50% *Leucaena*) for two weeks before treatments begin and have their urine checked again to confirm excretion of mimosine or DHP.

The treated group should receive the culture of DHP-degrading bacteria (currently an unnamed genus and species) via stomach tube into the rumen (50 ml to 200 ml) from an anaerobic tank pressurized with CO_2. The tube into the rumen should be flushed with CO_2 to reduce the risk of killing the bacteria by exposure to oxygen.

An alternative to using bacterial cultures in a broth medium is to use rumen fluid from a fistulated steer, if quarantine authorities will allow importation. The fluid in which the organisms are established comes from a steer grazing on *Leucaena*-dominant pasture. The rumen fluid is less sensitive to brief exposure to oxygen and is therefore more reliable. Control animals would receive an equal volume of drinking water into the rumen.

Since the beneficial effect of these microorganisms is dependent on increased feed inake, both groups should be offered feed freely and the quantities eaten recorded. Intake of mimosine can be calculated by analysis of the diet. Collection of urine on a daily basis with preservation of a proportion with acid will enable daily excretion of mimosine or DHP to be determined. In the absence of HPLC or other suitable instruments, the simple color test described earlier can be used on the acidified urine and a score of zero to five given for the intensity of the color reaction.

Animals should be weighed weekly to measure live-weight change. A decrease in excretion of mimosine or DHP is usually associated with an increase in serum thyroxine level, feed intake, and live weight of the treated group (Jones and Lowry 1984, Quirk et al. 1988). The controls may show a depressed appetite and may lose some hair. Sheep may shed their fleece.

When urine is free of DHP for a period of a month after infusion with the DHP-degrading bacteria, it can be assumed that the bacteria have become established in the rumens of the treated group. Usually five to seven days are required after infusion with rumen fluid for excretion of urinary DHP to stop (Jones and Lowry 1984, Wang et al. 1987b). When cultures of the DHP-degrading bacteria are used, it takes somewhat longer (Jones and Megarrity 1986, ILCA 1986). Large (44% to 98%) live-weight gain responses have been reported (Wang et al. 1987b, Quirk et al. 1988).

Critical Review

The use of rumen bacteria to degrade toxins and so improve forage utilization is new. The degradation of mimosine in *Leucaena* is the only example thus far, though it has been postulated that degradation of the hepatotoxin indospicine in the legume *Indigofera spicata* could be achieved by identifying suitable bacteria in African ruminants (Jones and Hegarty 1981). However, it can be assumed that most, if not all, plant toxins and other negative factors in herbage are broken down in the soil. Therefore, organisms must exist that, at least under aerobic conditions, are capable of using these compounds as substrates for growth. When there is no source of bacteria within ruminants, the soil beneath stands of plants containing toxic secondary plant compounds may well provide the necessary bacteria. Whether or not it is possible that facultative anaerobes capable of living in the rumen can be found remains to be seen. However, this seems unlikely because if they are present, ingestion of such bacteria in the dust and soil would be inevitable when these or associated plants are grazed, resulting in their colonization of the rumen. If this were the case, no problem would be identified.

The difficulty in exploiting this technology is two-fold: (1) identifying problem secondary plant compounds and (2) identifying specific rumen organisms to degrade these compounds. For the future, it seems probable that soil bacteria will need to be identified that can degrade problem secondary plant compounds and their capacity to do this incorporated into rumen bacteria. Unfortunately, the genetic systems of rumen bacteria are largely unknown (Smith and Hespell 1983), and until their genetic structure and biochemistry are better understood, alterations of these microbes by genetic engineering present a considerable challenge (Orpin 1988). Recombinant DNA technology is advancing so rapidly, however, that the goal of producing improved strains of rumen microorganisms to both degrade toxins and improve feed utilization now seems a reasonable one.

LITERATURE CITED

Anonymous. 1984. Leucaena: *promising forage and tree crop for the tropics*. Washington, DC: National Academy Press. 100 p.

Bamualim, A. 1981. *Nutritive value of some tropical browse species in the wet and dry seasons*. M Sc thesis, James Cook University of North Queensland, NZ. 103 p.

Bamualim, A. 1984. *The nutritive value of* Leucaena leucocephala *as a feed for ruminants*. Ph D thesis, James Cook University of North Queensland, NZ. 167 p.

Barry, T. N., and B. J. Blaney. 1987. Secondary compounds of forages. In: *The nutrition of herbivores* (J. B. Hacker and J. H. Ternouth, eds.). Sydney, Australia Academic Press pp 91–119.

Barry, T. N., and S. J. Duncan. 1984. The role of condensed tannins in the nutritional value of *Lotus pedunculatus* for sheep. I. Voluntary intake. *British Journal of Nutrition* 51: 485–491.

Bate-Smith, E. C. 1973. Haemanalysis of tannins: the concept of relative astringency. *Phytochemistry* 12: 907–912.

Coley, P. D., J. P. Bryant, and F. S. Chapin. 1985. Resouce availability and plant antiherbivore defense. *Science* 230: 895–899.

Elliott, R., and N. P. McMeniman. 1987. Supplementation of ruminant diets with forage. In: *The nutrition of herbivores* (J. B. Hacker and J. H. Ternouth, eds.). Sydney, Australia: Academic Press

Ffoulkes, D. 1986. Maximising the effective measurement of digestibility *in sacco*. In: *Forages in Southeast Asian and South Pacific agriculture*. ACIAR proceedings 12. Canberra: ACIAR. pp 124–127.

Freeland, W. J., P. H. Calcott, and D. P. Geiss. 1985. Allelochemicals, minerals and herbivore population size. *Biochemical Systematics and Ecology* 13: 195–206.

Gartner, R. J. W., and I. S. Hurwood. 1976. The tannin and oxalic acid content of *Acacia aneura* (mulga) and their possible effects on sulphur and calcium availability. *Australian Veterinary Journal* 52: 194–196.

Gartner, R. J. W., and D. R. Niven. 1978. Studies on the supplementary feeding of sheep consuming mulga (*Acacia aneura*). IV. Effect of sulphur on intake and digestibility and growth and sulphur content of wool. *Australian Journal of Experimental Agriculture and Animal Husbandry* 18: 768–772.

Gholz, H. L. (ed.). 1987. *Agroforestry: realities, possibilities and potentials*. International Council for Research in Agroforestry. Dordrecht, The Netherlands: Martinus Nijhoff. 227 p.

Goering, H. K., and P. J. Van Soest. 1970. *Forage fibre analysis*. Agricultural handbook 379. Washington, DC: Agricultural Research Service, USDA.

Goudberg, N. J. 1988. *The feeding ecology of three species of North Queensland upland rainforest ringtail possums*. Ph D thesis, James Cook University of North Queensland, NZ.

Hacker, J. B. 1974. Variation in oxalate, major cations, and dry matter digestibility of 47 introductions of the tropical grass setaria. *Tropical Grasslands* 8: 145–154.

Hagerman, A. E., and L. G. Butler. 1978. Protein precipitation method for the quantitative determination of tannins. *Journal of Agricultural Food Chemistry* 26: 809–812.

Harborne, J. B. 1973. A guide to modern techniques of plant analysis. In: *Phytochemical methods*. London: Chapman and Hall. pp 37–99.

Hegarty, M. P. 1982. Deleterious factors in forages affecting animal production. In: *Nutritional limits to animal production from pastures* (J. B. Hacker, ed.). Farnham Royal, UK: Commonwealth Agricultural Bureaux.

Hegarty, M. P., R. D. Court, and P. M. Thorne. 1964. Determination of mimosine and 3, 4-dihydroxypyridine in biological material. *Australian Journal of Agricultural Research* 15: 168–179.

Hegarty, M. P., C. P. Lee, G. S. Christie, R. D. Court, and K. P. Haydock. 1979. The goitrogen 3-hydroxy-4(1H)-pyridone, a ruminal metabolite from *Leucaena leucocephala*: effects in mice and rats. *Australian Journal of Biological Science* 32: 27–40.

Hegarty, M. P., and A. W. Pound. 1970. Indospicine, a hepatotoxic amino acid from *Indigofera spicata*: isolation, structure and biological studies. *Australian Journal of Biological Science* 23: 831–842.

ILCA (International Livestock Centre for Africa). 1986. ILCA Annual Report 1985/86, p 12. Addis Ababa. 88 p.

Jones, R. J. 1979. The value of *Leucaena leucocephala* as a feed for ruminants in the tropics. *World Animal Review* 31: 13–23.

Jones, R. J. 1985. *Leucaena* toxicity and the ruminal degradation of mimosine. In: *Plant toxicology* (A. A. Seawright, M. P. Hegarty, L. F. James, and R. F. Keeler, eds.). Melbourne: Dominion Press. pp 111–119.

Jones, R. J. 1988. The future for the grazing herbivore. Harry Stobbs Memorial Lecture for 1987. *Tropical Grasslands* 22: 97–115.

Jones, R. J., and M. P. Hegarty. 1981. Screening pasture plants for possible toxic effects on livestock. In: *Forage evaluation: concepts and techniques* (J. L. Wheeler and R. D. Mochrie, eds.). Townsville, Australia: CSIRO. pp 237–247.

Jones, R. J., and J. B. Lowry. 1984. Australian goats detoxify the goitrogen 3-hydroxy-4(1H) pyridone (DHP) after rumen infusion from an Indonesian goat. *Experientia* 40: 1435–1436.

Jones, R. J., and R. G. Megarrity. 1986. Successful transfer of DHP-degrading bacteria from Hawaiian goats to Australian ruminants to overcome the toxicity of *Leucaena*. *Australian Veterinary Journal* 63: 259–262.

Jones, R. J., and R. G. Megarrity. Unpublished data. CSIRO, Townsville, Queensland, Australia.

Juarini, E., H. Prasetyo, R. J. Petheram, and C. Liem. 1983. *Growth response to salt and sulphur in village sheep*. Research report, Project for Animal Research and Development at Balai Penelitian Ternak, Bogor, Indonesia. 47 p.

Lambourne, L. J., H. Bosman, and L. Reynolds. 1987. High concentrations of dihydroxypyridone (DHP) in urine of sheep and goats fed *Leucaena* in Ethiopia and Nigeria. In: *Herbivore nutrition research*. Sydney, Australia: Australian Society of Animal Production. pp 59–60.

Leche, T. F. 1977. Effects of a sodium supplement on lactating cows and their calves on tropical native pastures. *Papua New Guinea Agricultural Journal* 28: 11–17.

Levin, D. A., and B. M. York. 1978. The toxicity of plant alkaloids: an ecogeographic perspective. *Biochemical Systematics and Ecology* 6: 61–76.

Little, D. A., R. J. Petheram, and M. Boer. 1988. Observations on the mineral content and nutritive value of diets fed to village ruminants in the Indonesian districts of Bogor (West Java) and Pamekasan (Madura, East Java). *Tropical Agriculture* (Trinidad) 65: 213–218.

Lowry, J.B., N. Cook, and R.D. Wilson. 1984. Flavonol glycoside distribution in cultivars and hybrids of *Leucaena leucocephala*. *Journal of the Science of Food and Agriculture* 35: 401–407.

Lowry, J.B., Maryanto, and B. Tangendjaja. 1983. Autolysis of mimosine to 3-hydroxy-4 (H) pyridone in green tissues of *Leucaena leucocephala*. *Journal of the Science of Food and Agriculture* 34: 529–533.

Lowry, J.B., A. Thahar, T. Sitompul, and R.J. Petheram. 1983. Composition of some village ruminant feed-stuffs in West Java. *Proceedings of the 5th world animal production conference*. Tokyo. 599–600.

Lowry, J. B., and A. E. Sumpter. Problems with ytterbium precipitation as a method for estimation of plant phenols. *Journal of the Science of Food and Agriculture*. In press.

Lowry, J. B., and B. Tangendjaja. Unpublished research report. USAID supported project—Potential for leaf meal production from tree legumes in S. E. Asia. Balai Penelitian Ternak, Bogor, Indonesia.

Mahadevan, P. 1982. Patures and animal production. In: *Nutritional limits to animal production from pastures* (J. B. Hacker, ed.). Farnham Royal, UK: Commonwealth Agricultural Bureaux. pp 1–17.

Mahyuddin, P., D. A. Little, and J. B. Lowry. 1988. Drying treatment drastically affects feed evaluation and feed quality with certain tropical forage species. *Animal Feed Science and Technology* 22: 69–78.

McDowell, L. R., D.E. Morillo, J. Faria M., F.G. Martin, J.H. Conrad, and N. Wilkinson. 1987. Mineral status of beef cattle in specific regions of the Venezuelan Llanos. In: *Herbivore nutrition research*. Australian Society of Animal Production Occasional Publication. pp 209–212.

McDowell, R. E. 1980. The role of animals in developing countries. In: *Animals, feed, food and production* (R. L. Baldwin, ed.). Boulder, Colorado: Westview Press. pp 102–120.

McMeniman, N. P. 1976. Studies on the supplementary feeding of sheep consuming mulga (*Acacia aneura*). III. The provision of phosphorus, molasses and urea supplements under pen conditions. *Australian Journal of Experimental Agriculture and Animal Husbandry* 16: 818–822.

Megarrity, R. G. 1978. An automated colorimetric method for mimosine in *Leucaena* leaves. *Journal of Science, Food and Agriculture* 29: 182–186.

Megarrity, R. G. 1981. Rapid estimation of DHP in urine. *Leucaena Research Reports* 2: 16.

Norton, B. W. 1982. Differences between species in forage quality. In: *Nutritional limits to animal production from pastures* (J. B. Hacker, ed.). Farnham Royal, UK: Commonwealth Agricultural Bureaux.

Orpin, C. G., G. P. Hazlewood, and S. P. Mann. 1988. Possibilities of the use of recombinant-DNA techniques with rumen micro-organisms. *Animal Feed Science and Technology* 21: 161–174.

Pigden, W. J., C. C. Balch, and M. Graham (eds.). *Standardization of analytical methodology for feeds*. Ottawa: IDRC. 128 p.

Quirk, M. F., J. J. Bushell, R. J. Jones, R. G. Megarrity, and K. L. Butler. 1988. Liveweight gains on *Leucaena* and native grass pastures after dosing cattle with rumen bacteria capable of degrading DHP, a ruminal metabolite from *Leucaena*. *Journal of Agricultural Science* 111: 165–170.

Reed, J. D., P. J. Horvath, M. S. Allen, and P. J. Van Soest. 1985. Gravimetric determination of soluble phenolics including tannins from leaves by precipitation with trivalent Ytterbium. *Journal of Science, Food and Agriculture* 36: 255–261.

Ribereau-Gayon, P. 1972. *Plant phenolics*. New York: Hafner Publishing Co.

Rosenthal, G. A., and D. H. Janzen. 1979. *Herbivores: their interaction with secondary plant metabolites*. New York: Academic Press. 718 p.

Schultz, J. C., I. T. Baldwin, and P. J. Nothnagle. 1981. Hemoglobin as a binding substrate in the quantitative analysis of plant tannins. *Journal of Agricultural Food Chemistry* 29: 823–826.

Smith, C. J., and R. B. Hespell. 1983. Prospects for development and use of recombinant deoxyribonucleic acid techniques with ruminal bacteria. *Journal of Dairy Science* 66: 1536–1546.

Tejwani, K. G. 1987. Agroforestry practices in India. In: *Agroforestry: realities, possibilities and potentials* (H. L. Gholz, ed.). Dordrecht, The Netherlands: ICRAF and Martinus Nijhoff. pp 109–136.

Thahar, A., and R. J. Petheram. 1983. Ruminant feeding systems in West Java. *Agricultural Systems* 10: 87–97.

Van Soest, P. J., and J. B. Robertson. 1980. Systems of analysis for evaluating fibrous feeds. In: *Standardization of analytical methodology for feeds* (W. J. Pigden, C. C. Balch, and M. Graham, eds.). Ottawa: IDRC.

Wang, J., J. H. Yang, and T. S. Li. 1987a. Effects of feeding beef cattle with *Leucaena leucocephala* and toxicity observation. *Grassland of China* 4: 60–63.

Wang, J., J. H. Yang, and R. J. Jones. 1987b. Chinese cattle detoxify the leucaena toxin after Australian rumen fluid infusion. In: *Proceedings of the 4th annual conference of the Chinese Grassland Association*, November 12–17, 1987. Nanning, Guangxi.

Wilson, J. R., and D. J. Minson. 1980. Prospects for improving the digestibility and intake of tropical grasses. *Tropical Grasslands* 14: 253–259.

Yates, N. G. 1982. *Mineral supplements double growth rate of sheep fed* Leucaena. Research report, project for animal research and development at the Balai Penelitian Ternak, Ciawi, Indonesia. 43 p.

List of Acronyms

ACIAR	Australian Centre for International Agricultural Research, Canberra, Australia
AFRENA	Agroforestry Research Networks for Africa, ICRAF, Nairobi, Kenya
CAB	Commonwealth Agricultural Bureaux, Wallingford, UK
CAZRI	Central Arid Zone Research Institute, Jodhpur, India
CGIAR	Consultative Group on International Agricultural Research, Washington, DC
CIAT	Centro Internacional de Agriculture Tropical, Cali, Colombia
CRIDA	Central Research Institute for Dryland Agriculture, Hyderabad, India
CSIRO	Commonwealth Scientific and Industrial Research Organization, East Melbourne, Australia
FAO	Food and Agriculture Organization of the UN, Rome, Italy
IARC	International Agricultural Research Centres
IBP	International Biological Programme
IBPGR	International Board for Plant Genetics Resources, Rome, Italy
ICRAF	International Center for Research in Agroforestry, Nairobi, Kenya
ICRISAT	International Crops Research Institute for the Semiarid Tropics, Hyderabad, India
IDRC	International Development Research Centre, Ottawa, Canada
IITA	International Institute of Tropical Agriculture, Ibadan, Nigeria
ILCA	International Livestock Centre for Africa, Addis Ababa, Ethiopia
IRRI	International Rice Research Institute, Manila, Philippines
IUFRO	International Union of Forest Research, Vienna, Austria
ODI	Overseas Development Institute, London, England
ORSTROM	Office de la Recherche Scientifique et Technique d'Outre-Mer, France
RAPA	FAO's Regional Office for Asia and the Pacific, Bangkok, Thailand
USAID	US Agency for International Development, Washington, DC
USDA	US Department of Agriculture, Washington, DC

Author Index

Subject Index